COMPETITIVE NEMATOLOGY

The Author

Dr. N.G. Ravichandra, Professor of Plant Pathology and Scheme Head, All India Co-ordinated Research Project (Nematodes), Department of Plant Pathology, University of Agricultural Sciences, GKVK, Bengaluru, Karnataka, India, is involved in teaching, research and extension activities of Plant Nematology.

Dr. N.G. Ravichandra has 242 scientific research papers to his credit, published in National and International journals. He has authored six text books in Plant Pathology & Nematology, contributed chapters to eleven text books, edited five books and 12 chapters in Nematology and also authored ten books in Kannada.

Dr. N.G. Ravichandra is responsible for identifying and reporting a new species of root-knot nematode (*Meloidogyne enterolobii*) on Guava and a new host, Lac plant (*Flemingia semialalta* L.) for root-knot nematode (*Meloidogyne incognita*) from Karnataka. These are the new and first reports in India.

Dr. N.G. Ravichandra has been bestowed with the "Eminent Scientist Award" for the year 2018 for his research accomplishments and contributions. He has been recognized as the International Editor for the prestigious International journal 'Research Journal of Plant Pathology' published from California. He is also rewarded as the National Editor for the eminent journal 'Indian Phytopathology'. He is responsible for developing eight technologies for the management of root-knot nematodes infecting tomato, brinjal and rice, which have been included in the Package of Practices of University of Agricultural Sciences, Bengaluru, Karnataka, India.

Dr. N.G. Ravichandra has participated and presented original research papers on various aspects of plant diseases in National and International conferences. He has operated research projects funded by the Department of Bio-Technology & Indian Council of Agricultural Research, New Delhi, India as Principal Investigator and Co- Principal Investigator. He has undergone advanced training programs sponsored by the Department of Biotechnology & Indian Council of Agricultural Research, New Delhi, India. He is an active life member of professional societies including Nematological Society of India, Indian Phytopathological Society, Society of Mycology & Plant Pathology, National Environmental Science Academy and Institute of Agricultural Technology.

COMPETITIVE NEMATOLOGY

DR. N.G. RAVICHANDRA
Professor of Plant Pathology
& Head, AICRP (Nematodes)
Agricultural College, GKVK
University of Agricultural Sciences,
Bangalore-560 065, Karnataka

Published by
SCIENTIFIC PUBLISHERS (INDIA)
5 A, New Pali Road, P.O. Box 91
Jodhpur 342 001, INDIA
E-mail: info@scientificpub.com
Website: http://www.scientificpub.com

ISBN: 978-93-88812-85-6
eISBN: 978-93-88812-86-3

Printed in India

To

Sudesh Kumar and Tara

ACKNOWLEDGEMENT

My respectful thanks to,

Dr. S. Rajendra Prasad, Hon'ble Vice Chancellor, University of Agricultural Sciences, GKVK, Bengaluru

Dr. Y.G. Shadakshari, Director of Research, University of Agricultural Sciences, GKVK, Bengaluru

Dr. Umarao, Project Coordinator, AICRP (Nematodes), Indian Agricultural Research Institute, New Delhi.

I would like to express my deep sense of gratitude to my beloved parents Sri.N. Gurushankar and Smt. Parvathamma.

I wish to record my appreciation and affection to my wife, Ms. Deepthi and daughter, Ms. Anusha for their care, patience, keen interest and follow-up throughout the preparation of the manuscript. I am grateful to my in-laws for their affection and encouragement.

I remain ever grateful to my brother-in-law Mr. B. Sudesh Kumar and my sister Ms. N.G.Tara, to whom I fondly dedicate this book for their sustained moral support, affection and special care. I fondly acknowledge their keen interest and continued encouragement being showered on me.

My special thanks to Dr. T. Narendrappa, Professor & University Head, Dr.A.Nagaraja, Professor & Head, Dr. H.A.Prameela, Dr. N. Nagaraju, Dr.H.K.Ramappa, Dr. Y.M.Somashekhar, Dr. C.P. Manjula and Mr. B.R. Nagendra, Department of Plant Pathology, UAS, GKVK, Bengaluru.

I take this opportunity to thank M/s. Scientific Publishers, Jodhpur, Rajasthan, India for their help in processing the manuscript and getting the book published.

Foreword

Plant-parasitic nematodes are hidden enemies of crops. Nematodes, either alone or in combination with other pathogens, constitute an important constraint to world food production.

Plant Nematology is taught as a major subject in Undergraduate and Postgraduate degree programmes in Agricultural and Horticultural Universities. It is also a subject for various competitive examinations / entrance tests at All India Level including M. Sc. (Agri./Hort.), Ph.D.(Agri./Hort.), Agricultural Scientists Recruitment Board, Public Service Commission, Central Plant Quarantine, State Departments of Agriculture, Horticulture, Sericulture and Forestry apart from the recruitments in Pesticides, Seed and fertilizers industries. Students and candidates appearing for these examinations are in search of a book that fulfils their needs.

This book entitled "Competitive Nematology" covers the syllabus prescribed by the Indian Council of Agricultural and Research, New Delhi. It serves as a ready reckoner for various competitive examinations. Important aspects of Nematology courses from UG & PG Degree Programmes have been covered in the form of different question patterns including the multiple choice, true or false, matching and essay type. Answer key has been provided. In addition, a summary of the prominent historical developments in Nematology along with a detailed glossary of Nematological terms has been furnished. A list of pertinent reference books has also been included for further reading. As a supplement, appendices on major Nematological Journals and Societies, common names of predominant phytonematodes, conversion table and scoring procedure for root-knot and lesion nematodes has also been furnished for the benefit of the readers.

I am happy to note that this book is the first of its kind in India. I compliment Dr.N.G.Ravichandra for making an important contribution to the science of Nematology and congratulate him for his maiden efforts. I am sure, this book will be well received by all those preparing for competitive and entrance examinations.

(Dr. S. Rajendra Prasad)

CONTENTS

CHAPTER 1

Historical Developments in Nematology

ANCIENT BACKGROUND:

I. Major free living, animal, human and insect parasites:

2700 BC	Reference to 'Yellow Emperors' classic of internal medicine from China about *Ascaris*, the giant intestinal worm	
1553-50 BC	'Ebers papyrus'	*Ascaris lumbricides, Encyclostoma duodena* and *Dracunculus medinensis* (the guinea worm) in Egypt
1250 BC	Bible Nos. 21:6-9	Extraxtion of guinea worm by winding it around a stick from tissues of affected people which is still being practiced in many parts of North Africa and the Middle-East
430 BC	Hippocrates	'Aphorisms' by pin worm, *Enterobium vermicularis* in women and horses
53 BC-7 AD	Celsus	Distinguishes flat and round worms
200 AD	Columella	*Neoascaris vitulorum* from a calf
130-200 AD	Galen	Nematodes of fish
400 AD	Vegetius	Horse Ascarid, *Pareascaris equorum*
1200-1280 AD	Albertus	Nematodes from falcons
1519-1630	Caesalpinus	Giant dog kidney worm, *Dactophyma remale*
1547	Venergia	Filarid (*Spiruria*) from falcon
1623	Aldrovandus	Found dead grasshoppers with worms emerging from their bodies (in *De Animalibus insectis* 1623). He was the one who first used the word *Vermes*
1672	Lister	Described similar worms from a plant in his garden and compared his findings to that of Aldrovandus.
1742	Reaumur	Described a worm that was undoubtedly *Sphaerularia bombi* Dufour.
1771- 1832	Rudolphi	Included 350 species belonging to 11 genera in his works Entozoorum Synopsis. He also gave us the scientific name, Nematoidea.
1804-1885	SIebold	Introduced the concept of a life cycle involving different kinds of hosts, with different ways of penetration.
1656	Borellus	First free living nematode, vinegar eel-*Turbatrix aceti* in Europe

1667	Robert Hooke	*Ponegrellus redvivus*- 'eel in paste'
1663	Tysor	First to study the nemic anatomy and to describe a nematode egg
1684	Radi	Recorded pale cat, lion and fish as new hosts of nematodes
1786	Mullor	Free-living nematodes from fresh water
1817-1821	Bojanus	Anatomical studies of a variety of nematodes by free hand sections and dissections
1824	Cloquest	Anatomical studies of a variety of nematodes by free hand sections and dissections
1851	Leidy	First record of free-lving nematodes in USA
1851-52	Herbst	Transmission of *T.spinalis* from rats to swine
1858	Davainc	Life histories of *Ascaris lumbricoides* and *Trichuris trichiura*
1871	Fedtscheko	Cyclops as intermediate host for *Dacunculus medinensis* and disease transmission by drinkingwater contamination
1873	Butschli	First detailed descriptions of free-living nematodes
1876	Leukart	Extensive treatise onhuman Separate on & human helminthology
1878	Manson	Mosquitoes as intermediate host of human filarial worm, *Wucheraria bacrofti* (elephentiasis)
1880	Oerley	Described 202 nematodes in 27 genera
1887-88	Grassi	Experimental proof of direct infections of man by *Ascaris* eggs
1896-1911	Looss	Direct skin penetration by *Ancylostoma duodenloe* and complete life cycle
1900	Goldchmidtt	Nervous system of nematodes
1913	Seurat	Insects as intermediate hosts for Physocephalus and Ascaris
1923 & 1929	Steiner	Described *S. kraussei* and *S. glaseri*
1925	Micoletzky	Sporozoan parasites of nematodes
1929	Hobmaier	Earthewrm as intermediate host for pig lung worm
1930	Filipjev	Basis for current classification of Nemata
1930	Paramanov	Nemic relationships, evolution and phylogenesis
1951	Goubenov	*Placentomema gigantisma*

II. PLANT PARASITES:

International:

1743	Needham	First plant parasitic nematode, wheat gall nematode, *Anguina tritici*
1855	Berkeley	Root-knot nematode in cucumber
1857	Kuhn	*Ditylenchus dipsaci* infesting teasel heads
1859	Schacht	Sugar beet cyst nematode
1865	Bastien	Monograph on Anguillilidae

1871	Kuhn	CS_2 to control sugar beet cyst nematode
1873	Butscheli	Drawings and keys for identification
1884	De Man	Taxonomic monograph of soil and fresh water nematodes of the Netherlands. De man's formula for measurements
1884	Ritzemaboss	Biological races of *D. dipsaci*
1889	Neal	First record of root-knot nematode in USA
1891	Ritzemaboss	First foliar nematodes on strawberry, *Aphelenchoides frageriae*
1892	Atkinson	Association of root-knot with *Fusarium* in vascular wilt
1901	Hinger	Association of root-knot with bacterial wilt of tomato
1859-1932	N.A.Cobb	Coined the word 'Nema', described more than 100 nematodes. Father of American Nematology
1883-1921	Micoletzky	Monogram on free living soil nematodes
1889-1937	Filipjev	Manual of Agricultural Helminthology
1885-1953	T.Goodey	Monograph on soil and fresh water nematode, catalogued nematode diseases Helminthological Society of Washington started
1943	Carter	Nematological property of DD
1945	Christie	Nematological property of EDB
1949	Chitwood	Perineal pattern technique to differentiate species of *Meloidogyne*
1951	Christie and Perry	Pathogenicity of *Trichodorus christiei*
		First International Nematology Symposium, Rothemsted
1953	Mc Beth	Nematological property of DBCP
		European Society of Nematologists- First Journal, Nematologica Hewitt, Raski & Goheen
		Transmission of fan leaf of grapes virus by *Xiphinema index*
		Society of Nematologists (SON), USA
		The Organization of Tropical American Nematologists (OTAN)
		First issue of Journal of Nematology by SON
1973		First publication of *Nematologia Mediterannea* from Italy
1978		First Publication of *Revue de Nematologie* from France

B. India:

Even though the first ever authentic record of a phytonematode was made in 1901 by Barber with the discovery of root-knot nematode associated with tea, the discipline of Nematology can be considered to be only around 40 years old in India. The organized research on plant parasitic nematodes by full time researchers started towards the end of the fifties and early sixties. This organized research could be initiated only in 1958 with the identification of Molya disease of wheat and barley in Rajasthan State. The interest in the nematological research in the country increased with more and more researchers getting involved full time in this discipline. This was possible due to the training of several scientists/workers in Nematology within the country as well as abroad though short training courses or postgraduate/doctoral programmes.

Following are major contributions made in the field of Nematology in India after the first report by Barber in 1901.

Sl.No.	Contributor & Year	Contribution
1.	Butler, 1906	First report of Root-knot nematode on black pepper- Kerala
2.	Butler, 1913	First report of *Ditylenchus angustus* on rice-Bengal
3.	Hutchinson, 1917	First report of Tundu disease of wheat, *Anguina tritici*
4.	Ayyar, 1926,1933,1934	First report of root-knot nematode on vegetable & other crops –South India
5.	Dastur, 1936	First report of *Aphelenchoides besseyi*-Central province
6.	Thrumala Rao, 1956	First report of Root-knot nematode on Citrus
7.	Prasad, Mathur and Sehgal, 1959	First report of Heterodera avenae
8.	Siddiqui, 1961	Citrus Nematode, *Tylenchulus semipenetrans*
9.	Jones, 1961	*Globodera rostochiensis*-Nilgiris, Tamil Nadu
10.	Nair, Das and Menon, 1966	First report of *Radopholus similis* on Banana from Kerala
11.	Edward, J.C. and Misra, S.L., 1970	First Text Book on Nematology-"An Introduction to Plant Nematology"

Major Nematological Accomplishments in India:

- **1960:** Research initiation on nematodes of rice at the Central Rice Research Institute (CRRI), Cuttack
- **1961:** Section of Nematology established at the Department of Entomology, Agricultural College & Research Institute, Coimbatore
- **1961:** Nematology Units started functioning in the Divisions of Entomology/Plant Pathology at IARI, New Delhi
- **1961:** Nematoloy Units established at the Central Potato Research Institute (CPRI), Simla
- **1962:** Nematology Scheme at the College of Agriculture, Kanpur
- **1962:** ICAR Scheme on potato cyst nematode at Nilgiris, Tamil Nadu
- **1963:** Coordinated Scheme on Nematodes infesting cotton (ICAR) at Coimbatore and Bangalore
- **1964:** Research on Plant Parasitic Nematodes started at the G.B.Pant University of Agriculture & Technology, Pantnagar
- **1964:** Quarantine detection of potato cyst nematode in certified seed consignments from Scotland (100 tons)
- **1964:** FAO & Rockfeller Foundation organized an International Nematology Course at Indian Agricultural Research Institute (IARI), New Delhi under the supervision and guidance of Prof. D.J.Raski from California, USA, Dr.F.G.W. Jones and Dr.J.B.Goody from UK and many senior Nematologists from India
- **1965:** Coordinated Scheme on Nematodes of Fruits, Vegetables and Other Crops (ICAR) at Coimbatore, Hyderabad, Hisar and Aligarh
- **1966:** The establishment of full fledged Division of Nematology at IARI, New Delhi
- **1966-67:** Nematology units/departments started functioning in State Agricultural Universities and Central Agricultural Research Institutes
- **1967-68:** First South-East Asia Post-graduate Nematology courses of three months duration jointly organized by IARI, New Delhi and the Aligarh Muslim University, Aligarh, UP with the financial

support from the International Agricultural Centre, Wageningen (Netherlands), under the supervision of Dr.M.Oostenbrink and his colleagues

- **1968:** Initiation of full-fledged Post graduate teaching programme leading to M.Sc. and Ph.D degree Nematology at IARI, New Delhi and also at other Institutions and State Agricultural Universities
- **1968:** Nematology Unit set up in Department of Entomology at CRRI, Cuttack under PL-480 grant
- **1968:** Nematology Unit established in the Plant Quarantine Section, Division of Plant Introduction, IARI, New Delhi (upgraded in 1976 as the National Bureau of Plant Genetic Resources)
- **1968-69:** Second South-East Asia Nematology Course at Aligarh Muslim University, Aligarh and IARI, New Delhi
- **1969:** First All India Nematology Symposium at IARI, New Delhi
- **1969-70:** Third South-East Asia Nematology course at IARI, New Delhi
- **1970:** Nematology Unit established at the Indian Institute of Horticultural Research, Hessaraghatta, Bangalore
- **1971:** Establishment of Nematological Society of India (NSI) and publication of Indian Journal of Nematology (IJN) at IARI, New Delhi
- **1971:** First All India Nematology Workshop at IARI, New Delhi
- **1971:** Fourth South-East Asia Nematology course at IARI, New Delhi
- **1972:** Fifth South-East Asia Nematology course at AMU, Aligarh
- **1972:** Division of Nematology established at CPCRI, Regional Station, Kayangulum, mainly to work on *Radopholus similis* infesting coconut
- **1972-79:** UNDP Project on Plant Protection to strengthen Nematology at University of Agricultural Sciences, Bangalore
- **1973:** All India Short Term Course on Recent Advances in Nematological Techniques and control at UAS, Bangalore
- **1975:** Sixth South-East Asia Nematology course at IARI, New Delhi
- **1976:** Summer Institute on Phytonematology, Allahabad
- **1977:** Establishment of All India Coordinated Research Project on Nematode Pests of crops and their Management (AICRP) by Indian Council of Agricultural Research (ICAR), with its head quarter at IARI, New Delhi
- **1979:** Seventh South-East Asia Nematology Course at IARI, New Delhi
- **1979:** All India Nematology Workshop (AICRP) and Nematology Symposium at OUAT, Bhubaneswar, Orissa
- **1981:** Department of Nematology established at Tamil Nadu Agricultural University (TNAU), Coimbatore
- **1981:** Nematology Section established at CPCRI, Regional Research Station, Appangala, Karnataka for studies on nematodes infesting cardamom
- **1981:** All India Nematology Workshop (AICRP) and Nematology Symposium at TNAU, Coimbatore
- **1983:** All India Nematology Workshop (AICRP) and Nematology Symposium at HPKVV, Solan
- **1985:** All India Nematology Workshop (AICRP) and Nematology Symposium at Mohanlal Sukhada University, Udaipur, Rajasthan
- **1985:** Nematology Unit (Plant Quarantine Regional Station, NBPGR) started functioning at Hyderabad
- **1985:** Separate Scientific Panel fro Nematology constituted by ICAR

- **1986:** National Nematology Conference at IARI, New Delhi
- **1986:** Summer institute on "Nematode Diseases Problems and Their Management" at IARI, New Delhi
- **1986:** UAS, Bangalore was selected as a Centre of AICRP (Nematodes)
- **1987:** Group Discussions on "Nematological Problems of Plantation Crops" held at Sugarcane Breeding Institute, Coimbatore, Tamil Nadu
- **1987:** All India Nematology Workshop at Government Agricultural College, Pune, Maharashtra
- **1992:** Silver Jubilee Celebration of Division of Nematology, IARI, New Delhi
- **1992:** All India Nematology Workshop (AICRP) and Nematology Symposium at HAU, Hisar
- **1992:** Summer Institute on "Management of Plant Parasitic Nematodes in Different Crops" organized by ICAR, New Delhi at Haryana Agricultural University, Hisar Phytonematology" at Haryana Agricultural University, Hisar
- **1995:** All India Nematology Workshop (AICRP) and Nematology Symposium (Silver Jubilee Celebration of NSI) at IARI, New Delhi
- **1997:** Summer Institute on " Problems and Progress in Nematology during the past One decade" at IARI, New Delhi
- **1997:** National Nematology Symposium at GAU, Anand
- **1997:** Summer School on " Problems and Progress Nematology during the past one decade" at IARI, New Delhi
- **1997:** Biennial Group Meeting of AICRP (Nematodes) at University of Agricultural Sciences, Bangalore
- **1998:** Afro-Asian Nematology Conference at Coimbatore, Tamil Nadu
- **1999:** National Seminar on "Nematological Research in India :Challenges and Preparedness for the new millennium" and Biennial Group

Meeting of AICRP (Nematodes) at CSA University of Agriculture and Technology, Kanpur

- **1999:** Short-Term Refresher Course in Plant Nematology for the Plant Protection Officers, organized by Department of Plant Quarantine, Govt. of India
- **2000:** National Nematology Symposium on "Integrated Nematode Management" and Annual group Meeting of AICRP (Nematodes) at OUAT, Bhubaneswar
- **2000:** Winter School on "Physiological and Molecular Nematology" at IARI, New Delhi
- **2001:** National Congress on "Centenary of Nematology in India : Appraisal and Future Plans" and Biennial Group Meeting of the AICRP (Nematodes) at IARI, New Delhi

MULTIPLE CHOICE

I.

1. The number of lateral sectors present in a nematode body
   a. One b. Three c. Two d. Four
2. These are thickened wing-like extensions of cuticle
   a. Alae b. Papillae c. Rays d. Ocelli
3. Trilaminar outermost part of the cuticle, first layer to be laid down during moulting is known as
   a. Endocuticle b. Ectocuticle c. Epicuticle d. Mesocuticle
4. Adults and juveniles have the same cuticle structure in this family
   a. Longidoridae b. Trichodoridae c. Tylenchomorpha d. Heteroderidae
5. This number of layer of obliquely orientated and longitudinally aligned somatic muscle cells lies beneath the epidermis
   a. Single b. Double c. Triple d. Non-specific
6. The shape of the somatic muscle cell is mainly
   a. Spherical b. Spindle c. Rectangular d. Bulbous
7. The psuedocoelom contains some large amoeboid cells called
   a. Pseudocytes b. Pseudocoelomocytes c. Pseudocoelocytes d. Pseudocoelomocells
8. Nematode lips are of these patterns
   a. Two sub-dorsal, two sub-ventral and two lateral lips
   b. Two dorsal, two ventral and two lateral
   c. Two sub-dorsal, two ventral and two lateral
   d. Two dorsal, two sub-ventral and two lateral
9. In nematodes, number of primitively labial sensilla is
   a. Twelve b. Ten c. Six d. Fifteen
10. The aperture, the fovea, the canalis and the fusus or sensillar pouch are the components of
    a. Phasmid b. Amphid c. Deirid d. Hemizonid
11. Longidorids possess this number of stylet protractor muscles
    a. Eight b. Seven c. Six d. Four
12. In most virus vector trichodorid species, the entire lining of this organ acts as a retention site for tobraviruses
    a. Intestine b. Pharynx c. Mouth d. Stylet knobs
13. The intestine is entirely of this origin.
    a. Endodermal b. Ectodermal c. Mesodermal d. Epidermal
14. In *Meloidogyne*, these six large glands produce the gelatinous matrix in which the eggs are deposited
    a. Ventral oesophageal b. Dorsal oesophageal c. Rectal d. Lateral oesophageal
15. Germinal, telogonic and growth zones are connected with this organ of nematode
    a. Ovary b. Testes c. Vulva d. Intestine

16. The Z-organ is a characteristic feature of this genus
    a. *Radopholus* b. *Heterodera* c. *Globodera* d. *Xiphinema*
17. This layer in the egg is largely responsible for the impermeability of the eggshell
    a. Lipid layer b. Vitelline layer c. Chitinous layer d. Proteinaceous layer
18. Sensilla and dendrites are parts of this nematode organ
    a. Bursa b. Spicule c. Spermatheca d. Gubernaculum
19. Nematode sperm cells show amoeboid motility and are
    a. Non-flagellate, non-ciliate b. Flagellate and ciliate
    c. Flagellate and non-ciliate d. Non-flagellate and ciliate
20. Certain species of this genus have only three moults
    a. *Xiphinema* b. *Longidorus* c. *Helicotylecnhus* d. *Dolicodorus*
21. The majority of ectoparasitic nematodes remain motile whereas this genus becomes permanently attached to the root by the deeply embedded stylet
    a. *Hoplolaimus* b. *Xiphinema* c. *Cacopaurus* d. *Hemicycliophora*
22. In this genus, all juvenile stages, the male and the immature vermiform female are migratory, only the mature female being sedentary.
    a. *Nacobbus* b. *Meloidodera* c. *Meloidogyne* d. *Globodera*
23. The morphometric abbreviation, 'c' refers to
    a. Total body length divided by tail length
    b. Total body length divided by pharyngeal length
    c. Total body length divided by maximum body diameter
    d. Total body length
24. PCR is carried out for this number of cycles.
    a. 60-80 b. 10-20 c. 30-40 d. 0-10
25. *Ditylenchus dipsaci*, occurs as more than this number of biological races
    a. 20 b. 10 c. 31 d. 42
26. *Subanguina radicicola* causes this symptoms on roots of host plants
    a. Lesions b. Knots c. Cysts d. Stubby roots
27. This nematode deposits its eggs within the host
    a. *Meloidogyne graminicola* b. *M. incognita*
    c. *M. hapla* d. *M. indica*
28. This is frequently considered as being the most important factor for attracting phytonematodes.
    a. Ammonia b. Potassium c. Carbon dioxide d. Oxygen
29. Giant cells are most likely formed after repeated
    a. Endomitosis without cytokinesis b. Endo meiosis without cytokinesis
    c. Mitosis without cytokinesis d. Ectomitosis with cytokinesis
30. This nematode has only one generation per year
    a. *Meloidogyne arenaria* b. *M. naasi*
    c. *M. exigua* d. *M. batatiformis*

31. *Meloidogyne* species are active in soils with the following moisture field capacity
 a. 40-60% b. 10-20% c. 20-30% d. 30-40%
32. The majority of species of *Meloidogyne* reproduce by this method
 a. Bisexual b. Hermophroditism c. Parthenogenesis d. Apomixis
33. The majority of *Meloidogyne incognita* have 2n = this number of chromosomes
 a. 28-30 b. 12-14 c. 42–44 d. 35-38
34. In the following nematode, the eggshell is covered with small punctuations that resemble microvilli
 a. *Cactodera cacti* b. *Anguina tritici* c. *Pratylenchus* d. *Paratylenchus*
35. The thin-walled area on the vulval cone of a cyst nematode is known as
 a. Alae b. Papillae c. Ray d. Fenestra
36. *Globodera palida* can attack this crop also
 a. Chillli b. Tomato c. Okra d. Brinjal
37. *Pratylenchus brachyurus* is also widely associated with this crop
 a. Black pepper b. Tea c. Citrus d. Banana
38. *Radopholus bridge* infects this crop
 a. Turmeric b. Cardamom c. Coffee d. Cocoa
39. The burrowing nematode is more pathogenic to citrus in this soil type
 a. Clay b. Silt c. Sandy d. Loamy
40. 'Pisang Jari Buaya' and 'Yangambi km5' are two resistance sources for *R. similis* in this crop
 a. Citrus b. Banana c. Coffee d. Black pepper
41. *Ditylenchus africanus* is parasitic on this crop
 a. Onion b. Ground nut c. Garlic d. Cotton
42. *Ditylenchus* belongs to this subfamily
 a. Anguininae b. Radopholinae c. Ditylenchinae d. Heteroderinae
43. Heat-relaxed males of this genus assume a 'walking-stick' shape with the tail region curled ventrally
 a. *Hoplolaimus* b. *Helicotylenchus* c. *Aphelenchoides* d. *Xiphinema*
44. This nematode interacts with Xanthomonas begonia on Begonia
 a. *Aphelenchoides ritzemabosi* b. *A. frageriae*
 c. *A. besseyi* d. *A. begonia*
45. *Hoplolaimus and Helicotylenchus* come under this group
 a. Semi-ectoparasites b. Endoparasites
 c. Migratory endo d. Sedentary endo
46. *Criconemoides ornate* severly affects this crop
 a. Papaya b. Citrus c. Ground nut d. Corn
47. Meiotic parthenogenesis is also known as
 a. Apomixis b. Automixis c. Epimixis d. Epomixis

48. This is a type of hermaphroditism where sperm penetration activates oocyte development but the nuclei do not fuse.

a. Pseudophroidy b. Pseudogamy c. Pseudogymy d. Pseudotracy

49. During adverse conditions, the shed cuticles of this nematode are retained as sheaths to aid survival

a. *Rotylenchulus reniformis* b. *Meloidogyne hapla*

c. *M. arenaria* d. *Pratylenchus coffeae*

50. Lysine and phenylalanine are synthesized by this nematode

a. *Aphelenchoides* b. *Heterodera* c. *Meloidogyne* d. *Radopholus*

ANSWER KEY FOR I

Sl.No.	Ans.	Sl.No.	Ans.	Sl.No.	Ans.	Sl.No.	Ans.	Sl.No.	Ans.
1	c	11	a	21	c	31	a	41	b
2	a	12	b	22	a	32	c	42	a
3	c	13	a	23	a	33	c	43	c
4	a	14	c	24	c	34	a	44	a
5	a	15	a	25	a	35	d	45	a
6	a	16	d	26	b	36	b	46	c
7	a	17	a	27	a	37	c	47	b
8	a	18	b	28	c	38	a	48	b
9	a	19	a	29	a	39	c	49	a
10	b	20	a	30	b	40	b	50	a

II.

1. Nematodes possess a wide variety of sterols but cannot biosynthesize them de novo. These sterols are
   a. Tricyclic polyisoprenoids b. Tetracyclic polyisoprenoids
   c. Monocyclic polyisoprenoids d. Dicyclic polyisoprenoids
2. Chitin, a polymer of the following, is a major component of nematode eggshells
   a. *N*-acetylglucosamine b. Trehalose 6-phosphate synthetase
   c. 6-phosphate phosphatase d.Disaccharide trehalose
3. Male nematodes are heterogametic and are
   a. Sometimes XO, usually XY b. Always XO.
   c. Always XY d. Always XX
4. The chromosome number (haploid) in *Xiphinema index* is
   a. 8 b. 10 c. 6 d. 14
5. The major type of reproduction in *Pratylenchus brachyurus* is
   a. Amphimixis b. Parthenogenesis
   c. Mitotic parthenogenesis d.Meiotic parthenogenesis
6. The anal–lumbar commissures that link the pre-anal and lumbar ganglia are termed
   a. Caudalids b. Cephalids c. Cheilostomes d. Alae
7. Sex chromosomes are absent in the following genus and the sex ratio may be influenced by environmental factors, such as food availability and temperature.
   a. *Radopholus* b. *Globodra* c. *Anguina* d. *Hirschmaniella*
8. In following genus, the female nematode can be fertilized by several males, thus enhancing the genetic diversity of the offspring
   a. *Radopholus* b. *Meloidogyne* c. *Paratylenchus* d. *Trichodorus*
9. This genus exhibits both mitotic and meiotic parthenogenesis
   a. *Xiphinema* b. *Crconema* c. *Pratylenchus* d. *Helicotylenchus*
10. There are four bands of muscles in Nematodes, two are dorso-lateral and two
    a. Ventro-lateral b. Dorso-ventral c. Dorso-lateral d. Latero-ventral
11. Digestive system of nematodes have two openings and in form of
    a. Digestive tract b. Alimentary canal c. Neuron d. Muscules
12. The term 'Nematoda' actually refers to
    a. Pointed cells b. Pointed bodies c. Pointed ends d. Peaked heads
13. Undulating waves of contraction and relaxation of muscles helps nematodes in
    a. Excretion b. Digestion c. Locomotion d. Respiration
14. The nematode body has this number of lateral sectors.
    a. One b. Three c. Two d. Four
15. In addition to chemoreceptors and light receptors, this group of sense organ is also found in nematodes
    a. Tactoreceptors b. Hygroreceptors c. Textoreceptors d. Florareceptors

16. In the class Secernentea, amphidial openings are located in this region
    a. Oesophagous b. Ovary c. Lip d. Tail
17. Caudal papillae are normally situated at the
    a. Ventral side of male b. Dorsal side of male
    c. Dorsal side of female d. Lateral side of male
18. Ducts embedded in the lateral hypodermal cords are referred to as
    a. Excretory canals b. Oesophageal canals c. Hypodermal canals d. Nerve canals
19. The hollow stylet in most tylenchids exhibit this symmetry
    a. Diradiate b. Triradiate c. Monoradiate d. Polyradiate
20. The intestine in nematode is tube like and consists of this number of cell layers
    a. Four b. One c. Two d. Six
21. Most phytonematodes have this number of ovary and testis
    a. 2 ovaries, 1 testis b. 2 ovaries, 2 testes c. 1 ovary, 1 testis d. 1 ovary, 2 testes
22. The most common habitat of the vinegar eel worm, Turbatrix aceti is
    a. Fresh water b. Vinegar c. Marine d. Salt water
23. This nematode lives as an ectoparasite in the beginning of its life cycle, which later turns into endoparasitic
    a. *Radopholus similis* b. *Anguina tritici*
    c. *Meloidogyne enterolobii* d. *M.exigua*
24. For the extraction of the burrowing nematode from soil samples, optimum soil depth is up to
    a. 15 cm b. 25 cm c. 2 cm d. 35 cm
25. Basically soybean cyst nematode prefers this climate
    a. Tropical b. Temperate c. Subtropical d. a & c
26. In the plant damage function model for relating relative yield to the initial nematode population density $Y=M+(1-M)\ Z^{P-t}$, M stands for
    a. Maximum yield b. Minimum yield c. Medium density d. Maximum density
27. Ecdysteroids are polyhydroxylated derivatives of cholesterol containing a C-7double bond and a
    a. keto group at C-7 b. keto group at C-4 c. keto group at C-6 d. keto group at C-1
28. This organ is often referred to as sphincter
    a. Intestino-rectal valve b. Intestino-oesosphageall valve
    c. Testes-rectal valve d. Ovary-rectal valve
29. The cells produced by cleavage of the egg are referred to as
    a. Clastomeres b. Cleavomeres c. Stomatomeres d. Blastomeres
30. Avermectins are macrocyclic lactones derived from this part of Streptomyces avermitilis
    a. Mycelia b. Spores c. Cell walls d. Germ tube
31. This organ controls the passage of eggs from uterus to avoid blockage
    a. Oviduct b. Ovijector c. Microvilli d. Ovicardia

32. Males of this nematode release a pheromone in leaf tissue
    a. *Ditylenchus dipsaci* b. *Aphelenchoides ritzemabosi*
    c. *Aphelenchoides besseyi* d. *Ditylenchus angustus*
33. Generally chemoreceptors are present in this part of a nematode body
    a. Anterior b. Oesophageal glands c. Posterior d. Intestine
34. The somatic musculature in the early stage of the first stage juvenile is
    a. Platymyarian , circomyarian b. Platymyarian , celomyarian
    c. Platymyarian , meromyarian d. Polymyarian , meromyarian
35. During egg hatching process in some nematodes, the activity of this enzyme increases
    a. Leucine aminopeptidase b. Prolyl 4-hydroxylase
    c. Serine protease d. Zinc metalloproteases
36. The major feature in the hatching process of some nematodes is the
    a. Increase in the water content of juveniles b. Decrease in the water content of juveniles
    c. Increase in the temperature within the egg d. Decrease in the temperature within the egg
37. The H-shaped, unicellular muscle that controls the passage of faeces in nematodes is termed
    a. Supressor-ani muscle b. Progressive-ani muscle
    c. Depressor-ani muscle d. Excretory-ani muscle
38. This is the natural hatching factor for Heterodera glycines
    a. Glycinoeclepin -A b. Glycinolupin -A c. Glycinoeclepin-B d. Glycinoeclepin AB
39. The site of initial giant cell stimulation in sweet potato due to infection by root-knot nematodes is
    a. Vascular meristematic cells b. Vascular parenchyma cells
    c. Cortical tissues d. Epidermal tissues
40. Immediately after the insertion of stylet into a living cell, the cell wall becomes
    a. Thick b. Thin c. Multilayered d. Collapses
41. In juveniles of *Meloidogyne javanica*, this organ functions as receptor for hatching stimuli
    a. Amphid b. Phasmid c. Hemizonid d. Rectal gland
42. $Y=B_0 + B_1X_1 +$ Error refers to this model
    a. The Linear Regression Model b. The Multiple Point Model
    c. Duncan and Ferris Model d. The Scienhorst Model
43. There are many generations of root-knot nematodes per year, but in this species there is only one
    a. *Meloidogyne arenaria* b. *M. naasi*
    c. *M. exigua* d. *M. batatiformis*
44. Normally spermatozoa are stored in this structure in nematodes
    a. Seminal vesicle b. Spicule c. Uterus d. Gubernaculum
45. Farnesyl methyl ether influences molting in nematodes and is a
    a. Pheromone b. Fumigant c. Juvenile hormone d. Organophosphate
46. Modified host cells by *Nacobbus* spp., are referred to as
    a. Syncytia b. Giant cells c. Nurse cells d. Mother cells

47. Trichodorids can remove cytoplasm containing larger organelles because
    a. The wide lumen of the feeding tube
    b. The narrow lumen of the feeding tube
    c. Specialized oesophageal secretions
    d. Stout stylets
48. The galls formed at the root tips of rose by this nematode contain groups of highly active, enlarged multinucleate cells
    a. *Meloidogyne hapla*
    b. *M. arenaria*
    c. *M. brevicauda*
    d. *Xiphinema index*
49. This acts as a chemical stimulus to repel second stage juveniles of root-knot nematode
    a. Water
    b. Salt
    c. Sugar
    d. pH
50. Infective juveniles of cyst nematodes penetrate host in this manner
    a. Intercellular
    b. Stomatal opening
    c. Intracellular
    d. Extracellular

ANSWER KEY FOR II

Sl.No.	Ans.	Sl.No.	Ans.	Sl.No.	Ans.	Sl.No.	Ans.	Sl.No.	Ans.
1	b	11	b	21	a	31	b	41	c
2	a	12	c	22	a	32	a	42	a
3	a	13	c	23	b	33	a	43	b
4	b	14	c	24	a	34	c	44	a
5	c	15	a	25	b	35	a	45	c
6	a	16	c	26	b	36	a	46	a
7	b	17	a	27	c	37	c	47	a
8	a	18	a	28	a	38	a	48	d
9	c	19	b	29	d	39	b	49	b
10	a	20	b	30	a	40	a	50	c

III.

1. These novel compounds inhibit sterol metabolism in nematodes, hence can be used against nematodes
   a. Azasteroids b. Sterolins c. Chitosterol d. Azastrobins
2. A lipid layer is present in the egg shell of this nematode
   a. *Pratylenchus* b. *Radopholus* c. *Globodera* d. *Anguina*
3. One of the major changes found in the giant cell incited in tomato by Meloidogyne spp., is
   a. Numerous small vacuoles are displaced by a large vacuole
   b. Large vacuole is displaced by numerous small vacuoles
   c. Cytoplasm becomes non-granular
   d. Decrease in the number of cytoplasmic granules
4. This is the longest part in the male reproductive system
   a. Vas deferens b. Oocytes c. Sperm tail d. Testis
5. This nematode stage has maximum pheromone reception potential
   a. Adult male b. All juvenile stages c. Adult female d. Fourth stage juvenile
6. Grapevine chrome mosaic virus is transmitted by
   a. *Xiphinema americanum* b. *X. diversicaudatum*
   c. *X. index* d. *X. italiae*
7. The members of the following super family may have a single spicule or no spicule
   a. Trichuroidea b. Creconemoidea c. Dolichodoroidea d. Aphanomoidea
8. Avermectins usually affect this activity of nematodes
   a. Egg hatching b. Molting c. Migration d. Host penetration
9. One of the best characterized stimuli for nematodes is
   a. pH b. Soil solution c. Temperature d. Nitrogen in soil
10. In most nematodes, usually this stage releases pheromones
    a. Adult male b. Adult female
    c. Second stage juvenile d. All juvenile stages
11. The most competitive and toughest interaction among the following is between
    a. Sedentary endo & migratory ectoparasites b. Sedentary endo & sedentary endoparasites
    c. Migratory ecto & migratory endoparsites d. Migratory ecto & sedentary ectoparasites
12. Pheromone production in nematodes is normally maximum at this pH level
    a. 6.0 b. 7.0 c. 8.5 d. 3.0-5.0
13. The formula $dN / dt = rN \{(K-N)/K\}$ is related to this aspect of nematodes
    a. Population dynamics b. Ecology c. Epidemiology d. Damage model
14. This is the most commonly used nematode in laboratory for the bioassay studies
    a. *Seinura tempes* b. *Panagrellus redivivus*
    c. *Mononchus glidis* d. *Anguina agrostis*

15. *Ancyclostoma caninum* is a major nematode pest on
    a. Dog b. Whale c. Cat d. Horse
16. The cuticular thickening on the ventral wall of some of the strongylids is known as
    a. Telamon b. Guiding sheath c. X-organ d. Temolan
17. This is known as the 'age pigment' in nematode senescence
    a. Lipfuchcin b. Lipofuscin c. Lipomelanin d. Liponambin
18. POTATOPEST, a simulator model is concerned with this nematode pest on potato
    a. *Meloidogyne javanica* b. *M. hapla*
    c. *Pratylenchus penetrans* d. *M. indica*
19. The penetration of host by root-knot nematode juveniles is primarily
    a. Intercellular b. Intracellular c. Extracellular d. Hydathodes
20. The process of cell division in nematode embryology is known as
    a. Differentiation b. Cleavage c. Oogenesis d. Embryogenesis
21. Fluids within the eggs of *Globodera rostochiensis* contains
    a. Sucrose b. Galactose c. Trehalose d. Polysugars
22. Osmotic and ionic balance are important for the following genera, particularly for feeding and reproduction
    a. *Heterodera & Globodera* b. *Radopholus & Pratylenchus*
    c. *Trichodorus & Longidorus* d. *Hemicriconema & Criconemoides*
23. Somatic muscles in nematodes are arranged in longitudinal rows divided by the epidermal chords into the following number of symmetrical quadrants
    a. Two b. Three c. Four d. Six
24. Tryptophan is synthesized by this nematode
    a. *Pratylenchus* b. *Radopholus* c. *Meloidogyne* d. *Xiphinema*
25. The total lipid content of the plant-parasitic nematodes ranges from
    a. 11- 67% b. 5-25 % c. 26-48% d. 30-75%
26. Vaccenic acid, a major fatty acid group is found in this genus
    a. *Heterodera* b. *Ditylenchus* c. *Anguina* d. *Globodera*
27. Hatching in cyst nematodes is associated with the release of this compound
    a. Ribitol b. Inositol c. Glycerol, d. Trehalose
28. A decline in infectivity of Globodera rostochiensis has been correlated with a decline in the reserves of
    a. Lipid b. Carbohydrate c. Sugars d. Sterols
29. This genus retains its moulted cuticles to protect the infective stages
    a. *Radopholus similis* b. *Rotylenchulus reniformis*
    c. *Meloidogyne javanica* d. *M. hapla*
30. The first nematode resistance gene to be cloned, which confers resistance to Heterodera schachtii.
    a. *Hspto 1* b. *Hspro1* c. *Hspco 1* d. *Hspno 1*

31. The best-studied nematode resistance gene which confers resistance to three species of root-knot nematodes is

a. Tomato gene *Mj* b. Brinjal gene *Mi* c. Potato gene *Mi* d. Tomato gene *Mi*

32. The gene '*Hero*' *expresses* resistance to this genus

a. *Globodera* b. *Meloidogyne* c. *Heterodera* d. *Helicotylenchus*

33. Even at very low densities, this nematode can multiply fast enough to cause considerable damage very soon after the first infection

a. *Heterodera* b. *Meloidogyne* c. *Bursaphelenchus* d. *Pratylenchus*

34. This stage of juveniles of entomopathogenic nematodes attack insect larvae

a. J_2 b. J_3 c. J_4 d. J_1

35. Carrot disc culture is the best technique to culture this nematode

a. *Meloidogyne* b. *Heterodera* c. *Pratylenchus* d. *Ditylenchus*

36. These are fungivorous nematode genera

a. *Aphelenchus & Ditylenchus* b. *Aphelenchus & Nacobbus*
c. *Aphelenchoides & Tylenchorhynchus* d. *Aphelenchoides & Paratylenchus*

37. This nematode is of a major concern, particularly in South India

a. *Meloidogyne* b. *Heterodera* c. *Radopholus* d. *Pratylenchus*

38. Ocelli, in nematodes, generally act as

a. Light receptors b. Sound receptors c. Moisture receptors d. Host searching

39. Axial spear type is found in this genus

a. *Radopholus* b. *Meloidogyne* c. *Anguina* d. *Xiphinema*

40. Presence of a non- hollow mural tooth is a characteristic feature in

a. *Criconema* b. *Hemicycliophora* c. *Trichodorus* d. *Nacobbus*

41. The body cavity in this phylum is limited by a thin membrane devoid of nuclei, without any connection with epidermis

a. Gastrotricha b. Kinorhyncha c. Pseudocoelomata d. Rotifera

42. Body cavity lined by spaces within a compact peritoneum is termed

a. Schizocele b. Blastocele c. Epitheliocele d. Gymnocele

43. Common name of *Turbatrix aceti* is

a. *Dwarf nematode* b. *Vinegar nematode* c. *Stunt nematode* d. *Wool nematode*

44. Spring coiling movement is found in

a. *Helicotylenchus* b. *Seinura* c. *Mononchus* d. *Criconemoides*

45. Predatory nematode belongs to the order

a. Mononchida b. Dorylaimida c. Aphelenchida d. Tylenchida

46. Constricting ring are formed by this

a. *Dactylella bembicodes* b. D. *candida*
c. *Dactylaria keris* d. D. *anilae*

47. Tomato variety resistant to Meloidogyne species
    a. SL- 120 b. PNR-7 c. Hisar Lalit d. Pusa Rubi
48. Dogridge Grape root stock is resistant to this nematode
    a. Root knot nematode b. Dagger nematode c. Stunt nematode d. Lesion nematode
49. The citrus root stock resistant to Citrus nematode, *Tylenchulus semipenetrans*
    a. Coorg mandarin b. Nagpur lime c. Trifoliate orange d. Florida wonder
50. Barley varieties resistant to cereal cyst nematode, *Heterodera avenae*
    a. C-164 b. M-64 c. BH-75 d. Rajkiran

ANSWER KEY FOR III

Sl.No.	Ans.	Sl.No.	Ans.	Sl.No.	Ans.	Sl.No.	Ans.	Sl.No.	Ans.
1	a	11	b	21	c	31	d	41	d
2	c	12	a	22	a	32	a	42	a
3	b	13	a	23	c	33	c	43	b
4	a	14	b	24	c	34	b	44	a
5	a	15	a	25	a	35	c	45	a
6	c	16	a	26	b	36	a	46	b
7	a	17	b	27	d	37	c	47	c
8	a	18	c	28	a	38	a	48	b
9	c	19	a	29	b	39	d	49	c
10	b	20	b	30	b	40	c	50	d

IV.

1. This nematode generally inhabits soil at greater depths than most other nematodes
   a. *Meloidogyne javanica* b. *Heterodera avenae*
   c. *Anguina tritici* d. *Trichodorus christie*
2. Nematode movement is largely dependent on this
   a. Water b. pH c. Temperature d. Soil depth
3. Internal hydrostatic pressure in nematodes is important in this activity
   a. Respiration b. Circulation c. Movement d. Excretion
4. These phytonematodes can swim in water unlike most others
   a. *Meloidogyne & Heterodera* b. *Anguina & Trichodorus*
   c. *Globodera & Xiphinema* d. *Ditylenchus & Aphelenchoides*
5. Juveniles of this nematode can move towards heat produced by a light source
   a. *Nippostrongylus muris* b. *Placentonema*
   c. *Trichenella* d. *Toxoplasma*
6. Caudal alae may not be present in
   a. *Ditylenchus* b. *Hoplolaimus* c. *Pratylenchus* d. *Aphelenchoides*
7. Cuticular segmentation is very apparent in
   a. *Criconema* b. *Paratylenchus* c. *Meloidogyne* d. *Tylenchorhynchus*
8. The location of six lips in nematodes follows this pattern
   a. 2 sub dorsal, 2 lateral & 2 sub ventral b. 2 sub dorsal, 2 ventral & 2 sub lateral
   c. 2 sub ventral, 2 dorsal & 2 sub lateral d. 2 sub lateral, 2 ventral & 2 sub ventral
9. The anterior end of nematodes is generally of this type in basic symmetry
   a. Triradiate b. Hexaradiate c. Monoradiate d. Tetraradiate
10. Jute is a non-host for this serious nematode pest
    a. *Trichodrus christie* b. *Ditylenchus angustus*
    c. *Hirschmaniella oryzae* d. *Dolichodrus indicus*
11. *Hirschmaniella mangaloreansis* was reported on rice from
    a. West Bengal b. Tamil Nadu c. Orissa d. Karnataka
12. *Heterodera cajani* was reported for the first time from New Delhi, India on *Cajanus cajan* by
    a. Koshy (1967) b. Rao (1968) c. Prasad (1965) d. Siddiqui (1967)
13. This species of lesion nematode is widely distributed in chickpea growing areas
    a. *Pratylenchus zeae* b. *P. thornei* c. *P. coffeae* d. *P.mulchandi*
14. Root-knot nematode on potato tubers was reported for the first time from Shimla hills in 1951 by
    a. A.R. Seshadri b. C.L. Sethi c. M.J. Thirumalachar d. J.S. Gill
15. This the most common nematode infesting sugarcane
    a. *Helicotylenchus* b. *Tylenchorhynchus* c. *Meloidogyne* d. *Pratylenchus*

16. *Meloidogyne incognita* is responsible for breaking bacterial wilt resistance in this variety of brinjal
    a. Pusa Purple Cluster b. Pusa Purple Long c. Pusa Purple Round d. Pusa Green Long
17. This is commercially available egg parasite of phytonematodes
    a. *Pseudomonas fluorescens* b. *Pochonia chlamydosporia*
    c. *Purpureocillium lilacinum* d. *Pasteuria penetrans*
18. This is the major nematode pest on yam (Dioscorea spp.)
    a. *Hoplolaimus* b. *Hemicycliophora* c. *Trichodorus* d. *Scutellonema*
19. Egg-mass is found within the root tissue in this nematode
    a. *Meloidogyne graminicola* b. *M. hapla*
    c. *M. exigua* d. *M.africana*
20. 'Quarantine Act' has been imposed on this nematode in India
    a. *Radopholus similis* b. *Anguina tritici*
    c. *Globodera rostochiensis* d. *Heterodera avenae*
21. This nematode is yet to be reported from India
    a. *Criconema indica* b. *Rhadinaphelenchus cocophilus*
    c. *Scutellonema bradys* d. *Nacobbus spp.*
22. This is a common pest on palm trees
    a. *Radopholus similis* b. *Trichodorus christei*
    c. *Xiphinema index* d. *Dolichodorus spp.*
23. It is easy to culture this nematode on carrot discs and within the mesocarp of growing tender coconuts
    a. *Pratylenchus coffeae* b. *Rhadinaphelenchus cocophilus*
    c. *Radopholus similis* d. *Anguina tritici*
24. VTL-11and VTL-17 varieties of this crop are resistant to the burrowing nematode
    a. Coconut b. Arecanut c. Black pepper d. Citrus
25. *Radopholus similis* is economically important pest on this crop
    a. Betel vine b. Cardamom c. Yam d. Sweet potato
26. This nematode is of less importance on papaya
    a. *Meloidogyne javanica* b. *M. arenaria*
    c. *M. incognita* d. *Rotylenchulus reniformis*
27. Reniform nematode was reported on this crop for the first time from Hawaii
    a. Castor b. Cotton c. Black gram d. Cowpea
28. This is most commonly occurring spiral nematode in banana rhizosphere
    a. *Helicotylenchus dihystera* b. *H. multicinctus*
    c. *H. africanus* d. *H. erythinae*
29. *H. multicinctus* was reported for the first time in India during 1960 by Das from
    a. Kerala b. Hyderabad c. Kolkatta d. Haryana
30. Dry rot of yam is caused by
    a. *Scutellonema bradys* b. *P. coffeae*
    c. *Meloidogyne incognita* d. *Pratylenchus brachyurus*

31. Most economically important nematode pest on sweet potato is
    a. Potato rot nematode b. Reniform nematode c. Root-knot nematode d. Lesion nematode
32. *Rhyncohophorus palmorum* transmits this nematode
    a. *Radopholus similis* b. *Pratylenchus cocophilus*
    c. *Rhadinaphelenchus cocophilus* d. *Heterodera palida*
33. In India, root-knot nematodes were reported on jute for the first time in
    a. 1920 b. 1911 c. 1915 d. 1930
34. This is an example for a polyphagous nematode
    a. *Radopholus similis* b. *Heterodera avenae* c. *Xiphinema index* d. *Hoplolaimus indicus*
35. Most orders that contain endozoic fungal parasites of nematodes belong to
    a. Ascomycetes b. Phycomycetes c. Basidiomycetes d. Deuteromycetes
36. This is a predacious nematode genus
    a. *Caenorhabditis* b. *Tripyla* c. *Panagrellus* d. *Turbatrix*
37. This tardigrade can prey upon *Panagrellus redivivus*
    a. *Hypsibius myrops* b. *Discolaimus* c. *Butlerius* d. *Diplogasteer*
38. This is most commonly occurring nematode on coconut
    a. Lesion nematode b. Root-knot c. Spiral d. Burrowing
39. Posteriorly located guiding ring and the odotophore with a basal swelling is the characteristic feature of this genus
    a. *Longidorus* b. *Xiphinema* c. *Trichodorus* d. *Radopholus*
40. In De Man's formula of nematode measurements, 'c' stands for
    a. Cuticle length divided by body length b. Cuticle length
    c. Body length divided by tail length d. Body length divided by cuticle length
41. The outer surface of a nematode egg is generally
    a. Smooth b. With external ornamentation
    c. With Internal ornamentation d. With both external and internal ornamentations
42. A thin lipoidal membrane present within the nematode egg shell is termed
    a. Crystalline layer b. Vitelline layer
    c. Subvitelline layer d. Subcrystalline layer
43. *Tylenchorhynchus* female lays its eggs
    a. Within roots b. In gelatinous matrix c. In seed gall d. On root surface
44. The terms 'telogenic and hologenic' refers to this nematode organ
    a. Vulva b. Testis c. Ovary d. Spicule
45. In general, the size of males & females in nematodes
    a. Equal size b. Males are bigger
    c. Females are bigger d. May vary in different species
46. The circumoesophageal commissure is commonly called as
    a. Nerve ending b. Nerve ring c. Nerve cell d. a+b+c
47. Number of stylet knobs in nematodes is
    a. Two b. Four c. Not specific d. Three

48. A swollen portion of the oesophageal duct wwhich serves asa reservoir for oesophageal gland secretions is knowns
    a. Lumen b. Ampulla c. Renette d. Ray
49. Cardia refers to the
    a. Oesophago-intestinal valve b. Intestino-oesophageal valve
    c. Ovary duct d. Sperm duct
50. The secretions from this organ aids in the nematode moulting process
    a. Pseudocoelm b. Hypodermis c. Oesophagous d. Intestine

ANSWER KEY FOR IV

Sl.No.	Ans.	Sl.No.	Ans.	Sl.No.	Ans.	Sl.No.	Ans.	Sl.No.	Ans.
1	d	11	d	21	b	31	d	41	a
2	a	12	a	22	a	32	c	42	b
3	c	13	b	23	c	33	b	43	d
4	d	14	c	24	b	34	d	44	c
5	a	15	d	25	a	35	b	45	c
6	d	16	a	26	b	36	a	46	b
7	a	17	c	27	d	37	a	47	d
8	a	18	d	28	b	38	d	48	b
9	b	19	a	29	b	39	b	49	a
10	b	20	c	30	a	40	c	50	b

V.

1. The position of three oesophageal glands in nematodes is
   a. One dorsal, two subventral b. One ventral, two subdorsal
   c. One dorsal two lateral d. One dorsal two sublateral
2. Cuticle in nematodes is secreted by the
   a. Pseudocoelom b. Hypodermis c. Oessophagous d. Intestine
3. While developing the damage model for nematode diseases, '*Pi*' indicates the
   a. Initial population b. Population rate c. Population density d. internal population
4. While developing the damage model for nematode diseases, '*m*' indicates the
   a. Moisture b. Movement c. Maximum yield d. Minimum yield
5. While developing the damage model for nematode diseases, '*T*' indicates the
   a. Tolerance limit b. Minimum temperature
   c. Maximum temperatured. Threshold
6. In population dynamics, 'nematodes which grow large and have long life cycles with low rates of population increase' are referred to as
   a. K strategists b. 'r' strategists c. 'N' strategists d. 'L' strategists
7. The corky ring spot disease of potatoes is transmitted by
   a. *Longidorus elongatus* b. *Xiphinema index*
   c. *X. americanum* d. *Trichodorus chrestiei*
8. In males of the following genus, the stylet and oesophagous are often not as well developed as in the females
   a. *Paratylenchus* b. *Radopholus* c. *Helicotylenchus* d. *Pratylenchus*
9. The sperm producing organ is termed
   a. Spermatheca b. Spermagonium c. Spermatogonium d. Spermogam
10. The duration of life cycle of Aphelnchids is
    a. 7-10 days b. 15-20 days c. 30-35 days d. 40-60 days
11. This nematode feeds upon germinating seedlings causing pre-emergence death of seedlings
    a. *Meloidogyne* b. *Belonolaimus* c. *Anguina* d. *Heterodera*
12. This is an example for a migratory semi-endoparasite
    a. *Helicotylenchus* b. *Pratylenchus* c. *Hirschmaniella* d. *Paratylenchus*
13. The cuticle in nematode basically consists of
    a. Single layer b. Multiple layers c. Three layers d. Two layers
14. Following are responsible in the same order to prove that *Xiphinema index* transmits grapevine fan leaf virus
    a. Hewitt, Goheen and Cobb b. Hewitt, Raski and Cobb
    c. Hewitt, Raski and Goheen d. Hewitt, Cobb and Goheen

15. The presence of absence of this feature differentiates tylenchids from aphelenchids
    a. Phasmid b. Bursa
    c. Gubernaculum d. Dorsal oesophageal gland opening
16. Oesophagous in phytonematodes consists of
    a. Three parts b. Two parts c. Four parts d. Not specific
17. Nematodes are placed in this kingdom
    a. Eukaryota b. Prokaryota c. Animalia d. Zoomonia
18. At this pH, maximum hatching of nematode eggs takes place
    a. 7 b. 4-6 c. Less than 4 d. More than 7
19. Common name of *Dolichodorus* is
    a. Stunt nematode b. Awl nematode c. Cyst nematode d. Dagger nematode
20. Phytonematodes are generally confined to this top inch of soil
    a. 10 cm b. 5 cm c. 25 cm d. 40 cm
21. The perineal pattern of Meloidogyne are actually the modifications of
    a. Longitudinal markings b. Transverse markings
    c. Punctations d. Alae
22. The intestine in phytonematodes is referred to
    a. Heterocytous b. Myriocytous c. Oligocytous d. Polycytous
23. The oesophageal lumen in nematodes is of
    a. Radial symmetry b. Bilateral symmetry c. Triradiate symmetry d. Asymmetry
24. First molt in most nematodes occurs
    a. During hatching b. Before hatching
    c. After hatching d. After or during hatching
25. First Nematologist to propose that nematodes belong to the phylum Nemata
    a. J.N.Saser b. N.A.Cobb c. T.Goodey d. Needham
26. This is a typical parthenogenetic nematode genus
    a. *Heterodera* b. *Anguina* c. *Helicotylenchus* d. *Trichodorus*
27. These structures enhance the strength of cuticle
    a. Muscles b. Longitudinal markings
    c. Transeverse markings d. Punctations
28. This nematode reproduces once in a year
    a. *Pratylenchus coffeae* b. *Scutellonema brachyurus*
    c. *Xiphinema diversicaudatum* d. *Rhadinaphelenchus cocophilus*
29. The reproductive system in male nematodes joins the rectum to form a tube known as
    a. Anus b. Cloaca c. Microvilli d. Bursa
30. Rice root nematode is a
    a. Migratory ectoparasite b. Migratory endoparasite
    c. Sedentary endoparasite d. Sedentary enctoparasite

31. In general, soil and phytonematodes can withstand pressure of
    a. 10 atmosphere b. 4 c. 15 d. 2
32. This tool is normally employed to collect soil samples for nematode population estimation
    a. Hand showel b. Hand hoe c. Soil auger d. Spade
33. High population in this genus brings about sex reversal towards maleness
    a. *Radopholus* b. *Paratylenchus* c. *Pratylenchus* d. *Meloidogyne*
34. Cortical layer of most nematodes mainly consists
    a. Chitin b. Keratin c. Tannin d. Cutin
35. Pseudocoelom consists of the following systems
    a. Reproductory, excretory & nervous b. Reproductory, excretory & digestive
    c. Reproductory & excretory d. Excretory, nervous & digestive
36. Nematodes normally undergo anoxybiosis in
    a. Wet soils due to more CO_2 b. Wet soils due to less O_2
    c. Dry soils due to more CO_2 d. Variable
37. Cardia in nematodes aids mainly in preventing
    a. Digestion b. Absorption c. Regurgitation d. Excretion
38. Nematode distribution in top surface layer of sandy soil, in general, is
    a. Minimum b. Maximum c. Scattered d. Variable
39. During feeding by nematode, sudden necrosis in a host variety is the indication of its
    a. Resistance b. Toelrance c. Immunity d. Susceptibility
40. The most essential basic requirement for a Nematology laboratory is
    a. A set of sieves b. Refrigerator
    c. Washing chamber d. Stereo-binocular dissecting microscope
41. The phylum Nemata was divided into two classes by
    a. Paramanov b. Chitwood c. Thorne d. Peach
42. This phytonematode can survive well in heavy soils
    a. Root-knot b. Spiral c. Cyst d. Stem and bulb
43. This organ is very helpful in the formation of new cuticle during molting
    a. Hypodermis b. Pseudocoelom c. Oesophageal glands d. Rectal glands
44. Circum-oesophageal commissure is a part of this system in nematodes
    a. Digestive b. Nervous c. Excretory d. Reproductive
45. Soil samples are collected from a depth varying between
    a. 4”-6” b. 2”-4” c. 6”-10” d. 1”-4”
46. The median layer in nematodes comprises of
    a. Esterase b. Lipids c. Phenols d. Collagen
47. Root exudates from mustard contain
    a. Alpha terthenyl b. Allyl isothiocyanate c. Alpha diethienyl d. Thionemone

48. The immature stage of nematodes is actually called as

a. Larva b. Neonate c. Juvenile d. Nymph

49. Ventral overlapping of the oesophageal glands with the intestine is found in

a. Lesion nematode b. Rice root nematode c. Root-knot nematode d. a = b + c

50. These are very close to nematodes

a. Earthworms b. Rotifers c. Nematomorphs d. Priapulids

ANSWER KEY FOR V

Sl.No.	Ans.	Sl.No.	Ans.	Sl.No.	Ans.	Sl.No.	Ans.	Sl.No.	Ans.
1	a	11	b	21	a	31	a	41	[illegible]
2	b	12	a	22	c	32	c	42	d
3	c	13	c	23	c	33	d	43	c
4	d	14	c	24	b	34	b	44	b
5	a	15	d	25	b	35	a	45	a
6	a	16	a	26	a	36	b	46	d
7	d	17	c	27	d	37	c	47	b
8	a	18	b	28	c	38	a	48	c
9	b	19	b	29	b	39	a	49	d
10	a	20	c	30	b	40	d	50	c

VI.

1. The pre-hatch molt may be absent in this nematode type
   a. Marine b. Free living c. Animal d. Predaceous
2. This is responsible for the secretion of nematode cuticle
   a. Pseudocoelomatic fluid b. Gland secretions
   c. Hypodermis d. Enzymes
3. Highest distribution of cysts of Globodera rostochiensis is observed in the surface layer up to
   a. 10 cm b. 5 cm c. 18cm d. Less than 1 cm
4. The refractive index of the immersion oil used to view nematodes under 100x is
   a. 1.250 b. 1.500 c. 1.505 d. 1.515
5. This is an example for a predatory nematode
   a. *Mononchus* b. *Monacrosporium* c. *Legenidium* d. *Caloglyphus*
6. This nematode feeds on the spores/mycelia of *Rhizoctonia solani*
   a. *Ditylenchus dipsaci* b. *Dolichodorus annus* c. *Aphelenchus avenae* d. *Hoplolaimus indicus*
7. *Pseudomonas fluorescens* produces this toxin that inhibits nematodes
   a. Bulbiformin b. Butyric acid c. H_2S d. Phenazin
8. Optimum soil crumb size for movement of *Globodera rostochiensis* juveniles in sandy loam is
   a. 50-100 microns b. 150-250 microns c. 200-300 microns d. 100-150 microns
9. This stage in embryo formation may not be found in all nematode species
   a. Blastulla b. Coelogastrula c. Coeloblastula d. Gastrulla
10. In the following nematode type, all the three oesophageal glands open into the median bulb
    a. Aphelenchids b. Tylenchids c. Rhabditids d. Heteroderids
11. Soil sampling depth for deep rooted crops is
    a. 20-30 cm b. 30-40 cm c. 40-50 cm d. 15-20cm
12. This is the important factor influencing development of an endo-parasitic nematode
    a. Moisutre b. Temperature c. Host specificity d. Tissues
13. This nematode exhibits high activity at a moisture level just below the field capacity
    a. Reniform nematode b. Spiral nematode
    c. Ring nematode d. False root-knot nematode
14. A dead female nematode with eggs inside in Heteroderidae is termed
    a. Gall b. Cyst c. Dauer juvenile d. Egg mass
15. *Nacobbus batatiformis* is a parasite of this kind
    a. Semi-endo b. Sedentary ecto c. Migratory ecto d. Migratory endo
16. The term 'Adenophorea' was coined due to absence of
    a. Glandular type excretory system b. Caudal glands
    c. Phasmid d. Deirids

17. Hypodermis can store a large amount of this compound
    a. Lipids b. Amino acids c. Glycogen d. Chitin
18. Phytonematodes can navigate in soil pores of the following micrometer diameter
    a. 20-30 b. 5-10 c. 20-30 d. 50-55
19. To draw the nematodes accurately this is used
    a. Occular micrometer b. Camera lucida c. Measuring dish d. Stage micrometer
20. *Heterodera schactii* has more affinity towards
    a. CO_2 b. O_2 c. CO_2 & O_2 d. CO
21. The fluid sphere in which the cells in the blastula stage are arranged, is bounded by
    a. Double cell layer b. Triple cell layer c. Multi cell layer d. Single cell layer
22. Generally the effect of this soil factor on nematodes is indirect
    a. Aeration b. pH c. Moisture d. Temperature
23. Most phytonematodes under tylenchida contain this type of excretory system
    a. Rhabditid type b. Ascarid type
    c. Oxyurid type d. Asymmetrical tubular type
24. Normally, this temperature range is regarded as lethal to most nematodes
    a. Above 40° C b. 30- 40° C c. 20-30° C d. 15-20° C
25. Some species of soil nematodes may obtain sufficient oxygen from plant roots also
    a. Always b. Sometimes c. Never d. Controversial
26. Soil sample size is also determined based on the
    a. Spatial pattern of nematodes b. Soil moisture
    c. Type of host d. Soil type
27. Acid phosphatase has been frequently observed in this nematode
    a. Root-knot b. Lance c. Stubby root d. Cyst
28. In the following group, pre-hatch molt may be absent
    a. Animal parasitic b. Free living c. Marine d. Phytoparasitic
29. Lesion produced by *Pratylenchus penetrans* on peach roots are mainly due to
    a. Phnenolics b. Alfa glucosidase c. Amygdalin d. Benzaldehyde
30. In nematodes most muscle cells are located in this zone
    a. Body cavity b. Cuticular layer
    c. Matrix layer d. Hypodermis intercaudal zone
31. Muscle layers in nematodes are normally connected to each other through
    a. Meristematic linkages b. Ketonic linkages
    c. Glycoside linkages d. Cytoplasmic bridges
32. Generally nematodes stimulate the production of this compound in host plants
    a. Cytokinin b. Gibberlin c. IAA d. Ethylene

33. This hydrolytic enzyme is produced by the potato cyst nematode
    a. Invertase | b. Alkaline phosphatase
    c. Alpha galactosidase | d. Cellulase

34. The journal being published by the Nematological Society of India is
    a. Indian Journal of Nematology | b. Indian Nematology journal
    c. Nematological journal of India | d. Journal of Indian Nematology

35. This activity is aided by the muscle layer present in the outermost nematode body tube
    a. Secretion | b. Movement | c. Respiration | d. Excretion

36. The purpose of using a fixative in the preparation of mounts is to
    a. Clean nematodes | b. Kill nematodes | c. Activate nematodes | d. Preserve nematodes

37. For the extraction of 'full cysts' more suitable technique is
    a. Fenwick can | b. Fluidising column | c. 60-mesh sieve | d. 250 mesh sieve

38. The critical factor that influences nematode survival in storage is
    a. Moisutre and Temperature | b. Nematode genus & species
    c. Sannitation | d. Relative humidity

39. This layer offers resistance to nematode eggs under adverse climatic conditions
    a. Lipid layer | b. Vitelline layer | c. Chitinous layer | d. Protein layer

40. Japanese Nematological Society publishes this journal
    a. Japanese Nematology journal | b. Japanese journal of Nematology
    c. Journal of Japanese Nematology | d. Nematological journal of Japan

41. During aerobic starvation, normally nematodes utilize these stored nutrients
    a. Proteins | b. Lipids | c. Carbohydrates | d. Minerals

42. Tylenchids and aphelenchids differ most conspicuously by
    a. Presence or absence of bursa | b. Spicule shape
    c. Prominence of stylet knobs | d. Opening of dorsal oesophageal gland

43. This ectoparasite can induce galls on root tips
    a. *Meloidogyne arenaria* | b. *Trichodorus christei*
    c. *Xiphinema diversicaudatum* | d. *Tylenchorhynchus brevilineatus*

44. The most common fixative used in nematode mount preparation
    a. Glycerine | b. Formaldehyde | c. DPX mountant | d. Triethanol amine

45. Nematologica journal is being published by
    a. Nematological Society of India | b. Society of Nematologists, USA
    c. Society of European Nematologists | d. Afro-Asian Society of Nematologists

46. This is not a typical character of nematode musculature
    a. Oblique striation | b. Spindle shape
    c. Presence of Z-bundles | d. All of these

47. This part in nematodes contains chitin
   a. Stylet b. Cuticle c. Bursa d. Egg shell
48. According to latest publications, the pathogen of Molya disease of wheat is
   a. *Anguina tritici* b. *Heterodera avenae* c. *H.moli* d. *H.tritici*
49. 'Auger' in Nematology, refers to
   a. Extraction tool b. Mounting tool c. Sampling tool d. Staining tool
50. Decrease in the size of the body from larval stage to adult may be seen in
   a. *Meloidogyne hapla* b. *Radopholus similis*
   c. *Tylenchulus semipenetrans* d. *Heterodera zeae*

ANSWER KEY FOR VI

Sl.No.	Ans.	Sl.No.	Ans.	Sl.No.	Ans.	Sl.No.	Ans.	Sl.No.	Ans.
1	b	11	a	21	d	31	d	41	b
2	c	12	c	22	b	32	c	42	c
3	b	13	a	23	d	33	c	43	c
4	d	14	b	24	a	34	a	44	b
5	a	15	a	25	a	35	b	45	c
6	c	16	c	26	a	36	a	46	c
7	d	17	c	27	d	37	a	47	d
8	b	18	a	28	b	38	a	48	b
9	b	19	b	29	b	39	a	49	c
10	a	20	c	30	d	40	b	50	c

VII.

1. In general, soil sample at this depth range is collected for nematode population assay
   a. 5-15 cm b. 1-10 cm c. 10-15 cm d. More than 15 cm
2. The most essential factor for soil nematodes is
   a. CO_2 b. O_2 c. H_2O d. N
3. A 500 mesh sieve used for nematode extraction has
   a. 500 mesh /inch b. 500 mesh/cm c. 500 mesh/sq.inch d. 500 mesh/linear inch
4. This extraction technique is most suited for the extraction of ring nematodes
   a. Elutriation b. Incubation c. Waring blender d. Cobb's sieving
5. At this temperature soil and root samples are to be stored until further processing
   a. 10° C b. 15° C c. 5° C d. 20° C
6. In this nematode organ, muscles serve a 'ovijectors'
   a. Testis b. Ovary c. Spicules d. Vulva
7. This is the best fixative for nematodes
   a. Carnoy'fluid b. FAA c. TAF d. FAH
8. Members of this family possess more hemispherical shaped heads
   a. Tylenchidae b. Tylenchorhynchidae c. Hoplolaimidae d. Pratylenchidae
9. The modified host cells by citrus nematodes are termed
   a. Giant cells b. Nurse cells c. Transfer cells d. Syncytia
10. Nematodes extracted using technique are usually more active than those isolated from other techniques
    a. Maceration-filtration b. Mistifier
    c. Baermann's funnel d. Semi-automatic eluttriator
11. Fenwick can method of extraction is exclusively meant for the extraction of
    a. Cysts b. Juveniles c. Eggs d. Female nematodes
12. This is the common hatching agent used to hatch the eggs of Globodera rostochiensis
    a. Picrolinic acid b. Sodium metavandate
    c. Anhydrotertronic acid d. All these
13. White trap technique is used to extract this nematode form
    a. Free living b. Entomopathogenic c. Predacious d. Mycovorous
14. This chemical transmitter is responsible for the contraction of muscles in nematodes
    a. Lactose trinine b. Acetylecholine c. Diphosphate d. Chlorogenic acid
15. This is the common fixative used in fixing nematodes
    a. Potassium-iodine b. Sodium azide c. Formalin-glycerol d. NaOCl
16. Nematode stylet or spicule protrusion may be induced by using this compound
    a. Aqueous ammonia b. Glycerin c. Glycerol d. Acetic acid

17. When nematodes reverse their direction, the waves of bending pass from
    a. Anterior to posterior
    b. Anterior to lateral
    c. Posterior to anterior
    d. Lateral to posterior
18. The principle on which Cobb's decanting and sieving technique based is
    a. Nematode mobility
    b. Sieve pore size
    c. Hydrophilic nature of nematodes
    d. Gravitational force
19. The genome of *Meloidogyne incognita* comprises about this much DNA sequences
    a. 25 million
    b. 40 million
    c. 51 million
    d. 10 million
20. Three important steps of PCR technique are
    a. Annealing, denaturation and extension
    b. Denaturation, annealing and extension
    c. Extension, annealing and denaturation
    d. Denaturation, extension and annealing
21. The cuticle of *Heterodera cyst* has a very high accumulation of
    a. Tannin
    b. Keratin
    c. Quinone
    d. Polyphenols
22. The number of longitudinal incisures in Tylenchida ranges from
    a. 4-8
    b. 8-10
    c. 2-5
    d. 12-15
23. Race 'B' of *Rotylenchulus reniformis* multiplies on
    a. Cowpea
    b. Cotton
    c. Castor
    d. a + b + c
24. These cells have a strong neuromuscular relationship
    a. Circomyarian cells
    b. Coelomyarian cells
    c. Platymyarian cells
    d. Fibrillar cells
25. Evloved nematodes like *Meloidogyne* and *Heterodera* possess this type of muscle cells
    a. Platymyarian, meromyarian
    b. Platymyarian, coelomyarian
    c. Polymyarian, coelomyarian
    d. Polymyarian, circomyarian
26. Normally the size of the galls produced by *Meloidogyne arenaria* is
    a. Large
    b. Small
    c. Tiny
    d. Multiple
27. The lip region in nematodes is always
    a. Hexaradiate
    b. Triradiate
    c. Tetraradiate
    d. Polyradiate
28. This is the most common type of nematode movement
    a. Active
    b. Passive
    c. Undulatory
    d. Simulatory
29. Fenwick can technique was developed by
    a. Fenwick-1950
    b. Fenwick-1940
    c. Fenwick-1920
    d. Cobb-1940
30. Gelatinous matrix around the egg-mass in phytonematodes is secreted by
    a. Oesophageal glands
    b. Pseusocoelomic fluid
    c. Stylet
    d. Rectal glands
31. The bursa is relatively longer and subterminal in
    a. *Meloidogyne*
    b. *Globodera*
    c. *Radopholus*
    d. *Tylenchus*
32. Superimposed moulting occurs in J_2 and J_3 of
    a. *Anguina*
    b. *Criconema*
    c. *Hoplolaimus*
    d. *Meloidogyne*

33. In this nematode, both male and female gonads are present
a. *Meloidogyne* b. *Rhadinaphelenchus* c. *Caenorhabditis* d. *Bursaphelenchus*
34. 'Scutella' refers to the enlargement of this structure
a. Amphid b. Phasmid c. Hemizonid d. Lateral line
35. 'Endotoky" refers to the condition where
a. Egg development within the female body b. Penetration of nematodes into cortical cells
c. Spicule directing towards tail d. Alae protruding within cuticle
36. *Tylenchorhynchus* has this type of oesophageal glands
a. Rotating b. Abutting c. Spherical d. Bilobed
37. Capitulum region refers to
a. Ovary b. Spicules c. Stylet d. Bursa
38. *Trichodorus and Helicotylenchus* may be extracted using this technique
a. Elutriaiton b. Baermann Funnel
c. Cobb's decanting & sieving d. Centrifugal floatation
39. The lateral bilobed projections at the distal end of the gubernaculum are termed
a. Ocelli b. Carpillae c. Papillae d. Titillae
40. The stylet is strong and massive in
a. *Meloidogyne* b. *Radopholus* c. *Hoplolaimus* d. *Dolichodorus*
41. This genus possesses six large rectal glands in females and three in males
a. *Xiphinema* b. *Trichodorus* c. *Meloidogyne* d. *Tylenchus*
42. This genus has highly enlarged renette or excretory pore that occupies nearly 30% of the body cavity
a. *Tylenchulus* b. *Longidorus* c. *Hemicycliophora* d. *Nacobbus*
43. This is a better technique to extract nematodes from more number of soil samples in less time
a. Cobb's sieving b. Seinhorst elutriation c. Mist Chamber d. Fenwick can
44. In syncytia, intense synthesis of this compound is found
a. Only protein b. Only DNA c. Only RNA d. Both DNA and RNA
45. Number of ganglia attached to the nerve ring posteriorly is
a. 8 b. 4 c. 6 d. 2
46. The concept of growth stimulation at low densities is designated by
a. y=(c-)c-1)e) MP) b. y=(e-)c-1)c) MP) c. y=(c-)c-1)e) LP) d. y=(c-)c-1e KP)
47. These sense organs consist of a small elevation of the cuticle at the end of a nerve fiber
a. Phasmid b. Hemizonions c. Cephalic alae d. Caudalids
48. Extra corporeal digestion refers to the changes that occur inside the
a. Nematode during feeding b. Host plant during nematode feeding
c. Both the host and the nematode d. Outside the host
49. Three pairs of lips are arranged around the nematode mouth in this fashion
a. Triradiate b. Tetraradiate c. Monoradiate d. Hexaradiate

50. Deirids when present, are usually in this number

a. 2 b. 4 c. 6 d. 8

ANSWER KEY FOR VII

Sl.No.	Ans.	Sl.No.	Ans.	Sl.No.	Ans.	Sl.No.	Ans.	Sl.No.	Ans.
1	a	11	a	21	a	31	c	41	c
2	b	12	d	22	b	32	d	42	a
3	d	13	b	23	a	33	c	43	c
4	a	14	d	24	b	34	b	44	d
5	c	15	c	25	c	35	a	45	b
6	c	16	a	26	b	36	b	46	d
7	b	17	a	27	a	37	b	47	b
8	a	18	d	28	c	38	c	48	b
9	b	19	c	29	b	39	d	49	d
10	b	20	b	30	d	40	c	50	a

VIII.

1. These are not sense organs in nematodes
   a. Setae & papillae b. Phasmid & amphids
   c. Scales & spines d. Hemizonid & hemizonion
2. Ocelli are more common in this habitat of nematodes
   a. Soil b. Marine c. Animal d. Entomopathogenic
3. Carbon dioxide up to this concentration is generally suitable for nematode survival
   a. 2 % b. 15% c. 8% d. 20%
4. Alpha terthenyl compound present in the root exudates of this crop are toxic to nematodes
   a. Mustard b. Neem c. Tobacco d. Marigold
5. Allyl isothiocyanate is present in the root exudates of this crop
   a. Mustard b. Brinjal c. Sesame d. Tagetes spp.
6. Indole-3-acetic acid methylester (IAM) is detected in the juveniles of this nematode
   a. Reniform b. Stem and bulb c. Citrus d. False root knot
7. This stage undergoes dormancy in Ditylenchus dipsaci
   a. J_2 b. J_4 c. J_3 d. Adult
8. Taylor's 'Power Law' refers to
   a. Sampling scheme for nematode population estimation
   b. Nematode management optimization
   c. Community analysis of nematode populations
   d. Crop performance and nematode population
9. Most phytonematodes can survive in a solution of up to this much osmotic potential
   a. 20 atm. b. 35 atm. c. 10 atm d. 5 atm
10. For normal movement, nematodes depend on
    a. External hydrodynamic pressure b. Internal hydrodynamic pressure
    c. External hydrostatic pressure d. Internal hydrostatic pressure
11. The nerve ring encircles the middle part of this organ in tylenchids
    a. Intestine b. Pharynx c. Spicule d. Vulva
12. The major principle involved in extraction of nematodes by using Baermann's funnel technique is
    a. Gravitational force b. Nematode size c. More water d. Nematode mobility
13. The most common neurotransmitter present in nematodes is
    a. Acetylcholine b. Dopamine c. Adrenaline d. Noradrenaline
14. At this temperature range, *Meloidogyne javanica* can survive, while *Pratylenchus* cannot.
    a. 5-10ºC b. 15-20ºC c. 10-15ºC d. More than 20ºC
15. Prolonged temperature of 45-50^{0}C may cause sex reversal of females to males in
    a. *Radopholus* b. *Hemicriconemoides* c. *Tylenchorhynchus* d. *Meloidogye*

16. Number of thrusts made by the stylet of Pratylenchus on the host surface before penetration is
   a. 2-6 /sec. b. 5-10/sec c. 2-6/min d. 5-10/min
17. This is responsible for inducing sex reversal in nematodes
   a. Cytokinin b. Gibberlin c. Auxin d. Morphactin
18. This nematode can feed on several cells at one time by causing dissolution of cell walls
   a. Dagger b. Burrowing c. Root-knot d. Cyst
19. Nematode species with high rates of reproduction and shorter life span refers to
   a. s-selection b. r-selection c. m-selection d. n-selection
20. In nematodes, glandular type excretory system consists of this single specialized cell
   a. Giant cell b. Nurse cell c. Renette cell d. Mother cell
21. Most common enzymes secreted by nematodes are
   a. Cellulase, protease, amylase b. Chitinase, invertase, proteinase
   c. Pectinase, acid phosphatase, glucanase d. Esterase, polygalacturonase, phosphatase
22. Number of germinal cells present in genital primordium of burrowing nematode
   a. Four b. Twelve c. Three d. Two
23. This is a migratory ectoparasite
   a. *Trichodorus* b. *Nacobbus* c. *Pratylenchus* d. *Tylenchorhynchus*
24. Sexual dimorphism actually means
   a. Didelphic females b. Diorchic males
   c. Morpholigically dissimilar adults d. Morpolgiclaly similar adults
25. The stylet of this nematode is very long measuring 100-300 um.
   a. *Meloidogyne* b. *Globodera* c. *Heterodera* d. *Longidorus*
26. These are regarded as highly specialized phytonematodes
   a. Sedentary ectoparasitesb.Sedentary endoparasites
   c. Migratory endoparasites d.Migratory ectoparaites
27. Mitotic parthenogensis is exhibited by this nematode
   a. Root-knot b. Burrowing c. Cyst d. Spiral
28. Race 'A' of *Rotylenchulus reniformis* multiplies on
   a. Cowpea b. Castor c. Cotton d. All these
29. Oosternbrink elutriator apparatus is used to extract nematodes from
   a. Roots b. Foliage c. Stem d. Soil
30. This group is known as the most efficient osmoregulator
   a. Phytonematodes b. Animal parasitic nematodes
   c. Marine nematodes d. Human forms
31. The enzyme beta-glucosidase secreted by this nematode is responsible for root necrosis
   a. *Radopholus* b. *Pratylenchus* c. *Rhadinaphelenchus* d. *Helicotylenchus*

32. Oocytes, if produced at distal end, then such ovary is referred to
    a. Polygenic b. Hologenic c. Telogenic d. Delphagenic
33. Feeding by this nematode on phloem parenchyma causes roots to become flaccid
    a. *Meloidogyne* b. *Globodera* c. *Hoplolaimus* d. *Longidorus*
34. This sense organ is a biconvex structure forming a semicircle on the ventral side between lateral fields
    a. Hemizonid b. Papillae c. Cephalid d. Amphid
35. Spreading decline differs from slow decline by exhibiting this symptom
    a. Wilt b. Die back c. Defoliation d. Twisted leaves
36. Seinhorst's models are grouped under this group of models
    a. Critical point model b. Multiple point model c. Regression model d. Single point model
37. Spear is normally long with rounded basal knobs in
    a. *Trichodorus* b. *Dolichodorus* c. *Belonolaimus* d. *Criconema*
38. Following nematode can synthesize all amino acids essential for them, while most nematodes cannot.
    a. *Nacobbus* b. *Helicotylenchus* c. *Ditylenchus* d. *Tylenchorhynchus*
39. This genus is migratory and both ecto-and endoparasitic in nature
    a. *Meloidogyne* b. *Pratylenchus* c. *Anguina* d. *Hoplolaimus*
40. The terminal galls formed by dagger nematode are due to
    a. Hypertrophy b. Hyperplasia c. Hypotrophy d. Hypoplasia
41. A peloderan bursa covers
    a. Entire tail b. Only tail tip c. Three-thirds of tail d. Two-thirds of tail
42. In de Man formula, 'P' refers to
    a. Position of phasmid b. Size of phasmid c. Size of papilla d. Position of papilla
43. This equipment is used to measure the oxygen consumption by intact nematodes specimens
    a. Tensiorespirometry b. Thermorespirometry c. Microrespirometry d. Hygorespirometry
44. *Tylenchulus semipenetrans* initiates trophic nurse cells consisting of
    a. Epidermal cells b. Mesodermal cells c. Pericycle cells d. Cortical cells
45. Cellulase and pectinase enzymes are released by this nematode that cause dissolution of the cell wall
    a. *Ditylenchus* b. *Anguina* c. *Radopholus* d. *Aphelenchoidesi*
46. If the tip of the ovary is curved back, it is termed as
    a. Reflexed ovary b. Outstretched ovary c. Condensed ovary d. Backtip ovary
47. This nematode category possess highly evolved parasitism
    a. Sedentary ectoparsitic b. Migratory ecto
    c. Migratory endo d. Sedentary endo
48. *Rhadinaphelenchus cocophilus* is transmitted by
    a. *Monochamus alternatus* b. *M. carolinensis*
    c. *Rhynchophorus palmorum* d. *R. cocophiulm*

49. During the early phase of nematode infection, transpiration rate of the host plant

    a. Increases but later decreases b. Increases

    c. Decreases but later increases d. Decreases

50. *Xiphinema diversicaudatum* reproduces once in

    a. 6 months b. 12 months c. 2 months d. 30 days

ANSWER KEY FOR VIII

Sl.No.	Ans.	Sl.No.	Ans.	Sl.No.	Ans.	Sl.No.	Ans.	Sl.No.	Ans.
1	c	11	b	21	a	31	b	41	b
2	b	12	d	22	d	32	c	42	[illegible]
3	c	13	a	23	a	33	c	43	c
4	d	14	a	24	c	34	a	44	d
5	a	15	d	25	d	35	b	45	c
6	b	16	a	26	b	36	c	46	a
7	b	17	d	27	a	37	b	47	d
8	a	18	b	28	d	38	c	48	c
9	c	19	c	29	d	39	c	49	a
10	a	20	d	30	a	40	b	50	b

IX

1. White tip rice nematode was identified as *Aphelenchoides besseyi* by
   a. Dastur b. Allen c. Perry d. Thorne
2. Broad and short spicules are characteristic features of
   a. *Anguina* b. *Ditylenchus* c. *Pratylenchus* d. *Meloidogyne*
3. Burrowing nematode can survive in fallow soil for a period of
   a. 1 year b. 2 years c. 3-6 months d. 8-10 months
4. The two flask technique employed to extract nematodes is based on the principle of
   a. Centrifugal force b. Sedimentation
   c. Counter flow of water d. Size of nematodes
5. Rapid reproduction rate in root-knot nematode is because the body length-ovary ratio is
   a. More than three b. Less than three c. More than four d. Less than four
6. Due to this nature of nematodes, it is not very easy to frame uniform models
   a. Small size b. Polyspecific community
   c. Wide host range d. Slow movement
7. Adults of this nematode can undergo quiescence
   a. *Radopholus similis* b. *Meloidogyne hapla*
   c. *Aphelenchoides besseyi* d. *Pratylenchus coffeae*
8. Nematode population on a host increases up to a certain limit with respect to time, which is called
   a. Peak level b. Saturation lelvel c. Roof level d. Ceiling level
9. This is a common predator on phytonematodes occurring in natural soils
   a. *Monacrosporium drechselri* b. *Pochonia chlamydosporia*
   c. *Lagenidium caudatum* d. *Tyrophagus similis*
10. Mustard root exudates contain following compound that inhibits nematodes
    a. Alpha terthienyl b. Allyl isothiocyanate c. Allyl thiocyanate d. Thionemone
11. The posterior part of the stoma is surrounded by anterior oesophagous as it lies embedded in the oesophageal tissue and this state is termed
    a. Oesophagealstoma b. Oesophagous stoma c. Oesophagostome d. Stomaphgous
12. This genus has a strong and massive stylet
    a. *Radopholus* b. *Meloidogyne* c. *Hoplolaimus* d. *Pratylenchus*
13. This is the largest part of the alimentary canal of nematodes
    a. Oesophagous b. Cuticle c. Intestine d. Stoma
14. Nematode activity generally, is high at a field capacity of
    a. 40-60 % b. 30-40 % c. 60-80% d. 10-20%
15. Maximum egg production in root-knot nematodes occurs at the temperature range of
    a. 35-37°C b. 15-30°C c. 25-32°C d. 33-40°C

16. Citrus race of *Radopholus similis* is parasitic on
    a. Citrus b. Banana c. Banana and citrus d. Coconut
17. The genus *Aphelenchoides* was created by
    a. M.Fischer b. Chitwood c. J.Ritzemaboss d. N.A.Cobb
18. Index of similarity is employed to compare this aspect of nematodes
    a. Population b. Density c. Diversity d. Community
19. This is the best technique to extract sluggish nematodes from soil
    a. Centrifugal floatation b. Elutriation
    c. Sieving +Flocculation d. Baermann's Funnel
20. This stage undergoes quiescence in Anguina
    a. J4 b. J3 d. J1 d. J2
21. Sinus cell and renette are the organs of this system
    a. Nervous b. Excretory c. Reproductory d. Digestive
22. In this genus, intestine overlaps oesophageal glands in a long lobe
    a. *Bursaphelenchus* b. *Rhadinaphelenchus* c. *Aphelenchus* d. *Aphelenchoides*
23. *Aphelenchoides ritzemabosi* and *Ditylenchus dipsaci* enter the host through
    a. Wounds b. Hyadathodes c. Lenticells d. Stomata
24. Predisposition caused by biotic factors is known as
    a. Biotic disposition b. Biodisposition c. Bio predisposition d. Biotic predisposition
25. Stylet is not present in males of this genus
    a. *Belonolaimus* b. *Xiphinema* c. *Hemicycliophora* d. *Criconemoides*
26. Maximum number of body annules is found in
    a. *Criconemoides* b. *Criconemella* c. *Heterodera* d. *Longidorus*
27. For most nematodes this is the natural environment
    a. Plant tissues b. Terrestrial c. Aquatic d. Meadow
28. Banana race of *Radopholus similis* is parasitic on
    a. Citrus b. Banana c. Banana and citrus d. Coconut
29. This is the most important available soil water for nematode activity
    a. Hygroscopic water b. Capillary water c. Gravitational water d. Not specific
30. Paratylenchidae falls under this super family
    a. Criconematoidea b. Diphtherophoroidea c. Dorylaimoidea d. Tylenchoidea
31. The terms 'R-strategy and K-strategy' refer to the following aspect of phytonematodes
    a. Population b. Modeling c. Reproduction d. Management
32. The first phytonematode *Anguina tritici* was named as *Vibrio tritici* in 1799 by
    a. Steinbuch b. Filipjev c. Milne d. Rhode
33. Relative density +relative frequency+ relative dominance accounts to
    a. Importance value b. Prominence value c. Index of similarity d. Absolute biomass

34. This is the smallest nematode
 a. *Xiphinema* b. *Tylenchorhynchus* c. *Anguina* d. *Trichodorus*
35. This is a non-cyst forming genus
 a. *Meloidodera* b. *Ephippoidera* c. *Punctodera* d. *Circodera*
36. *Hirschmaniella* induces
 a. Interveinal necrosis b. Interveinal chlorosis c. Intraveinal chlorosis d. Intraveinal necrosis
37. This nematode is associated with swamps, marshes and wet locations
 a. *Globodera* b. *Tylenchorhynchus* c. *Dolichodorus* d. *Criconema*
38. Bacterial feeding nematodes are found in
 a. Diplogasteridae b. Tylenchidae c. Mononchida d. Plectide
39. When nematodes reverse their direction, the waves of bending pass from
 a. Lateral to posterior b. Anterior to lateral c. Anterior to posterior d. Posterior to anterior
40. J.N.Sasser started this project
 a. International *Globodera* Project b. International *Meloidogyne* Project
 c. Worldwide *Meloidogyne* Project d. Continental *Meloidogyne* Project
41. *Heterodera goettingiana* was first reported on pea in 1892 by
 a. Liebscher b. T.Goodey c. Poinar d. Steiner
42. Spreading decline of citrus is induced by
 a. *Radopholus similis* b. *Tylenchulus semipenetrans*
 c. *Rotylenchulus reniformis* d. *Pratylenchus indica*
43. The purpose of using sugar solution in centrifugal flotation technique is
 a. Easily soluble in waterb. Lower specific gravity
 c. Higher specific gravityd. Easily available
44. Generally, soil inhabiting nematodes need large quantities of
 a. Oxygen b. Carbon dioxide c. Hydrogen d. Carbon monoxide
45. In cereal cyst nematode, the cysts are
 a. Afenestrate b. Circumfenestrate c. Ambifenestrate d. Bifenestrate
46. *Criconema* is the synonym of this genus
 a. *Mulveyotus* b. *Dolichodorus* c. *Basiriolaimus* d. *Hoplolaimus*
47. The person who was responsible to recognize Nematology as a separate discipline
 a. J.N.Sasser b. B.G.Chitwood c. F.G.W.Jones d. N.A.Cobb
48. In nematodes various openings are located on
 a. Lateral side b. Ventral side c. Dorsal side d. Not specific
49. Among all factors this factor affects the duration of life cycle in nematodes
 a. Soil type b. Soil moisture c. Soil temperature d. Host type
50. Sheath nematodes belongs to the family
 a. *Criconematoidea* b. *Heteroderoidea* c. *Hoplolaimoidea* d. *Longidoroidea*

ANSWER KEY FOR IX

Sl.No.	Ans.	Sl.No.	Ans.	Sl.No.	Ans.	Sl.No.	Ans.	Sl.No.	Ans.
1	a	11	c	21	b	31	a	41	a
2	a	12	c	22	b	32	a	42	a
3	c	13	c	23	d	33	a	43	c
4	c	14	a	24	b	34	d	44	a
5	a	15	c	25	c	35	b	45	b
6	b	16	c	26	c	36	a	46	c
7	c	17	a	27	b	37	[illegible]	47	d
8	d	18	d	28	b	38	a	48	b
9	a	19	a	29	b	39	c	49	c
10	b	20	d	30	d	40	b	50	c

X.

1. A large egg sac found at the end of the posterior end of female is the characteristic of
   a. *Heterodera cajani* b. *H.avenae* c. *H.sorghi* d. *H.zeae*
2. This nematode can sense and get more attracted to carbon dioxide
   a. *Meloidogyne javanica* b. *Anguina tritici*
   c. *Ditylenchus dipsaci* d. *Radopholus similis*
3. Suspended animation of nematodes due to desiccation is termed
   a. Anabiosis b. Anhydrobiosis c. Anoxybiosis d. Cryptobiosis
4. Slow decline of citrus is caused by
   a. *Radopholus similis* b. *Tylenchulus semipenetrans*
   c. *Rotylenchulus reniformis* d. *Pratylenchus indica*
5. The appropriate way of estimating nematode population density in soil is the number of nematodes per
   a. gram dried soil b. cc soil c. ml soil d. g wt soil
6. Necrosis of cells of localized areas is by
   a. *Radopholus* b. *Helicotylenchus* c. *Pratylenchus* d. *Globodera*
7. This genus can retain virus particles for shorter time compared to other genera
   a. *Xiphinema* b. *Longidours* c. *Paratrichodorus* d. *Paralongidours*
8. Sugar beet cyst nematode was reported by
   a. Weischer b. Kuhn c. Schacht d. Schneider
9. Criconematids are named after this character
   a. Creconemoid type of pharynx b. Presence of spines and scales
   c. Slow and sluggish movement d. Coarse ring-like annules
10. This genus can grow steadily throughout life to a maximum size
    a. *Radopholus* b. *Meloidogyne* c. *Ditylenchus* d. *Heterodera*
11. This solution can make nematodes shrink
    a. Alkaline b. Salt c. Hypertonic d. Hypotonic
12. Too high temperature actually inhibits this activity of nematodes
    a. Invasion b. Movement c. Hatching d. Reproduction
13. In *Globodera* cysts are
    a. Spherical b. Pear shaped c. Lemon shaped d. Squarish
14. In histopathological techniques, mounted slides are finally cleared in a mixture of alcohol, xylol and
    a. Olive oil b. Balsam oil c. Canada balsom oil d. Clove oil
15. This condition can inhibit the hatching and development of nematodes
    a. High osmotic pressure b. High moisture
    c. Low osmotic pressure d. Low moisture
16. Normally most dormant stages of nematodes are resistant to
    a. Nematicides b. Insecticides c. Fungicides d. Herbicides

17. Top leaves of rice are twisted due to the attack of
    a. *Ditylenchus destructor*
    b. *D.angustus*
    c. *Aphelenchoides besseyi*
    d. *D.dipsaci*
18. Nematodes in soil are attached to roots in gradients of electrical potential due to
    a. Sodium thionite
    b. Sodium thionate
    c. Sodium dithionide
    d. Sodium dithionite
19. The number of *Aphelenchoides composticola* needed to kill mushroom mycelium in 100 g compost
    a. 1000
    b. 100
    c. 10,000
    d. 2000
20. The specific gravity of nematodes is
    a. 1.10
    b. 1.15
    c. 1.00
    d. 1.05
21. Population reduction of a phytonematode in fallow soil follows
    a. Symmetric curve
    b. Normal curve
    c. Hypobolic curve
    d. Hyperbolic curve
22. Sex differentiation in nematodes is mostly controlled by
    a. Physiological mechanism
    b. Genetic mechanism
    c. Environmental conditions
    d. Host
23. The resemblance among nematode species which is not due to inheritance from common ancestry is known as
    a. Homology
    b. Homoplane
    c. Homoplasty
    d. Homolysis
24. This genus has smallest sized female nematode
    a. *Heterodera*
    b. *Meloidogyne*
    c. *Nacobbus*
    d. *Criconema*
25. Labial framework is strongly sclerotized in
    a. *Tylenchulus*
    b. *Helicotylenchus*
    c. *Criconema*
    d. *Rotylenchulus*
26. This is the major factor to decides the development of a nematode inside the host
    a. Temperature
    b. Moisture
    c. Host specificity
    d. Soil factors
27. This predaceous fungus attacks nematodes infesting sugarcane
    a. *Catenaria vermicola*
    b. *C.bemicoides*
    c. *Dactylella cinophaga*
    d. *D.candida*
28. Large numbers of phytonematodes can be propagated in vitro by using this technique
    a. Tissue culture
    b. Callus culture
    c. Meristem culture
    d. Sterile root culture
29. *Rhadinaphelenchus cocophilus* along with infested tissues is transmitted by
    a. *Aphis gossypii*
    b. *Nephotettix virecens*
    c. *Bemisia taabaci*
    d. *Rhynchophorus palmorum*
30. In the feeding sites of the following nematode, high frequency of plasmodesmata is seen
    a. *Heterodera*
    b. *Radopholus*
    c. *Pratylenchus*
    d. *Anguina*
31. This is the longest nematode
    a. *Longidorus elongates*
    b. *Rhadinaphelenchus cocophilus*
    c. *Belonolaimus longicaudatus*
    d. *Paralongidorus maximus*
32. The gonads of Meloidogyne and *Tylenchulus semipenetrans* can be distinguished in this juvenile stage
    a. J1
    b. J2
    c. J3
    d. J4

33. *Radopholus similis* was first noticed in 1890 by
    a. *de Man* b. *N.A.Cobb* c. *Thorne* d. *Atkinson*
34. Perennial pattern technique is related to this genus
    a. *Heterodera* b. *Meloidogyne* c. *Radopholus* d. *Anguina*
35. Trichodorus and Xiphinema are differentiated by this parameter
    a. Relaxation posture b. Body length c. Body width d. Stylet size
36. Twisting and malformation of basal leaves in onion is due to
    a. *Ditylenchus dipsaci* b. *D.angustus* c. *D.destructor* d. *D.myceliophagus*
37. The varied adopted nematode populations are referred to
    a. Biotypes b. Ecotypes c. Pathotypes d. Races
38. The egg production per cyst at high densities,
    a. Increases b. Decreases c. Varies d. Neutral
39. 'Onion bloat' is due to
    a. *Ditylenchus destructor* b. *D.dipsaci*
    c. *D. allii* d. *D. globus*
40. The genus Trichodorus was created by Cobb in
    a. 1910 b. 1921 c. 1903 d. 1913
41. Double staining in histology consists of
    a. Lactophenol and cotton blue b. Cotton blue and fast green
    c. Safranin and Fast green d. Safranin and Cotton blue
42. This bioagent possess nerve poison to kill phytonematodes
    a. *Mononchus* b. *Strongylus* c. *Seinura* d. *Chamberiella*
43. *Heterodera avenae* is an inhabitant of
    a. Warm climate b. Cool climate c. High humid climate d. Not specific
44. This anhydrobiotic nematode has been used in US Space Programmes for the study of the effect of high energy particles in space
    a. *Mononchus* b. *Caenorhabditis elegans*
    c. *Ditylenchis dipsaci* d. *Panagrellus redividua*
45. After reaching a high level, the nematode population varies around equilibrium, which is termed as
    a. Equilibrium effect b. Static effect c. Constant effect d. Ceiling effect
46. This topic covers the factors which prevent a nematode population from growing at its full potential rate
    a. Ecological potential b. Environmental potential
    c. Environmental limit d. Ecological limit
47. Homoeostasis refers to the ability to maintain constant
    a. Body temperature b. Body fluids c. Resting period d. Movement
48. The ecology of nematodes in relation to the individual species refers to
    a. Auteoecology b. Proecology c. Synecology d. Deltaecology

49. This biotype of potato cyst nematode is absent in India

a. Ro1 b. Ro5 c. Pa1 d. Pa3

50. Soil incorporation with saw dust may not be very effective in this soil type

a. Heavy sandy b. Clayey c. Porous c. Silty

ANSWER KEY FOR X

Sl.No.	Ans.	Sl.No.	Ans.	Sl.No.	Ans.	Sl.No.	Ans.	Sl.No.	Ans.
1	a	11	c	21	d	31	d	41	a
2	c	12	b	22	c	32	b	42	c
3	a	13	a	23	a	33	b	43	b
4	b	14	c	24	b	34	b	44	b
5	b	15	b	25	b	35	b	45	d
6	b	16	a	26	c	36	a	46	b
7	d	17	b	27	a	37	b	47	a
8	c	18	c	28	b	38	b	48	a
9	d	19	a	29	d	39	b	49	a
10	c	20	d	30	a	40	d	50	b

XI.

1. The first record of nematodes from birds, namely falcons was provided by
   a. Columella b. Albertus magnus c. Vegetinus d. Caesalpinus
2. The free living nematode commonly referred as Vinegar eel was first discovered by
   a. Borellus b. Owen c. Leady d. Tyson
3. The nematode was viewed under microscope for the first time by
   a. Bojanus b. Cliquet c. Bastian d. Borellus
4. The first to study nemic anatomy and describe a nematode egg
   a. Redi b. Vircho c. Tyson d. Bastian
5. Free living nematodes from fress water was first reported by
   a. Herbse b. Muller c. Schneider d. Dujardin
6. Anatomic studies of varieties of nemas by free hand section dissection was conducted by
   a. Bojanus b. Cloquet c. Deising d. Both a and b
7. The detailed description of free living nematodes using camera lucida was first described by
   a. Butschli b. Stewart c. Paramanov d. None of the above
8. The nervous system of nematodes was first described by
   a. Needham b. Atkinson c. Goldschmidt d. T. Goody
9. Hypothesis concerning nemic relationship, evolution and phylogenics was formulated by
   a. Paramanov b. Christie c. Carter d. Oostenbrink
10. Nematode as plant parasites was reported for the first time in
    a. N. A. Cobb b. T. Needham c. Julius kuhn d. A. Schmidt
11. The first plant parasitic nematode to be noticed and reported by T. Needham in 1743 was
    a. *Heterodera schachtii* b. *Meloidogyne incognita*
    c. *Ditylenchus dipsaci* d. *Anguina tritici*
12. The first plant parasitin nematode (wheat seedgall nematode) was named as Vibrio tritici in 1799 by
    a. Filipjev b. Milne c. Steinbuch d. None of the above
13. The present name of wheat seedgall nematode, anguoina tritici was given by
    a. Milne (1919) b. N. A. Cob (1913) c. Steinbunch (1799) d. Filipjev (1936)
14. The life cycle of wheat seedgall nematode was studied by
    a. T. Needham b. J. G. Steinbunch
    c. C. Devaine d. M. J. Berkeley
15. Father of Nematology
    a. Cobb b. Bastian c. Jones d. Barber
16. Occurrence of root-knot nematode in cucumber was first reported in 1855 by
    a. M. J. Berkley b. C. Liebscher c. Barber d. E. A. Bessey
17. Stem and bulb nematodes *Ditylenchus dipsaci* infecting the heads of teasel was first discovered by
    a. W. Carter (1943) b. Atkinson (1892) c. Neal (1894) d. Julius Kuhn(1857)

18. Sugarbeet cyst nematode was reported in 1859 by
   a. Kuhn b. Schneider c. H. Schacht d. Weischer
19. *Heterodera schachtii* was described and named by
   a. A. Schmidt b. E. A. Goeldi c. J. Ritzemaboss d. E. A. Bessy
20. The first specific mention of root knot nematode, *Anguiillula marioni* was described in 1879 by
   a. E. A. Goeldi b. Schacht c. M. Cornu d. Needham
21. The lesion nematode was first described in 1880 by
   a. Atkinson b. De man c. Hedwig hirschmann d. None of the above
22. Detailed description of *meloidogyne incognita* (root knot of coffee) was published in 1887 by
   a. M. Cornu b. E. S. Bessy c. G. F. Atkinson d. E. A. Goeldi
23. The detailed life cycle of *Heterodera schchtii* was published in 1888 by
   a. H. Schacht b. Van Gundy c. A. Strubell d. Seinhorst
24. Name of the scientist who used CS_2 (Carbon disulphide) as soil fumigant for the control of sugarbeet cyst nematode
   a. Julius Kuhn b. Schmidt c. Schacht d. Hue
25. The existence of physiological races in plant parasitic nematodes was reported as early as 1888 by
   a. Carter b. Mountain c. Sturhan d. J. Ritzemaboss
26. Foliar nematode, *Aphelenchoides fragariae* on straw berry was first recorded in 1891 by
   a. J.f dastur b. J. Ritzemaboss c. Mankau d. Duddington
27. Cyst nematode, *Heterodera goettingiana* on pea roots was discovered in 1892 by
   a. C. Liebscher b. T. Goodey c. Steiner d. Poinar
28. The first record of involvement of root knot nematode and Fusarium complex in vascular wilt of cotton was discovered in 1892 by
   a. Oostenbrink b. Carter c. Atkinson d. Christie
29. Taxonomic monograph of soil and fresh water nematodes of the Netherlands was authored by
   a. Filipjev b. De man c. T. Goodey d. Oostenbrink
30. Name of the scientis(s) who independently published on root knot nematode in United States
   a. Atkinson b. Neal c. N. A. Cob d. Both a and b
31. Association of root knot nematode with bacterial wilt of tomato was first reported by
   a. Atkinson b. Hewitt c. Hunger d. Goheen
32. The term "Nematology" was coined by
   a. N. A. Cobb b. Kuhn c. Needham d. schacht
33. Father of American Nematology
   a. G. Thorne b. N. A. Cobb c. F.g.w. Jones d. Kuhn
34. Cobb sieving technique was developed by
   a. Mcbath b. Chitwood c. Hue d. N. A. Cobb

35. The generic name nacobbus is coined in the honour of eminent nematologist

a. N. A. Cobb b. Oostenbrink c. Christie d. carter

36. The burrowing nematode was first observed in 1890 by

a. Thorne b. Linford c. N. A. Cobb d. Oliveira

37. A detailed account on morphology and life cycle of citrus nematode, *Tylenchulus semipenetrans* was published in 1913 by

a. N. A. Cobb b. Filipjev c. Stewart d. Paramanov

38. The genus *Tylenchorynchus* and *Trichodorus* was created by cobb in

a. 1911 b. 1912 c. 1913 d. 1914

39. The genus *Radopholus* was described by

a. N. A. Cobb b. Thorne c. Linford d. Both a and b

40. The idea of hot treatment of plant materials was given by

a. Julis kuhn b. Christie c. Carter d. Marcinowski

41. Goodey was known for his contribution in the field of nematode

a. Ecology b. Morphology c. Taxonomy d. Diseases

42. Nomenclature of golden nematode of potato, *Globodera rostochiensis* was proposed by

a. Wollenweber b. Massee c. Zimmerman d. Thorne

43. Potato cyst nematode was first discovered in Nassau country, long Islands, USA in

a. 1913 b. 1931 c. 1914 d. 1946

44. The book "plant parasitic nematode and the diseases they cause"is written by

a. T. Goodey b. Perry c. Stekhoven d. Dastur

45. Monograph on soil and fresh water nematode by

a. Thorne b. Parvatha reddy c. Siddiqui d. T. Goodey

46. "Nematodes that are Importantace for Agriculture" is written by

a. Raski b. Filipjev c. Goheen d. Hewitt

47. The book "Nematodes that are important for agriculture" was translated from Russian to English under the title "A Manual of Agricultural Helminthological" by

a. J. H. Schuurman stekhovan b. Van Gundy

c. Holdeman d. Graham

48. The survival mechanisms of nematodes was studied by

a. Triantaphyllou b. Van gundy c. Caveness d. Oostenbrink

49. Genetics of plant nematode relationship was first reported by

a. Raski b. Norton c. N. A. Cobb d. Triantaphyllou

50. First person to demonstrate that bacteria are carried through digestive tract of nematodes

a. Kalinenko b. Jensen c. Stewart d. Dchindler

ANSWER KEY FOR XI

Sl.No.	Ans.	Sl.No.	Ans.	Sl.No.	Ans.	Sl.No.	Ans.	Sl.No.	Ans.
1	b	11	d	21	b	31	c	41	c
2	a	12	c	22	d	32	a	42	a
3	d	13	d	23	c	33	b	43	c
4	c	14	c	24	a	34	d	44	a
5	b	15	b	25	d	35	a	45	d
6	d	16	a	26	b	36	c	46	a
7	a	17	d	27	a	37	a	47	b
8	c	18	c	28	c	38	c	48	a
9	a	19	a	29	b	39	b	49	b
10	b	20	c	30	d	40	d	50	c

XII.

1. Norton was known for his contribution in this aspect of nematodes

   a. Management b. Morphology c. Taxonomy d. Ecology

2. The nematicidal value of DD (dichloropropane dichloropropene) was discovered in 1943 by

   a. Perry b. Christie c. Carter d. Dastur

3. The nematicidal value of EDB (ethylene dibromide) was discovered in 1945 by

   a. Christie b. Perry c. Chitwood d. N. A. Cobb

4. The world's first formal university cource in Nematology was thought by

   a. D. W. Fenwick b. Allen c. M. T. Franklin d. R.S Picther

5. White tip nematode of rice was identified as *Aphelencoides bessyi* by

   a. Allen b. Dastur c. Perry d. Fischer

6. The book "the potato nematode, a dangerous parasite to potato monoculture was written by

   a. F.J.W Jones b. B.G.Peters c. Oostenbrink d. Pitcher

7. Name the scientists who discovered the role of ectoparasites as plant pathogens

   a. Christie b. Perry c. Van gundy c. Both a and b

8. Augmentation of *Fusarium* wilt of cotton by *Belanolaimus longicaudatus* was first reported by

   a. Christie and perry b. Holdemanand graham

   c. Barron and duddington d. Cobb

9. The fact that wounding of roots by *Meloidogyne* facilitated infection of tobacco roots by *Pseudomonas solanacearum* was demonstrated by

   a. Lucas *et al* b. Hewitt *et al* c. Pitcher *et al* d. Siddiqui *et.al.*,

10. The first report of virus transmission by a planr parasitic nematode was made by

    a. Johnson, 1945 b. Christie, 1951

    c. Harrison& Cadman, 1959 d. Hewitt, Raski and Goheen, 1958

11. Association of *Aphelencoides ritzemabosyi* with the bacterium *Corynebacterium fascians* was demonstrated by

    a. *Hewitt et al* b. *Lucas et al* c. *Pitcher and Crosse* d. *Carter et.al.,*

12. The genus Rotylenchulus was recorded on cowpea for the first time in 1940 by

    a. Linford b. Oliveira c. Mankau d. Both a and b

13. The nematicidal value of DBCP (dibromochloro propane) was discovered by

    a. Ritzemaboss b. Hue c. Mcbath d. Martin

14. Classification of different species of meloidogyne by perennial pattern of annulations was discovered by

    a. T. Goodey b. Chitwood c. Ritzemaboss d. G. Thorne

15. The genus Meloidogyne was separated Heterodera by

    a. Chitwood b. Muller c. Goeldi d. Cornu

16. An insight on the biological control of plant parasitic nematodes was provided by
    a. Barron and duddington b. Mankau and linford
    c. Sayre and Zukerman d. All of the above
17. Biological control of insects with nematodes was demonstrated by
    a. Nickel b. Poinar c. Steiner d. All of the three
18. The fundamental information on developmental biology and genetics of *Caenorhabditis elegans* was provided by
    a. Brenner b. Dougherty c. Nicholas d. All of the three
19. Vapour heat treatment and hot air treatments for nematode control was introduced by
    a. T.R Hewitt b. R. Latta c. K. Marcinowski d. All the three
20. The role of burrowing nematode, *Radopholus similis* in causing spreading decline of citrus was proved by
    a. Suit b. Ducharme c. Oliveira d. Both a and b
21. The generic name *Hirschmanniella* is coined in honour of an eminent US Nematologist
    a. Hedwig Hirschmann b. N. A. Cobb c. Chitwood d. Sasser
22. Rice root nematode, *Hirschmanniella* oryzae was assigned by
    a. Luc b. Goodey c. Both a and b d. Oostenbrink
23. Root knot nematode on tea in south india was first reported by
    a. Siddiqi b. Butler c. Ayyar d. Barber
24. Occurance of root knot nematode on vegetables in india was first reported by
    a. Thirumala rao b. Ayyar c. Butler d. Goheen
25. Root knot nematode on black pepper in kerala was first reported by
    a. Butler b. L. C. Coleman c. Nair d. Menon
26. Ufra disease on rice (*Ditylenchus angustus*) was first reported from Bengal by
    a. Milne b. Prasad c. Butler d. Mathur
27. Presence of wheat seed gall nematode was first reported from Bengal by
    a. Nair b. Milne c. Siddiqi d. Ayyar
28. Lesion nematode was first reported in 1933 from India by
    a. Mayne b. Subramanian c. Dastur d. Both a and b
29. Root knot nematode on citrus in India was first reported by
    a. Thirumala rao b. Duster c. Barber d. Koshy
30. Occurance of white tip of rice *Aphelenchoides besseyi* was reported from central India by
    a. Prasad b. Duster c. Menon d. Edward
31. Potato cyst nematode in India was first reported in 1961 by
    a. Berkely b. N.A. Cob c. Jones d. Siddiqi
32. Occurance of cerial cyst nematode was first reported from Rajasthan by
    a. Nair *et al* b. Vasudeva c. Hewitt *et al* d. Sengupta

33. Molya disease on barley in India was first reported from Rajasthan by Prasad et al., in the year
    a. 1959 b. 1979 c. 1969 d. 1989
34. Citrus nematode, *Tylenchulus semipenetrans* was first reported from Uttar Pradesh by
    a. Nair b. Menon c. Siddiqi d. Das
35. The genus *Trichodorus* was split into *Trchodorus* and *Paratrichodorus* by
    a. N.A. Cob b. Siddiqi c. Jones d. None of thsee
36. Reniform nematode, *Rotylenchus reniformis* on cotton was first reported in India in 1963 by
    a. Seshadri b. Sivakumar c. Siddiqi d. Both a and b
37. The first person to report the presence of *Trichodorus* nematode in India
    a. Sivakumar b. Siddiqi c. Prasad d. Both a and b
38. Occurrence of burrowing nematode on banana was reported from kerala by
    a. Nair b. Das c. Menon d. a + b + c
39. European Society of Nematologist was founded in the year
    a. 1955 b. 1956 c. 1961 d. 1969
40. One of the this is first international journal devoted for Nematology. This is one of the first international journal devoted for Nematology
    a. Organization of Tropical American Nematologists
    b. European Journal of Nematologists
    c. Nematological Society of India
    d. Nematologica
41. In this year Society of Nematologists was founded in US
    a. 1961 b. 1959 c. 1966 d. 1967
42. Organization of tropical American nematologists was founded in
    a. 1965 b. 1967 c. 1969 d. 1971
43. Journal of Nematology is published from
    a. Organization of tropical American Nematologists
    b. European society of Nematologists
    c. Society of Nematologists, USA
    d. Nematological society of India
44. Nematology newsletter is published from
    a. Organization of Tropical American Nematologists
    b. European Society of Nematologists
    c. Nematological Society of India
    d. Society of Nematologists, USA
45. Nematropica journal is published from
    a. Nematological Society of India
    b. Society of Nematologists, USA
    c. European Society of Nematologists
    d. Organization of Tropical American Nematologists

46. The journal of Nematologia Mediterranea is published from
    a. Italy b. Europe c. USA d. France
47. The journal Revue de nematologie is published from
    a. Australia b. France c. Germany d. Italy
48. Nematological society of India was founded in the year
    a. 1965 b. 1967 c. 1969 d. 1971
49. Journal of Nematology published in India
    a. Indian journal of Nematology b. Journal of Nematology
    c. Nematologica d. Plant Nematology
50. In India for the first time Deparment of Nematology was established in 1977 at
    a. G.B. Pant University of Agricultural and Technology
    b. Central Agricultural University
    c. Haryana Agricultural University, Hissar
    d. Tamil Nadu Agricultural University

ANWER KEY FOR XII

Sl.No.	Ans.	Sl.No.	Ans.	Sl.No.	Ans.	Sl.No.	Ans.	Sl.No.	Ans.
1	d	11	c	21	a	31	b	41	a
2	c	12	d	22	c	32	a	42	b
3	a	13	c	23	d	33	c	43	c
4	b	14	b	24	a	34	b	44	d
5	a	15	a	25	c	35	d	45	d
6	c	16	d	26	b	36	b	46	a
7	d	17	d	27	a	37	d	47	b
8	b	18	d	28	a	38	a	48	c
9	a	19	d	29	b	39	c	49	a
10	d	20	d	30	c	40	b	50	c

XIII.

1. The first Indian university to introduce postgraduate programme in Plant Nematology in 1979 was
   a. Marathwada Agricultural University b. Punjab Agricultural University
   c. Kerala Agricultural University d. Tamil Nadu Agricultural University
2. The division of Nematology was established at IARI in the year
   a. 1963 b. 1966 c. 1969 d. 1971
3. Laboratory for potato cyst nematode research was established in the year 1963 at
   a. Nilgiri hills b. Palani hills
   c. Uthagamandalam d. Both a and b
4. AICRP on nematode pests of crops and their control as started in the year
   a. 1977 b. 1979 c. 1981 d. 1982
5. International Meloidogyne project was led by
   a. Jones b. J.N. Sasser c. Zimmermann d. Wollenweber
6. The abstracting service of world literature on Nematology is provided by
   a. Nematological Society of India
   b. Commonwealth Agricultural Bureau, England
   c. European Society of Nematologists
   d. Organization of Tropical American Nematologists
7. 'Fungicide and Nematicides Tests' is being published from
   a. Organization of Tropical American Nematologists
   b. American Phytopathological Society
   c. Society of Nematologists, USA
   d. Nematological Society of India
8. The bibliography on plant, soil and fresh water nematodes is a card title service published by
   a. E. J. Brill b. T. Goodey c. N. A. Cobb d. W.S.Gill
9. The headquarter of International *Meloidogyne* Project was
   a. Rothamsted Experimental Station, UK b. University of California, USA
   c. University of Florida, USA d. North Carolina State University, USA
10. Who has contributed most for the strengthening of nematology in India?
    a. F.G.W. Jones b. J. N. Sasser c. M. Oostenbrink d. D. J. Raski
11. Nematology as a separate discipline was first advocated by
    a. B. G. Chitwood b. N. A. Cobb c. J. N. Sasser d. F. G. W. Jones
12. The book 'An Introduction to Nematology' was authored by
    a. B. G. Chitwood b. M. B. Chitwood c. Both a and b d. N. A. Cobb
13. Author of "Text Book on Introductory Plant Nematology" published by the ICAR,New Delhi
    a. Raman. K. Walia b. Harish k Bajaj c. P. Parvatha reddy d. Both a and b

14. The book 'Horticultural Nematology' is authored by
    a. N.G.Ravichandra b. R. S. Singh c. K. Sitaramaiah d. E. I. Jonathan
15. 'Plant Pathogens: The Plant Parasitic Nematodes' is written by
    a. R. S. Singh b. K. Sitaramaiah c. Harish k Bajaj d. Both a and b
16. The book "A Treatise on Phytonematology" is penned by
    a. G. Rajendran b. S. Kumar c. K. Devarajan d. P. Parvatha Reddy
17. Awad instituted by Nematological Society of India for research work based on thesis work
    a. Raski award b. Shah award c. D' souza award d. Best thesis award
18. Awad instituted by Nematological Society of India for contributer in the field of Nematology during the proceding five years.
    a. D' souza award b. Shah award c. Raski award d. Best thesis award
19. The most accepted theory of chemotaxis based upon interactions between molecules emanating from host and nematode was proposed by
    a. Kuhn, 1959 b. Klink et al., 1984
    c. Zukerman & Jenson, 1984 d. Van Gundy & Rackman, 1961
20. Nematode implicated in the peach replant problem in Canada
    a. *Pratylenchus vulnus* b. *P. pratensis* c. *Radopholus similis* d. *Pratylenchus penetrans*
21. This nematode race attacks cotton
    a. *Meloidogyne incognita race 3* b. *M. incognita race 1*
    c. *M. javanica race 1* d. *M. arenaria race 2*
22. Cluster bean may be used to differentiate host races of
    a. *Heterodera glycines* b. *H.ciceri* c. *H.cajani* d. *H.trifoli*
23. This tissue is involved in the induction of nurse cells by Tylenchulus semipenetrans
    a. Pericycle b. Cortical parenchyma c. Endodermis d. Vascular tissues
24. This root stock is resistant to *Pratylenchus coffeae*
    a. *Coffea arabica* b. *C.robusta* c. *C.liberica* d. *C.canephora*
25. The most commonly used RKI scale is
    a. 0-4 b. 0-8 c. 1-5 d. 0-5
26. Resistance to cyst nematodes is based mainly on the inhibition of
    a. Molting b. Hatching c. Penetration d. Movement
27. Presence of these in host plants analogous to antibodies in animals responds to nematode infection
    a. Collagen b. Agglutin c. Collagenin d. Agglutinins
28. Normally, nematode resistant varieties contain more of
    a. Polyphenol oxidase b. Phenolic oxidase c. Free polyphenol d. Conjugated polyphenol
29. The major sterol traced in nematodes
    a. Lathosterol b. Cholesterol c. Campesterol d. Cholestanol
30. This bioagent secretes nematode cuticle dissolving enzymes
    a. Fungus b. Protozoa c. Mycorrhiza d. Bacterium

31. Soil incorporation with the chopped leaves of this crop minimizes root-knot nematode population
    a. Sapota b. Mango c. Guava d. Pineppal
32. Early planting of this crop leads to escape of nematode attack
    a. Tomato b. Banana c. Potato d. sugar beet
33. Crop rotation with this crop can reduce the incidence of *Meloidogyne hapla* on groundnut
    a. Linseed b. Sunflower c. Maize d. Sesamum
34. Brown ring of daffodils and Hyacinth is due to
    a. *Ditylenchus destructor* b. *D. dipsaci*
    c. *D.galliae* d. *D.angustus*
35. Telogenic ovary is
    a. Oocytes are produced throughout its length b. Oocytes are produced at its proximal end
    c. Oocytes are produced at its distal end d. Oocytes preoduced Iin lateral end
36. Meloidogyne has rapid reproduction rate because
    a. Ovary-body length ratio is more than three b. Ovary-body length ratio is less than three
    c. Ovary-body length ratio is more than one d. Ovary-body length ratio is less than one
37. Organs of copulation found in male nematode are
    a. "Z" organ b. Phasmids c. Bursa d. Spicules
38. A peloderan bursa covers
    a. Only tail tip b. Only cloacal opening c. Encircle tail upto tip d. Two thirds of tail
39. Nematodes with one testis are called
    a. Polyorchic b. Diorchic c. Monarchic d. Monorchic
40. In diorchic condition the number of testis present in
    a. One b. Two c. Three d. None of the above
41. A sphincter muscle is associated with
    a. Cloaca b. Gubernaculums c. Ovary d. Anus
42. A cuticularied structure lying beneath the pair of spicule is known as
    a. Bursa b. Vas deferens c. Cloaca d. Gubernaculums
43. In the uterus sperms normally move in this fashion
    a. Amoeboid b. Swarming c. Crawling d. Irregular
44. *Partylenchus penetrans*, *Partylenchus vulni* and *Partylenchus coffeae* are bisexual nematodes with
    a. Parthenogenic reproduction b. Amphimetric reproduction
    c. Both a and b d. Intersex
45. An intersex is an individual which exhibits a blending of
    a. Male characters b. Female characters c. Only male characters d. both a and b
46. Intersex is found in genera
    a. *Meloidogyne* b. *Ditylenchus* c. *Xiphinema* d. *Both a and b*
47. Sex reversal is observed in
    a. *Globodera* b. *Ditylenchus* c. *Meloidogyne* d. *Radopholus*
48. The following induces sex reversal in nematodes
    a. IAA b. Morphactin c. Cytokinins d. Ethylene

49. The genital primordium consists of thid number of germinal cells in Radopholus similis

a. 2 b. 4 c. 3 d. 6

50. The highest number of chromosomes (3n=54) occurs in

a. *Meloidogyne arenaria* b. *Heterodera*

c. *M. incognita* d. *Radopholus similis*

ANSWER KEY FOR XIII

Sl.No.	Ans.	Sl.No.	Ans.	Sl.No.	Ans.	Sl.No.	Ans.	Sl.No.	Ans.
1	d	11	b	21	a	31	d	41	a
2	b	12	c	22	c	32	c	42	d
3	c	13	d	23	b	33	c	43	c
4	a	14	a	24	d	34	b	44	b
5	b	15	d	25	d	35	b	45	d
6	b	16	d	26	b	36	a	46	d
7	c	17	a	27	a	37	d	47	c
8	a	18	b	28	b	38	a	48	b
9	d	19	c	29	b	39	b	49	a
10	a	20	d	30	d	40	d	50	b

XIV

1. Nematodes have this symmetry
   a. Bilateral b. Lateral c. Radial d. Trilateral
2. 'Pseudocoelom' denotes
   a. Body cavity line by mesodermal epithelium on both sides
   b. Body cavity formed as a result of gastrulation
   c. Body cavity filled with fluid
   d. Body cavity not lined by mesodermal epithelium on both sides
3. This structure helps in maintaining the shape of nematodes
   a. Pseudocoelom b. Cuticle and muscle c. Pseudocoelomic fluid d. Annulations of cuticle
4. Nematodes are
   a. Diploblastic b. Triploblastic c. Tetrablastic d. Monoblastic
5. Nematode body shape is typically
   a. Vermiform b. Pleomorphic c. Round d. Rod
6. Nematodes are also known as
   a. Flat worms b. Earth worms c. Eel worms d. Tape worms
7. Females of root knot nematodes are
   a. Spherical shaped b. Melon shaped c. Tubular shaped d. Kidney shaped
8. Cyst nematode females are
   a. Kidney shaped b. Melon shaped c. Tubular shaped d. Spherical shaped
9. Females of reniform nematode, *Rotylenchus reniformis* are of
   a. Round shaped b. Tubular shaped c. Melon shaped d. Kidney shaped
10. Citrus nematode females are
    a. Vermiform b. Tubular shaped c. Spiral shaped d. Cocci shaped
11. The length of the nematode may
    a. 0.01 mm to 10.0 mm b. 0.01mm to 11.0 mm
    c. 0.2 mm to 10.0 mm d. 0.3 mm to 11.0 mm
12. The body width of nematode may vary from
    a. 0.01 mm to 0.5 mm b. 0.1 mm to 0.7 mm c. 0.2 mm to 0.9 mm d. 0.3 mm to 1.0 mm
13. Smallest nematode is
    a. *Paralongidorus maximus* b. *Paratylenchus hamatus*
    c. *Pratylenchus coffeae* d. *Longidorus elongatus*
14. Longest nematode is
    a. *Belanolaimus longicaudatus* b. *Longidorus elingatus*
    c. *Radopholus similis* d. *Paralongidorus maximus*
15. The nematodes lack organs for
    a. Circulation b. Respiration c. Both a and b d. Excretory

16. The nematode lack
    a. Digestive and excretory system
    b. Circulatory and respiratory system
    c. Reproductive and circulatory system
    d. Nervous and muscular system
17. All the natural openings of nematodes are present on
    a. Ventral side
    b. Dorsal side
    c. Laterel side
    d. Not confined to any particular side
18. Exchange of gases takes place through
    a. Hypodermis b. Cuticle c. Microvilli d. Rectum
19. The feeding stylet and copulatory spicules are formed from
    a. Hypodermis b. Outer body tube c. Cuticle d. Muscle
20. Under high pressure the nematode body maintains its shape with the aid of
    a. Cuticle b. Fluid filled interior c. Muscles d. Stylet
21. The structures which helps in strengthening of the cuticle
    a. Punctuations
    b. Transverse striations
    c. Longitudinal markings
    d. Caudal alae
22. Punctations are visible in
    a. Rugose wall pattern of *Heterodera*
    b. Perineal patern of *Meloidogyne hapla*
    c. Scales of *Criconematids*
    d. Both a and b
23. Transverse striations pronounced in
    a. Tylenchidae b. Heteroderidae c. Criconematicidae d. All
24. The perineal pattern in *Meloidogyne* and rogues wall pattern of *Heterodera* are
    a. Modification of punctuations
    b. Modification of longitudinal markings
    c. Modification of transverse
    d. Modificaitons of lateral lines
25. The following aids taxonomists in species identification
    a. Caudal alae b. Ridges c. Cervical alae d. Longitudinal alae
26. The presence alae in the anterior part of the body around the neck is known as
    a. Cervical alae b. Caudal alae c. Longitudinal alae d. Transvverse alae
27. Caudal alae which are restricted to males as coupulatory bursa are present in the
    a. Anterior region b. Ventral region c. Lateral region d. Posterior region
28. The nematode cuticle is basically
    a. 1-layerd structure b. 2-layerd structure c. 3- layerd structure d. 4-layered
29. Cortical layer of nematode comprises of
    a. Keratin b. Dermin c. Tannin d. Paraffin
30. Chemically median layer consists of
    a. Lipids b. Collagen c. Esterase d. Fats
31. The nematode cuticle is secreted by
    a. Matrix b. Fibrilar layer c. Cortex d. Hypodermis

32. Lorge amount of glycogen is stored in
    a. Cuticle b. Cloaca c. Hypodermis d. Buccal cavity
33. Nematode muscle cells are situated in the
    a. Intercaudal zone of hypodermis b. Body cavity
    c. Adjacent to cuticle d. Matrix layer
34. The main function of hypodermis is
    a. Serves as exoskeleton for attachment of muscles
    b. Secretion of cuticle
    c. To accommodate longitudinal nerves and excretory canals
    d. Storage of reserve food material
35. Hypodermal glands act as
    a. Osmotic regulators b. Ionic regulators
    c. Both a and b d. Either a or b
36. The caudal glands of the hypodermis which secrete adhesive substances for anchoring of nematodes are found in
    a. Lateral region b. Ventral region c. Tail region d. Dorsal region
37. Sensory organs like amphids, phasmids and deirids are associated with the
    a. Hypodermal glands b. Caudal glands
    c. Oesophgeal glands d. Rectal glands
38. In nematodes the longitudinal running hypodermal chords are
    a. One b. Three c. Four d. Five
39. Which is not a unique feature of nematode muscles?
    a. Spindle shape
    b. Oblique striation
    c. Innervations processes joining nerve chords
    d. Presence of z-bundles
40. Protein predominantly present in thick mycofilaments of nematodes
    a. Tropomyosin b. Myosin c. Actin d. Paramyosin
41. Protein predominantly present in thin mycofilaments of nematodes
    a. Tropomyosin b. Paramyosin c. Actin d. Myosin
42. In the outer body tube of the nematode the muscle layer is responsible for
    a. Locomotion b. Secretion c. Respiration d. Digestion
43. On the basis of arrangement of muscle cells the following types are identified
    a. Holomyarian b. Meromyarian c. Polymyarian d. All
44. Holomyarian consists of
    a. 2 muscle cells b. 2-5 muscle cells
    c. More than 5 muscle cells d. 1 muscle cell

45. Muscle cells present in meromyarian

a. 1 muscle cells
b. 2 muscle cells
c. 2-5 muscle cells
d. More than 5 muscle cells

46. Polymyarian consists of

a. 2 muscle cells
b. 3 muscle cells
c. 2-5 muscle cells
d. More than 5 muscle cells

47. Types of muscle cells based on shape

a. Platymyarian b. Coelomyarian c. Circomyarian d. All

48. Platymyarian cells are usually

a. Flat shaope b. U shaped c. Round d. Rod shaped

49. Platymyarian oells arc usually

a. Circomyarian b. Coelomyarian c. Meromyarian d. Polymyarian

50. The shape of coelomyarian is

a. Round b. U shaped c. Rods d. Pleomorphic

ANSWER KEY FOR XIV

Sl.No.	Ans.	Sl.No.	Ans.	Sl.No.	Ans.	Sl.No.	Ans.	Sl.No.	Ans.
1	a	11	c	21	a	31	d	41	c
2	d	12	a	22	b	32	c	42	a
3	c	13	b	23	d	33	a	43	d
4	b	14	d	24	c	34	b	44	a
5	a	15	c	25	b	35	d	45	c
6	c	16	b	26	a	36	c	46	d
7	b	17	a	27	d	37	a	47	d
8	d	18	b	28	c	38	c	48	a
9	d	19	c	29	a	39	d	49	c
10	b	20	a	30	b	40	b	50	b

XV.

1. Coelomyarian cells are usually
   a. Meromyarian b. Coelomyarian c. Circomyarian d. Polymyarian
2. The coelomyarian cell is virtually a storage depot for
   a. Glycogen b. Lipids c. Proteins d. Starch
3. Muscle cells are connected to each other by means of
   a. Alpa- 1, 4 glycosidic linkage b. Peptide linkage
   c. Cytoplasmic bridges d. Phenolic linkages
4. Muscles serving as ovijectors are found in this nematode organ
   a. Vulva b. Spicules c. Testis d. Uterus
5. This is the chemical transmitter, which contracts muscles in nematodes
   a. Acetylcholine b. Chlorogenic acid
   c. Pentose monophosphate d. Lactose dinine
6. The stomodeum includes
   a. Buccal cavity b. Pharynx c. Mouth and lips d. All
7. All known plant parasitic nematodes possess
   a. Amphids b. Phasmids c. Stylet d. Spicule
8. In Secernentea the stylet is derived from the fusion of stomatal linig known as
   a. Stomatostylet b. Odontostylet c. Onchiostylet d. Ontarostylet
9. In adenophorea yhe stylet is thought to be derived from a tooth which is termed as
   a. Onchiostylet b. Odontostylet c. Stomatostylet d. a + b + c
10. Stomatostylets are present in
    a. Dorylaimids b. Rhabditids c. Trchodorids d. Tylenchids
11. In dorylaimids stylets develop from
    a. Oesophagus b. Buccal cavity c. Mouth d. Median bulb
12. Nematodes penetrate plant tissues with the aid of
    a. Saliva b. Labial papilla c. Stylet d. Cervical
13. Stylet knobs are tulip shaped in
    a. *Hoplolaimus* b. *Trichodorus* c. *Xiphinema* d. *Meloidogyne*
14. Anchor sahped stylet knob is present in
    a. *Longidorus* b. *Xiphinema* c. *Hoplolaimus* d. *Criconematids*
15. Plant parasitic nematodes possess
    a. Two-part oesophagus b.Three-part oesophagus
    c. Four-part oesophagus d. Single part
16. Oesophagus is the largest part of stomodeum and found between
    a. Stoma and intestine b. Stoma and rectum
    c. Intestine and rectum d. Cardia and median bulb

17. Lumen in nematode is divided into
 a. Four parts b. Two parts c. Three parts d. No parting
18. The oesopageal lumen in nematodes is
 a. Asymmetry b. Triradiate symmetry c. Bilaterally symmetry d. Radial symmetry
19. In dorylaimodea the oesophagus is divided into
 a. Anterior procorpus b. Posterior corpus c. Isthmus d. Both a and b
20. In tylenchida the oesophagus is divided into
 a. Procorpus b. Posterior bulb c. Isthmus d. a + b + c
21. Median bulb is large and square shaped in
 a. Mononchids b. Criconematids c. Aphelenchids d. Hoplolaimus
22. Median bulb is a
 a. Muscular pumping organ b. Glandular
 c. Globular d. Squarish
23. Nematode oesophagus shows
 a. Asymmetry b. Radial symmetry c. Bilateral symmetry d. Triradiate symmetry
24. The lumen of the oesophagus consist of one or more these muscular or globular swellings
 a. Bulbs b. Glands c. Both a and b d. Villi
25. A chamber functioning as efficient pump in case of tylenchids
 a. Valve apparatus b. Oesophagus c. Stylet-knobs d. Intestine
26. The oesophageal intestine is called
 a. Butterfly valve b. Phasmids c. Cardia d. Amphids
27. Cardia helps in preventing
 a. Absorption b. Passage c. Digestion d. Regurgitation
28. The finger like projections of the plama membrane which increase the surface area of the intestine is known as
 a. Microvilli b. Macrovilli c. Both a and b d. Semivilli
29. Tylenchids and Aphelenchids differ most conspicuously by
 a. Prominence of stylet knobs b. Shape of spicule
 c. Presence/absence of bursa d. Opening of dorsal oesophageal gland
30. The intestinal cells are richg in
 a. Glycogen b. Lipids c. Protein d. a + b + c
31. When the number of cells is less than 128 the intestine is termed as
 a. Polycytous b. Oligicytous c. Heterocytous d. Mycriocytous
32. If the number of cells exceeds 128, the intestine is knownn as
 a. Heterocytous b. Oligicytous c. Polycytous d. Nevirocytous
33. The plant parasitic nematodes the intestine is
 a. Oligicytous b. Mycriocytous c. Polycytous d. Heterocytous

34. Intestinal cells dissimilar in size and shape
    a. Polycytous b. Oligicytous c. Anisocytous d. Isocytous
35. Proctodeum comprises of
    a. Rectum b. Anus c. Intestine d. Both a and b
36. The titillate small projections are commonly associated with
    a. Stylet b. Rectum c. Gubernaculum d. Cloaca
37. In case of males the reproductive system joins the rectum to form a common tube called
    a. Cloaca b. Gubernaculums c. Microvilli d. Anus
38. During moulting new cuticle formation is greatly helped by this structure
    a. Psedocoel b. Oesophageal glands
    c. Hypodermis d. Intestine
39. The oesophagel or pharyngeal glands enter
    a. Intestine b. Proctodeum c. Stomodeum d. Nerve ring
40. Oesophageal glands overlap the intestine ventrally in
    a. *Pratylenchus* b. *Hirscmanniella* c. *Meloidogyne* d. a + b + c
41. All the three oesophageal glands open into median bulb in
    a. Tylenchids b. Aphelenchids c. Heteroderids d. Rhabditids
42. Oesophageal glands overlapsthe dorsal side of the intestine
    a. *Hoplolaimus and Radopholus* b. *Radopholus and Hirchmanniella*
    c. *Hoplolaimus and Pratylenchus* d. *Hoplolaimus and Anguina*
43. Intestine in this genus overlaps oesophageal glands in a long lobe
    a. *Aphelenchus* b. *Aphelenchoides* c. *Bursaphelenchus* d. *Rhadinaphelenchus*
44. A butting type of oesophageal gland is present in
    a. *Rotylenchulus* b. *Tylenchulus* c. *Tylenchorhynchus* d. *Hoplolaimus*
45. In rootknot nematodes gelatinous matrix is secreted by
    a. Rectum b. Rectal glands c. Oesophageal gland d. Posterior bulb
46. The feeding phase of nematode is known as
    a. Injection phase b. Ingestion phase c. Digestion phase d. a and b
47. The chemical changes taking place inside the plant tissues in response to nematode feeding as
    a. Extra corporeal digestion b. Predigestion
    c. Biochemical digestion d. Hydrolysis
48. Large fixed cells in pseudocoel are termed as
    a. Pseudocoelocytes b. Pseudocyes c. Coelomycets d. Psedocyelite
49. Pseudocoelom consists of
    a. Nervous system b. Excretory system c. Reproductive system d. a + b + c
50. Nervous system in plant parasitic nematodes is
    a. Peripheral nervous system b. Central nervous system
    c. Reproductive system d. Lateral type

ANSWER KEY FOR XV

Sl.No.	Ans.	Sl.No.	Ans.	Sl.No.	Ans.	Sl.No.	Ans.	Sl.No.	Ans.
1	d	11	a	21	c	31	b	41	b
2	a	12	c	22	a	32	c	42	a
3	c	13	c	23	d	33	a	43	d
4	b	14	d	24	c	34	c	44	c
5	b	15	b	25	b	35	d	45	b
6	d	16	a	26	c	36	b	46	d
7	b	17	c	27	d	37	a	47	a
8	a	18	b	28	a	38	b	48	c
9	b	19	d	29	d	39	c	49	d
10	d	20	d	30	d	40	d	50	b

XVI.

1. Circumoesophageal commissure is a part of
   a. Nervous system b. Digestive system c. Circulatory system d. Excretory system
2. The sensory structure found in nervous system of nematodes are
   a. Hemizonoids b. Caudalids c. Cephalids d. a + b + c
3. Chemo sensory organs located at the anterior end of the nematodes are called
   a. Mouth b. Phasmids c. Stylet d. Amphids
4. Paired lateral cephalic sence organs of nematodes are
   a. Phasmids b. Amphids c. Caudalid d. Hemizonid
5. The nematode organ whiuch has important sensory function is
   a. Amphid b. Ampulla c. Bursa d. Dierid
6. Caudal sensory organs in nematodes are called
   a. Stylet b. Spicules c. Phasmids d. Caudalids
7. The enlargement of phasmids in few nematodes is known as
   a. Bursa b. Scutella c. Cephalic papillae d. Caudalids
8. The caudal sensory organs associated with reproduction exhibit
   a. Parthenogenesis b. Intersex c. Sexual dimorphism d. Hermophroditism
9. A sense organ consisting of of a small elevation of the cuticle at the end of a nerve fiber
   a. Papilla b. Phasmid c. Ampulla d. Hemizonioms
10. These pouch like sense organs are combinations of nerve endings and gland cells
    a. Hemizonids b. Amphids c. Ampullae d. Papillae
11. These are not related to nervous system in nematodes
    a. Scales and spines b. Setae and papillae
    c. Amphids and phasmids d. Hemizonid and hemizonion
12. In Tylenchida the nerve ring encircles the middle part of
    a. Intestine b. Oesophagus c. Pharynx d. Intestine
13. In Dorylaimida nerve ring is present around the
    a. Ventral region of oesophagus b. Posterior of oesophagus
    c. Narrow anterior of oesophagus d. Dorsal region of pharynx
14. Neurotransmitter widely present in nematodes
    a. Dopamine b. Adrenaline c. Noradrenaline d. Acetylcholine
15. Excretory system present in nematode
    a. Canalicular b. Glandular c. Tubular d. a and b
16. Excretory pore in nematode is located in
    a. Cephalic end b. Caudal end
    c. The anterior one third d. The posterior one third

17. In *meloidogyne javanica* the excretory pore is located near the
    a. Posterior region
    b. Basal region of median bulb
    c. Ventral region
    d. Anterior region
18. In *Tylenchulus semipenetrans* the excretory pore is located in the
    a. Posterior region
    b. Dorsal region
    c. Ventral region
    d. Lateral region
19. Tubular type of excretory system is
    a. Tylenchid type (asymmetrical)
    b. Ascarid type (inverted U shape)
    c. Rhabditid type
    d. a + b + c
20. The excretory system present in majority of the plant parasitic which fall under the order Tylenchida is
    a. Asymmetrical tubular type
    b. Rhabditid type
    c. Oxyurid type
    d. Ascarid type
21. Glandular type consists of a single specialized cell known as
    a. Nurse cell
    b. Renette cell
    c. Giant cell
    d. Syncytium
22. Glandular type of excretory system is present in
    a. Rhabditida
    b. Secernentea
    c. Adenophorea
    d. Tylenchida
23. The main function of excretory system is
    a. Excretion of toxic substances
    b. Secretion of certain chemicals
    c. Osmoregulation
    d. a + b + c
24. These are very efficient Osmoregulators
    a. Animal parasitic nematodes
    b. Plant parasitic nematodes
    c. Marine nematodes
    d. Free living nematodes
25. Majority of nematodes are
    a. Dioecious
    b. Monoecious
    c. Unisexual
    d. Varies
26. Sexual dimorphism means
    a. Males and females are separated
    b. Females possess didelphic ovaries
    c. Males and females are morphologically dissimilar
    d. Males are monorchic
27. Sexual dimorphism is seen in
    a. *Meloidogyne*
    b. *Heterodera*
    c. *Globodera*
    d. a + b + c
28. Sexual dimorphism is exhibited by
    a. *Rotylenchulus*
    b. *Tylenchulus*
    c. *Nacobbus*
    d. a + b + c
29. Nematodes reproduce mainly through
    a. Viviparity
    b. Parthenogenesis
    c. Amphimixis
    d. Polyembryony
30. Mitotic parthenogenesis is found in this nematode
    a. *Meloidogyne*
    b. *Radopholus*
    c. *Heterodera*
    d. *Rotylenchulus*
31. Parthenogenesis is very common in
    a. *Heterodera*
    b. *Meloidpgyne*
    c. *Hemicycliophora*
    d. a + b + c

32. Parthenogenic nematode are grouped under
    a. Hoplolaimidae b. Tylenchidae c. Rhabditidae d. Heteroderidae
33. Male nematodes differ with female nematodes in
    a. Male nematodes have nervous system
    b. Male nematodes are cylindrical and filiform while females are bigger in size
    c. Male possesses stylets and females possess spicules
    d. Absence of sense organs
34. Spermatheca is found in the nematode body of
    a. Adult female b. Adult Male c. Both d. Juveniles
35. The females of all nematodes can be distinguished by the presence of
    a. Valva b. Vagina c. Spermatheca d. a + b + c
36. In monodelphic condition, always the number of ovaries found is
    a. One b. Two c. Three d. Four
37. Monodelphic ovaries are found in
    a. *Ditylenchus* b. *Pratylenchus* c. *Hoplolaimus* d. a and b
38. In monodelphic prodelphic the ovary is
    a. Anteriorly directed b. Posteriorly directed
    c. One anterior and one posterior d. a and b
39. In monodelphic ophistodelphic the ovary is
    a. Anteriorly directed b. One anterior and one posterior
    c. Posteriorly directed d. a and b
40. Monodelphic ophistodelphic is found in
    a. *Meloidogyne* b. *Xiphinema* c. *Radopholus* d. *Ditylenchus*
41. Nematodes with two ovaries are known as
    a. Didelphic b. Ophistodelphic c. Monodelphic d. Diorchic
42. In didelphic prodelphic both ovaries are
    a. Posteriorly directed b. Anteriorly directed
    c. One anterior and one posterior d. Varies
43. Didelphic ovaries are found in
    a. *Meloidogyne* b. *Heterodera* c. *Globodera* d. a + b + c
44. Amphidelphic means nematodes with two ovaries
    a. Anteriorly directed b. Posteriorly directed
    c. One anterior and one posterior d. Varies
45. Examples for amphidelphic ovaries
    a. *Tylenchorhynchus* b. *Hoplolaimus* c. *Helicotylenchus* d. a + b + c
46. When both ovaries are posteriorly directed the condition is known as
    a. Ophistodelphic b. Amphidelphic c. Prodelphic d. Kinodelphic

47. The ovary which is always straight and does not curve back in most of the plant parasitic nematodes is known as

a. Reflexed ovary b. Outstretched ovary c. Condenced ovary d. Straight ovary

48. Outstretched ovaries are found in

a. *Tylenchorhynchus* b. *Radopholus* c. *Hirschmanniella* d. a + b + c

49. In dorylaimid nematodes tip of the ovary is curved back. Such ovary is known as

a. Reflexed ovary b. Condenced ovary c. Outstretched ovary d. Vurved ovary

50. Hologenic ovary is

a. Oocytes are produced at its distal end

b. Oocytes are produced throughout its length

c. a and b

d. Oocytes are produced laterally

ANSWER KEY FOR XVI

Sl.No.	Ans.	Sl.No.	Ans.	Sl.No.	Ans.	Sl.No.	Ans.	Sl.No.	Ans.
1	a	11	a	21	b	31	d	41	b
2	d	12	b	22	c	32	c	42	a
3	d	13	c	23	d	33	b	43	b
4	b	14	d	24	b	34	a	44	d
5	a	15	d	25	a	35	d	45	c
6	c	16	c	26	c	36	a	46	d
7	b	17	b	27	d	37	d	47	a
8	c	18	a	28	d	38	a	48	b
9	d	19	d	29	b	39	c	49	d
10	c	20	a	30	a	40	b	50	a

XVII.

1. In a typical phytonematode life cycle, number of stages are
   a. 2 b. 4 C. 6 d. 8
2. The number of larval stages found in a nematode is
   a. One b. Two c. Three d. Four
3. Nematode moult
   a. Once b. Twice c. Thrice d. Four times
4. Nematodes usually undergo first moult
   a. Before hatching b. After hatching c. During hatching d. Varies
5. The pre-hatch moult may also be absent in some nematodes
   a. Parasitic b. Saprophytic c. Marine d. Free living
6. Larvae of *Xiphinema index* are reported to emerge from the egg before
   a. Second moult b. First moult c. Thitd moult d. Not specific
7. In blastula stage the cells are arranged to form a fluid sphere which is bounded by
   a. Single layer of cells b. Two layer of cells c. Three layer of cells d. Absence of cells
8. In gastrula stage the early embryo consists of an open mouthed sac like body with
   a. No layer of cells b. Single layer of cells c. Two layer of cells d. Four layer of cells
9. The layer present in typical nematode egg shell
   a. Lipid layer b. Chitinuous layer c. Vitelline layer d. Crystalline layer
10. The layer which offers resistance to nematode eggs under adverse environmental conditions
    a. Chitinuous layer b. Lipid layer c. Vitelline layer d. Micro layer
11. Chitin is present only in
    a. Eggshell b. Cuticle c. Spicule d. Stylet
12. In plant parasitic nematodes embryonic development has been studied in
    a. *Radopholus similis* b. *Ditylenchus dipsaci*
    c. *Meloidogyne incognita* d. *Both a and b*
13. The process of pushing ventral surface of nematode embryo into the interior is known as
    a. Epiboly b. Endoly c. Ectoboly d. Ecoboly
14. Artificial hatching agent to hatch potato cyst nematode is
    a. Anhydrotetronic acid b. Picrolinic acid
    c. Sodium metavandate d. a + b + c
15. Following is the hatching factor from root exudates of soybean
    a. $H_{18}C_{20}O_6$ b. $H_{14}C_{24}O_8$ c. $H_{18}C_{24}O_6$ d. $H_{10}C_{20}O_{12}$
16. There is a definite relationship between the onset of moulting and this aspect of nematode
    a. Feeding habit b. Genus c. Size d. Parasitism
17. The most characteristic feature of post embrytonic development in plant parasitic nematodes is

a. Ecdysis b. Moulting c. Both a and b d. Giant cell

18. Pre-hatch moulting has not been noticed in
   a. *Tylenchulus semipenetrans* b. *Meloidogyne incognita*
   c. *Rotylenchulus reniformis* d. *Aphelencoides besseyi*
19. The larva ensheathed by the cuticles of previous stage is called
   a. Cyst b. Neonate larva c. Old larva d. Dauer larva
20. A mass of resistant desiccated larva frequently found on bulb crops is
   a. Dauer larva b. Firefly blast c. Cyst d. Desiccated nematode
21. The immature stage of nematodfe are called as
   a. Larva b. Moults c. Juveniles d. Nymphs
22. The infective stage in all endoparasitic nematodes is
   a. J1 stage b. J2 stage c. J3 stage d. Adult
23. This stage of embryo formation is not found in all species of nematodes
   a. Blastula b. Psedoblatula c. Gastrula d. Coelograstula
24. Eggs are laid in gelatinous matrix in
   a. *Meloidogyne incognita* b. *Globodera*
   c. *Heterodera* d. *Anguina*
25. The rate of egg laying in *Radopholus similis* and *Aphelenchoides ritzemabosi* is
   a. 4 & 2 b. 2 & 4 c. 4 & 8 d. 6 & 4
26. The fecundity ratevof ectoparasites and migratory endoparasites ranges between
   a. 5-10 b. 10-25 c. 25-35 d. 30-45
27. The fecundity rate of sedentary endoparasites and semi endoparasites ranges between
   a. 50-100 b. 100-250 c. 50-500 d. 300-450
28. The cyct nematode feeding stimulates the development of a typical cell system called
   a. Host cells b. Syncytium c. Nurse cells d. Feeding tubes
29. Development of giant cell involves this mechanism
   a. Repeated mitosis without cytokinesis b. Repeated mitosis with cytokinesis
   c. Repeated meiosis without cytokinesis d. Repeated meiosis with cytokinesis
30. Growing root knot nematodes get nourishment from
   a. Giant cells b. Nurse cells c. Feeding cells d. Host cells
31. This genus of nematodes has the special types of feeding cell systems which serve as source of continuous supply of nutrients to the nematodes
   a. *Meloidogyne* b. *Heterodera* c. *Globodera* d. a + b + c
32. The second stage juveniles infective in
   a. Cyst nematode b. Root knot nematode
   c. Stem and bulb nematode d. a + b
33. Infective stage in all species of Meloidogyne is
   a. Second stage juvenile b. Pre adult female

c. Fourth stage juvenile d. First stage juvenile

34. The dead heteroderid nematodes containing viable eggs are called

a. Dauer larva b. Cysts c. Egg sac d. Z organ

35. In this genus due to quinine tanning, presence of white to brown cysts projecting on the root surface is a characteristic symptom

a. *Longidorus* b. *Trichodorus* c. *Meloidogyne* d. *Globodera*

36. After removal from the host cysts turn into this colour

a. Dark brown b. Black c. Light brown d. Orange brown

37. Acid phosphatase has been frequently reported from the following nematode

a. Root knot b. Cyst c. Dagger d. Lesion

38. The hydrolytic enzyme produced by potato cyst nematode

a. Cellulose b. Invertase c. Alkaline phosphatase d. Alpa-galactosidase

39. Not all species increase in size as they pass through larval stage to maturity. There may even be a decrease as in

a. Citrus nematode b. Cyst nematode c. Root knot nematode d. Lesion nematode

40. The cells formed due to the feeding of *Tylenchulus semipenetrans*

a. Transfer cells b. Gaint cells c. Nurse cells d. Syncytium

41. "nurse cells" are the modification of these tissues

a. Cortex mesophyll b. Cortex-epithelial c. Cortex-parenchyma d. Vascular

42. These cells are absent in nematode

a. Hypodermal b. Flame cells c. Mesodermal cells d. Endodermal cells

43. A large egg sac at the posterior end of female is present in the cyst nematode

a. *Heterodera avenae* b. *H. sorghi* c. *H. cajani* d. *H. zeae*

44. Plant parasitic nematodes generally contain more

a. Proteins b. Enzymes c. Polyphenols d. Lipids

45. The major energy reserve in plant parasitic nematode is

a. Lipids b. Carbohydrates c. Proteins d. Triacylglycerols

46. The mechanism of host finding in nematodes was proposed by

a. Zukerman b. Jenson c. Linford d. a and b

47. The most accepted theory of chemotaxis based upon interactions between molecules emanating from host and nematode was proposed by

a. Kuhn, 1959 b. Klink et al., 1970

c. Zukerman & Jenson, 1984 d. Van gundy & Rackman, 1961

48. In general, soil-inhabiting nematodes require ample supply of

a. CO_2 b. O_2 c. H_2O d. N

49. Oxygen consumption by intact specimens of several species of plant parasitic nematodes are measured by

a. Thermometer b. Tensiometer c. Microrespirometer d. Tensiorespirator

50. 'Some nematode species that occur in soil obtain sufficient O_2 from roots of plants'. This statement is

a. False b. Partly true c. Controversial d. True

ANSWERS KEY FOR XVII

Sl.No.	Ans.	Sl.No.	Ans.	Sl.No.	Ans.	Sl.No.	Ans.	Sl.No.	Ans.
1	c	11	a	21	c	31	d	41	a
2	d	12	c	22	b	32	d	42	b
3	d	13	a	23	d	33	a	43	c
4	a	14	d	24	a	34	b	44	d
5	d	15	b	25	a	35	d	45	a
6	b	16	c	26	b	36	c	46	d
7	a	17	b	27	[illegible]	37	b	47	c
8	c	18	c	28	d	38	d	48	b
9	d	19	b	29	b	39	a	49	c
10	b	20	c	30	a	40	c	50	b

XVIII.

1. Sometimes delay in hatching of cysts is mainly due to accumulation of
   a. CO_2 b. O_2 c. H_2O d. H_2S
2. This nematode can sense and get more repelled to CO_2
   a. Radopholus *similis* b. *Ditylenchus dipsaci*
   c. *Meloidogyne incognita* d. None of these
3. During aerobic starvation usually nematodes utilize these stored nutrients
   a. Carbohydrates b. Proteins c. Lipids d. Fats
4. The optimum temperature required for growth and development of plant parasitic nematodes is in the range of
   a. 5-30°C b. 10-25°C c. 15-35°C d. 20-35°C
5. The upper temperature limit for development of *meloidogyne* species and *globodera rostochiensis*
   a. 5-15°C b. 15-30°C c. 20-30°C d. 25-35°C
6. The optimum temperature required for egg production in root knot nematodes varies between
   a. 15-20°C b. 25-32°C c. 35-37°C d. Above 37°C
7. Globodera rostochiensis develops and produces eggs most rtapidly at
   a. 10°C b. 14°C c. 18°C d. 21°C
8. Natural environment of nematode is
   a. Terrestrial b. Aquatic c. Plant tissue d. Marine
9. Nematodes inhabiting the above ground parts of plants as well as soil inhabiting nematodes are resistant to
   a. Desiccation b. Moisture c. Temperature d. Alkaline soil
10. Suspended animation of nematode due to desiccation is called
    a. Anhydrobiosis b. Anoxybiosis c. Cryptobiosis d. Anabiosios
11. Nematodes normally shrink in this solution
    a. Hypotonic b. Hypertonic c. Alkaline d. Salt
12. Nematodes quickly becomes turgid and active in this solution
    a. Hypotonic b. Isotonic c. Alkaline d. Both a and b
13. The larvae of *Anguina tritici* in wheat galls can survive up to
    a. 28 years under dry conditions only b. 9-14 years under moist conditions only
    c. Both a and b d. Only few months
14. *Ditylenchus dipsaci* have been revived from infected plant material after
    a. 23 years of storage b. 25 years of storage c. 28 years of storage d. 32 years of storage
15. Hatching and development of nematodes is inhibited under
    a. High osmotic pressure b. Low osmotic pressure
    c. High moisture d. Low moisture

16. Following found that at pH 6 there is maximum hatching of Globodera rostochiensis
    a. Duggan b. Seshadri c. Williams d. Ellenby
17. Ammonium ions are responsible for hatching of Meloidogyne incognita eggs, was reported by
    a. Oteifa b. Ross c. Castaner d. Kimpinski
18. The stage at which *Anguina tritici* survives dormancy is
    a. First stage larva b. Second stage larva c. Pre adult d. Adults
19. In Ditylenchus and Aphelenchoides the resistant stages are
    a. Second stage larva b. Pre adult c. Adults d. Both b and c
20. Dormant stages of nematodes are particularly resistant to the action of
    a. Insecticides b. Nematicides c. Fungicides d. Bioagents
21. The reduction of population of plant parasitic nematodes in fallow soil follows
    a. Hyperbolic curve b. Symmetric curve c. Normal curve d. Asymmetric curve
22. Sex determination in nematodes is governed by
    a. Genetic mechanism b. Nutritional mechanism
    c. Physiological mechanism d. Both a and b
23. The most obvious factor influencing development of a nematode within the host is
    a. Soil b. Temperature c. Host specificity d. Moisture
24. The most promising method of propagating large quantities of plant parasitic nematodes in laboratory is
    a. Meristem tip culture b. Callus culture c. Tissue culture d. Sterile root culture
25. *Heterodera* and *Globodera* the gonads are clearly distinguished in
    a. First larval stage b. Second larval stage c. Thrid larval stage d. Adult stage only
26. *Meloidogyne* and Tylenchulus semipenetrans gonads can be distinguished in
    a. First larval stage b. Second larval stage c. Thrid larval stage d. Forth larval stage
27. *Meloidogyne hapla* and *Meloidogyne javanica* are strongly attracted to
    a. Low redox potential b. High redox potential
    c. Extreme redox potential d. None of the above
28. *Ditylenchus dipsaci* is reported to be attracted by
    a. Oxidizing agents b. Reducing agents c. Both a and b d. Either a or b
29. Various reducing agents which attract some species of nematodes
    a. Sodium dithionite b. Glutathione c. Cysteine d. a + b + c
30. Many species of nematodes aggregate at the source of oxidizing agents like
    a. Tyrosine b. Potassium permanganate
    c. Oxygen d. a + b + c
31. The apical 2 mm of excised root tips and the next 6 mm have these respective effects on the nematodes
    a. Repel, attractive b. Attract, repellent
    c. Attract, neutral effect d. Repel, neutral effect

32. The chemorecptor sense organ which plays an important role in the orientation of larvae to the source of chemicals

a. Amphids b. Phasmids c. Dierids d. Caudalids

33. Enzymes such as Amylase, Invertase and Pectinase have been detected in

a. *Globodera rostochiensis* b. *Heterodera schachtii*
c. *Ditylenchus destructor* d. a + b + c

34. Long distance spread of nematodes is usually by means of

a. Wind b. Seed
c. Water d. Infected planting material

35. Meloidogyne species can enter the host in wet soil through

a. Hydathodes b. Lenticels c. Stomata d. Nectarines

36. *Ditylenchus dipsaci and Aphelenchoides ritzemabosi* enter the host mainly through

a. Stomata b. Wounds c. Hydathodes d. Lenticells

37. Quiescence due to desiccation has been reported in

a. *Ditylenchus dipsaci* b. *Anguina agrostis*
c. *Tylenchus balsamophilus* d. *Aphelenchoides ritzemabosi*

38. In *Anguina* species the following stage undergoes quiescence

a. First stage larva b. Second stage larva c. Pre adult d. Adult

39. In *Ditylenchus dipsaci* which stage undergoes quiescence?

a. Second stage larva b. Third stage larva c. Fourth stage larva d. Adult

40. In *Aphelenchoides ritzemabosi* quiescence is undergone by

a. Third stage larva b. Fourth stage larva c. Pre adult d. Adults

41. Low temperature induced quiescence is reported in

a. *Ditylenchus dipsaci* b. *Ditylenchus destructor*
c. *Anguina tritici* d. *Aphelenchoides ritzemabosi*

42. Survival time of Anguina tritici in fallow soil

a. 2 to 17 weeks b. 1 year c. 1.5 to 2 year d. 3 years

43. *Aphelenchoides ritzemabosi* survives in fallow soil

a. 12 to 17 weeks b. 6 to 9 months c. 11 to 16 months d. 1 year

44. The survival time of *Globodera rostochiensis* larvae and cysts in fallow soil

a. 3 months and 2 years b. 6 months and 5 years
c. 9 months and 7 years d. 12 months and 7 years

45. *Meloidogyne* species survives without moisture for

a. 2 to 7 months b. 3 to 9 months c. 5 to 11 months d. 6 to 16 months

46. *Radopholus similis* survive in fallow soil for a period of

a. 3 to 6 months b. 9 to 12 months c. 1 year d. 1.5 to 2 years

47. The survival time of *Tylenchulus semipenetrans* in fallow soil is

a. 2 years b. 5 years c. 7 years d. 10 years

48. This group of nematodes develops very fast compared to other groups
    a. Phytonematodes   b. Animal parasites   c. Marine nematodes   d. Bacterial feeders
49. When nematode reverses its direction the waves of bending pass from
    a. Posterior to anterior   b. Anterior to posterior   c. Lateral to posterior   d. Anterior to lateral
50. For normal movement nematodes depend on
    a. External hydrodynamic pressure   b. External hydrostatic pressure
    c. Internal hydrodynamic pressure   d. Internal hydrostatic pressure

ANSWERS KEY FOR XVIII

Sl.No.	Ans.	Sl.No.	Ans.	Sl.No.	Ans.	Sl.No.	Ans.	Sl.No.	Ans.
1	c	11	b	21	a	31	b	41	a
2	d	12	d	22	d	32	a	42	b
3	a	13	c	23	c	33	d	43	a
4	b	14	a	24	b	34	c	44	c
5	c	15	c	25	c	35	b	45	d
6	a	16	d	26	b	36	a	46	a
7	c	17	a	27	a	37	a	47	d
8	b	18	b	28	c	38	b	48	c
9	a	19	d	29	d	39	c	49	b
10	d	20	b	30	d	40	d	50	a

XIX.

1. A homonym means
   a. Same name given to two different taxa
   b. Two different names given to same taxon
   c. Both a and b above
   d. Same name given to three different taxa
2. A lectotype is
   a. One of the syntypes selected subsequently to serve as a 'type'
   b. A specimen on which original description of the species is based
   c. Specimen other than holotype available at the time of original description
   d. A paratype opposite in sex to holotype
3. Phytonematology is the study of
   a. Nematodes
   b. Eel worms
   c. Plant parasitic nematodes
   d. Free living forms
4. The classification of plant parasitic nematodes is mainly based on
   a. Biological characters
   b. Cultural characters
   c. Physiological characters
   d. Morphological characters
5. Nematodes are placed within the kingdom
   a. Prokaryotes
   b. Animalia
   c. Eukaryotes
   d. Monera
6. The concept that nematodes belong to phylum Nemata was first proposed by
   a. N.A. Cobb
   b. Berkeley
   c. Goodey
   d. Paramanov
7. The concept that nematodes belong to phylum Nemata proposed by Cobb was strongly supported by
   a. Marcinowski
   b. Gubanov
   c. Triantaphyllou
   d. Maggenti
8. The phylum Nematoda or Nemata was reinststed in 1958 by
   a. Atkinson
   b. Raski
   c. Chitwood
   d. Oostenbrink
9. The latest classification of Nematoda was given by
   a. Siddiqi
   b. Hooper
   c. Golden
   d. a + b + c
10. The phylum Nemata is divided into two classes
    a. Secernentea /Phasmidia
    b. Adenophorea / Aphasmidia
    c. Both a and b
    d. Monochida /Dorylaimida
11. The phylum Nemata was divided into Secernentea and Adenophorea by
    a. Thorne (1961)
    b. Paramanov (1963)
    c. Barnes (1968)
    d. Chitwood (1974)
12. This group is most closely related to nematodes
    a. Priapulids
    b. Nematomorphs
    c. Earth worms
    d. Rotifers
13. The term 'adenophorea' is coined after
    a. Post labial amphids
    b. Glandular type of excretory system
    c. Absence of caudal glands
    d. Absence of phasmids
14. Most of the plant parasitic nematodes belong to order

a. Rhabditida b. Dorylamida c. Tylenchida d. Aphelenchida

15. In tylenchids stylet develop from

a. Mouth b. Buccal wall c. Oesophagus d. Median bulb

16. Stomatostylets are present in

a. Tylenchids b. Rhabditids c. Trichodorids d. Dorylaimids

17. In Tylenchida the Oesophagus is divided into

a. Procorpus b. Median bulb c. Isthmus d. a + b + c

18. The excretory system present in majority of the plant parasitic nematodes which fall under the order Tylenchida is

a. Asymmetrical tubular type b. Rhabditid type
c. Oxyurid type d. Ascarid type

19. The order tylenchida is divided into two sub-order

a. Tylenchina b. Aphelenchina c. Dorylaimina d. a and b

20. The suborder Tylenchina is divided into superfamily

a. Tylenchoidea b. Heteroderoidea c. Criconematoidea d. a + b + c

21. The superfamilies of suborder Tylenchina

a. Neotylenchoidea and Mesotylenchoides b. Atylenchoidea and Platytylenchoidea
c. Neotylenchoidea and Atylenchoidea d. Platytylenchoidea and Zeotylenchoidea

22. Tylenchids and aphelenchids differ most conspicuously by

a. Opening of dorsal oesophageal gland b. Prominence of stylet knobs
c. Presence/ absence of bursa d. Shape of spicule

23. Nematode members of this family have more hemispherical shaped heads

a. Tylenchidae b. Tylenchorhynchidae
c. Hoplolaimidae d. Pratylenchidae

24. The number of longitudinal incisures in Tylenchida varies from

a. 2-6 b. 4-8 c. 10-12 d. 8-10

25. Tylenchidae was earlier named as

a. Leptotylenchidae b. Nacobbidae c. Antarctidae d. Platylenchidae

26. Stem and bulb nematode *Ditylenchus dipsaci* infesting the heads of teasel was first discovered by

a. W. Carter (1943) b. Atkinson (1892) c. Neal (1894) d. Julius Kuhn (1857)

27. *Ditylenchus* is categorized under this family

a. Dolichodoridae b. Anguinidae c. Hoplolaimidae d. Neotylenchidae

28. This genus can grow steadily throughout life to maximum size

a. *Meloidogyne* b. *Heterodera* c. *Ditylenchus* d. *Anguina*

29. "Potato race, tulip race, onion race and daffodil race" refer to this nematode

a. Rot nematode b. Stem and bulb nematode
c. Foliar nematode d. Cyst nematode

30. Nematode wool is characteristic in

a. *Radopholus* b. *Hoplolaimus* c. *Meloidogyne* d. *Ditylenchus*

31. Ufra disease on Rice was first reported from Bengal by
   a. Milne b. Prasad c. Butler d. David
32. The first plant parasitic nematode to be noticed and reported by T. Needham in 1743 was
   a. *Anguina tritici* b. *Meloidogyne incognita*
   c. *Ditylenchus dipsaci* d. *Heterodera schachtii*
33. The first plant parasitic nematode was named as *Vibrio tritici* in 1799 by
   a. Filipjev b. Milne c. Steinbuch d. Needham
34. The present name of wheat seed gall nematode, *Anguina tritici* was given by
   a. Milne (1919) b. N.A. Cobb (1913) c. Steinbuch (1799) d. Filipjev (1936)
35. The life cycle of wheat seed gall nematode was studied by
   a. T. Needham (1743) b. J.G. Steinbuch (1799)c. C. Devaine (1857) d. M.J. Berkeley (1855)
36. Spicules are broad and short in case of
   a. *Anguina* b. *Ditylenchus* c. *Pratylenchus* d. a + b + c
37. *Tylenchorhynchus* belongs to super family
   a. Aphelenchoidea b. Tylenchoidea c. Dorylaimoidea d. Heteroderoidea
38. Abutting type of oesophageal gland is present in
   a. *Meloidogyne* b. *Tylenchulus* c. *Tylenchorhynchus* d. *Hoplolaimus*
39. Stunt or stylet nematode is
   a. *Criconema* b. *Anguina* c. *Xiphinema* d. *Tylenchorhynchus*
40. The common name of Tylenchorhynchus (stylet nematode) was given by
   a. *G. Steiner* b. *Filipjev* c. *N.A. Cobb* d. *Nicholas*
41. Awl nematode is
   a. *Paralongidorus* b. *Paratylenchus* c. *Dolichodorus* d. *Meloidodera*
42. Spear is usually long with rounded basal knobs in case of
   a. *Belonolaimus* b. *Dolichodorus* c. *Trichodorus* d. *Meloidogyne*
43. *Belanolaimus* is also referred to as
   a. Sting nematode b. Needle nematode c. Pin nematode d. Ring nematode
44. Family Pratylenchidae consists of subfamily
   a. Pratylenchinae b. Radopholinae c. Nacobbinae d. a + b + c
45. The genus *Pratylenchus* was created by
   a. Filipjev, 1936 b. T. Goodey, 1946 c. Chitwood, 1953 d. Cobb, 1937
46. Maximum contribution to taxonomyof the genus Pratylenchus is by
   a. Sher and Allen b. Café Filho and Huang
   c. Handoo and Golden d. a + b + c
47. Oesophageal glands overlap the intestine ventrally in
   a. *Trichodorus* b. *Ditylenchus* c. *Pratylenchus* d. a + b + c
48. All the 6 lip areas form a cephalic plate in this genus

a. *Xiphinema* b. *Pratylenchus* c. *Longidorus* d. *Tylenchus*

49. *Pratylenchus coffeae* was first described by

a. Golden b. Handoo c. Zimmermann d. Filho

50. *Pratylenchus indicus* infesting tomato and brinjal was described in 1960 by

a. Prasad b. Das c. Loof d. Allen

ANSWERS KEY FOR XIX

Sl.No.	Ans.	Sl.No.	Ans.	Sl.No.	Ans.	Sl.No.	Ans.	Sl.No.	Ans.
1	a	11	a	21	c	31	c	41	c
2	b	12	b	22	b	32	a	42	b
3	c	13	d	23	a	33	c	43	a
4	d	14	c	24	d	34	d	44	d
5	b	15	b	25	a	35	c	45	a
6	a	16	a	26	d	36	a	46	d
7	d	17	d	27	b	37	b	47	c
8	c	18	a	28	c	38	c	48	b
9	d	19	d	29	b	39	d	49	c
10	c	20	d	30	d	40	a	50	b

XX.

1. The lesion nematode was first described in 1880 by
   a. Atkinson b. Filipjev c. Hedwig Hirschmann d. de Man
2. Root lesion nematode is commonly known is
   a. *Pratylenchus* b. *Hoplolaimus* c. *Pratylenchus* d. *Helicotylenchus*
3. Lesion nematode is also called
   a. Needle nematode b. Meadow nematode c. Pin nematode d. Reniform nematode
4. The burrowing nematode was first observed in 1890 by
   a. Thorne b. Linford c. N.A. Cobb d. Oliveira
5. Genus *Radopholus* was first described in 1893 on the basis of a male specimen obtained from necrotic roots of which of the following crops growing in Fiji Island
   a. Coconut b. Sugar beet c. Coconut d. Banana
6. The genus *Radopholus* was described by
   a. Thorne b. N.A. Cobb c. Linford d. a and b
7. Oesophageal glands overlap the intestine dorsally in
   a. *Criconema* b. *Radopholus* c. *Tylenchulus* d. *Meloidogyne*
8. Spear is strong with well developed basal knobs in the genus
   a. *Radopholus* b. *Pratylenchus* c. *Ditylenchus* d. *Anguina*
9. Shethoid nematode is
   a. *Nacobbus* b. *Rotylenchulus* c. *Longidorus* d. *Hemicriconemoides*
10. The generic name *Hirschmanniella* is coined in honour of an eminent US Nematologist
    a. Martin Hirschmann b. Johnson Hirschmann c. Hedwig Hirschmann d. Lewis Hirschmann
11. Rice root nematode was discovered on rice roots in Indonesia and described as Tylenchus oryzae by
    a. J. Von Breda de Haan b. Bergman c. Vecht d. G. Thorne
12. Rice root nematode, *Hirschmanniella oryzae* was assigned by
    a. Luc b. Goodey c. Both a and b d. Oostenhrink
13. Oesophageal glands overlap the intestive ventrally in
    a. *Scutellonema* b. *Hoplolaimus* c. *Globodera* d. *Hirschmanniella*
14. Nematode genus Hopolaimus daday was first recorded from an island min the Paraguay river in the year
    a. 1900 b. 1905 c. 1910 d. 1915
15. Stylet knob is tulip shaped in
    a. *Hoplolaimus* b. *Xiphinema* c. *Criconema* d. *Tylenchulus*
16. Family Hoplolaimidae includes
    a. *Hoplolaimus* b. *Scutellonema* c. *Trichodorus* d. *Both a and b*

17. Position of phasmids in the *Hoplolaimus*
    a. Anterior to Vulva
    b. Posterior to vulva
    c. One anterior and one posterior to vulva
    d. Lateral to vulva
18. A strong cephalic framework is the characteristic of this family
    a. Belanolaimidae
    b. Hoploaimidae
    c. Nacobbidae
    d. Aphelenchoididae
19. Select the correct statement
    a. *Pratylenchus* has small spear
    b. *Pratylenchus* has medium spear
    c. *Hoplolaimus* has more stout spear
    d. *Hoplolaimus* has less stout spear
20. Neddle nematode is
    a. *Trichodorus*
    b. *Paralongidorus*
    c. *Longidorus*
    d. b + c
21. This genus has enlarged phasmid but situated not opposite to each other
    a. *Scutellonema*
    b. *Rotylenchulus*
    c. *Hoplolaimus*
    d. *Helicotylenchus*
22. The genus *Rotylenchulus* was recorded on cowpea for the first time in 1940 by
    a. Linford
    b. Oliveira
    c. Mankau
    d. a and b
23. The term 'adanal' refers to
    a. Proximity of the anus
    b. Proximity of the anal glands
    c. Within the anus
    d. Cross section of the anus
24. "Brown ring' disease is caused by this nematode on daffodils
    a. *Radopholus citrophilus*
    b. *Pratylenchus penetrans*
    c. *Ditylenchus dipsaci*
    d. *Prartrichodorus indicus*
25. Labial framework is strongly sclerotized in the genus
    a. *Rotylenchulus*
    b. *Helicotylenchulus*
    c. *Tylenchulus*
    d. *Criconema*
26. Criconemtidas are named after
    a. Slow and sluggish movement
    b. Presence of spines and scales
    c. Criconemoid type of oesophagus
    d. Coarse ring-like annules
27. Criconemtidae includes this genus
    a. *Criconemoides*
    b. *Hemicriconema*
    c. *Criconema*
    d. *Nothocriconema*
28. Presence of coarse annules is the distict feature of this family
    a. Criconemtidae
    b. Hoploaimidae
    c. Pratylenchidae
    d. Tylenchidae
29. Earlier name of this genua is Criconemoides
    a. *Macroposthonia*
    b. *Bakernema*
    c. *Criconema*
    d. *Hemicriconema*
30. *Basiriolaimus* is the synonym of this genus
    a. *Mulveyotus*
    b. *Hoplolaimus*
    c. *Dolichodorus*
    d. *Criconema*
31. Anchor shaped stylet knob is present in
    a. Criconemtids
    b. Xiphinema
    c. Hoploaimus
    d. Longidorus
32. The head is asymmetrical of this genus
    a. *Tylenchocriconema*
    b. *Criconema*
    c. *Criconemoides*
    d. *Hemicycliophora*

33. Maximum number of body annules is present in this genus
   a. *Criconemella* b. *Hoploaimus* c. *Heterodera* d. *Criconemoides*
34. The genus *Criconemoides* was discovered by
   a. de man b. Hofmanner c. Taylor d. Manzel
35. The genus *Criconema* was discovered by
   a. Hofmanner b. Manzel c. Chitwood d. a and b
36. Common name of *Criconema* is
   a. Spine nematode b. Spiral nematode c. Sessile nematode d. Sheath nematode
37. *Hemicycliophora* belongs to the super family
   a. Heteroderoidae b. Hoplolaimoidea c. Criconematoidea d. Rotylenohoidea
38. This nematode posses a short stylet
   a. *Xiphinema* b. *Belonolaimus* c. *Hemicycliophora* d. *Longidorus*
39. The term 'bell' also refers to this organ in nematodes
   a. Ovary b. Bursa c. Spicule d. Spermatheca
40. Stylet is absent in males in
   a. *Xiphinema* b. *Belonolaimus* c. *Hemicycliophora* d. *Criconemoides*
41. The genus Hemicycliophora was discovered by
   a. de Man b. Chitwood c. Birchfield d. Thorne
42. The sheatoid nematode, Hemicirconemoides was discovered by
   a. Chitwood b. Birchfield c. de Man d. Both a and b
43. A detailed account on morphology and life cycle of *Tylenchulus semipenetrans* was published in 1931 by
   a. N. A. Cobb b. Filipjev c. Stewart d. Paramanov
44. Citrus nematode is unique due to
   a. Degenerate feeding apparatus of males b. Mono-prodelphic reproduction system
   c. Elongate irregular neckd.Posterior location of excretory pore
45. The term 'apicad' means
   a. Within the apex b. Away from the apex c. Toward the apex d. Overlapping the apex
46. The generic name *Meloidogyne* is derived from
   a. Latin b. Greek c. Spanish d. German
47. Occurrence of root knot nematode in cucumber was first reported in 1855 by
   a. Barber b. C. Liebscher c. E.A. Goeldi d. A. Strubell
48. The first specific mention of root knot nematode, Anguillula marioni was described in 1879 by
   a. M. Cornu b. Schacht c. E. A. Goeldi d. Needham
49. Detailed description of *Meloidogyne exigua* was published in 1887 by
   a. M.Cornu b. E.S.Bessey c. G.F.Atkinson d. E.A.Goeldi

50. Classification of different species of *Meloidogyne* by perinneal pattern of annulations was discovered by

a. T.Goodey b. Chitwood c. Ritzemaboss d. G.Thorne

ANSWERS KEY FOR XX

Sl.No.	Ans.	Sl.No.	Ans.	Sl.No.	Ans.	Sl.No.	Ans.	Sl.No.	Ans.
1	d	11	a	21	a	31	a	41	a
2	a	12	c	22	d	32	b	42	d
3	b	13	d	23	a	33	d	43	a
4	c	14	b	24	c	34	c	44	b
5	d	15	a	25	b	35	d	45	c
6	a	16	d	26	d	36	a	46	b
7	b	17	c	27	c	37	b	47	c
8	a	18	b	28	a	38	c	48	a
9	d	19	c	29	b	39	b	49	d
10	c	20	d	30	d	40	c	50	b

XXI.

1. The genus Meloidogyne was separated from Heterodera by
   a. Chitwood b. Muller c. Goeldi d. Cornu
2. The name '*Meloidogyne*' on the basis of shape was given by
   a. Chitwood, 1949 b. White, 1919 c. Neal, 1949 d. Goeldi, 1887
3. 'Cabbagy plant' a disese symptom of teazle is incited by
   a. *Anguina tritici* b. *Dolichodorus indicus* c. *Ditylenchus dipsasi* d. *Heterodera teasle*
4. The terminal cell of the ovary is known as
   a. Cup cell b. Cap ooll c. Apical cell d. Tip cell
5. The common species of *Meloidogyne* can be most accurately characterized by
   a. Perineal patterns b. En-face views
   c. Esterase enzyme profiles d. Host differential tests
6. The nematode genus which induces feeding cell formation in the host
   a. *Heterodera* b. *Meloidogyne* c. *Globodera* d. a + b + c
7. Female of this genus of nematode is smallest
   a. *Heterodera* b. *Anguina* c. *Meloidogyne* d. *Nacobbus*
8. 'Ring shape' is also referred to
   a. Circonate b. Circinate c. Sphericate d. Circynate
9. The anal opening is technically known as
   a. Anal orifice b. Cloacal orifice c. Cloacal aperture d. Anal aperture
10. Cystoid nematode is
    a. *Meloidodera spp* b. *Heterodera spp* c. *Globodera spp* d. *Meloidogyne spp*
11. The generic name *Nacobbus* is coined in the honour of an eminent Nematologist
    a. Christie b. Oostenbrink c. N.A. Cobb d. Carter
12. The term 'crenation' refers to a
    a. Straight projection b. Pointed projection
    c. Rod shaped projection d. Rounded projection
13. These are the cuticular protruberabces in nematodes
    a. Probolae b. Enaboles c. Creaboles d. Prebolae
14. *Heterodera* belongs to super family
    a. Atylenchoidea b. Neotylenchoidea c. Tylenchoidea d. Heteroderoidea
15. Sugar beet cyst nematode was reported in 1859 by
    a. Kuhn b. Schneider c. H.Schacht d. Weischer
16. *Heterodera schachtii* was described and named by
    a. E.A.Goeldi b. A.Schmidt c. J. Ritzemaboss d. E.A.Bessey
17. Nomenclature of *Globodera rostochiensis* was first proposed by

a. Wollen Weber b. Massee c. Zimmerman d. Sasser

18. Potato cyst nematode was first discovered in Nassau county, Long Islands, USA in the year

a. 1913 b. 1931 c. 1941 d. 1946

19. The detailed life cycle of *Heterodera schachti* was published in 1888 by

a. H. Schacht b. A. Strubell c. Van Gundy d. Seinhorst

20. Cyst nematode, *Heterodera goettingiana* on pea roots was discovered in 1892 by

a. Poinar b. T. Goodey c. Steiner d. C. Liebscher

21. The five pathotypes of *Heterodera avenae* was reported by

a. Swarup *et al.*, 1980 b. Mathur 1974 c. Yadav 1985 d. Koshg 197

22. Lemon shaped cysts are common feature in

a. *Heterodera* b. *Globodera* c. *Meloidogyne* d. a + b

23. Lateral tactoreceptors located near the nerve ring are termed

a. Phasmids b. Cephalids c. Deirids d. Caudalids

24. In *Heterodera* fenestration is usually

a. Ambifenestrate b. Bifenestrate c. Circumfenestrate d. a + b + c

25. In *Heterodera avenae* the cysts are

a. Ambifenestrate b. Afenestrate c. Circumfenestrate d. Bifenestrate

26. Members of Heteroderidae have this type of stylet

a. Long b. Short c. Saccate d. Varies

27. The name *Globodera* was coined because of

a. Global distribution of nematode b. Globe-like shape of eggs

c. Global cultivation of host crop d. Globe-like shape of females/cysts

28. In *Globodera* cysts are

a. Round b. Pear shaped c. Lemon shaped d. Squarish

29. In Globodera the fenestration is

a. Ambifenestrate b. Afenestrate c. Circumfenestrate d. Bifenestrate

30. This is a non-cyst forming genus

a. *Ephippoidera* b. *Meloidodera* c. *Punctodera* d. *Globodera*

31. Many members of this genus inhabit tunnels of wood bearing beetles where they feed on fungi

a. *Rhadinaphelenchus* b. *Bursaphelenchus* c. *Hemicycliophora* d. *Ditylenchus*

32. Intestine in this genus overlaps oesophageal glands in a long lobe

a. *Aphelenchus* b. *Aphelenchoides* c. *Bursaphelenchus* d. *Rhadinaphelenchus*

33. 'Digitiform' means

a. Shape of a finger b. Uneven distribution c. Uniform distribution d. Shape of a pin

34. Paratylenchidae comes under super family

a. Diptherophoroidea b. Criconematoidea c. Dorylaimoidea d. Tylenchoidea

35. The genus *Paratylenchus* was discovered by

a. Bastian b. Fischer c. Micoletzky d. a + b

36. The process of hatching from the egg is also known as

a. Ecdysis b. Eclosion c. Ecylotion d. Desltion

37. Ectophore means

a. Nerve ring pore b. Cuticular pore c. Oesophageal pore d. Excretory pore

38. The genus Aphelenchoides is include under the super family

a. Dorylaimoidea b. Aphelenchoidea c. Tylenchoidea d. Aphelenchoidodea

39. The first species of *Aphelenchoides* was described by

a. H.C.Bastian b. M.Fischer c. Sayre d. Allen

40. The genus *Aphelenchoides* was created by

a. Chitwood b. J.Ritzemaboss c. M.Fischer d. Sayre

41. Phasmids are excessively small in the genus

a. *Belonolaimus* b. *Hoplolaimus* c. *Aphelenchoides* d. *Radopholus*

42. *Aphelenchoides fragariae* on strawberry was first recorded in 1891 by

a. J.F.Dastur b. J.Ritzemaboss c. Mankau d. Duddington

43. White tip nematode of rice was identified as *Aphelenchoides besseyi* in 1951 by

a. Allen b. Dastur c. Perry d. None of the above

44. Odontostylet is present in

a. Aphelenchoidea b. Tylenchoidea c. Criconematoidea d. Dorlaimoidea

45. The number of suborders in the order Dorlaimoida is

a. Two b. Six c. Four d. Three

46. Suborder Dorlaimina includes the super family

a. Dorlaimoidea b. Actinolaimoidea c. Leptonchoidea d. Belondiroidea

47. Super family Dorlaimoidea includes the family

a. Longidoridae b. Trichodoridae c. Paralongidoridae d. a and b

48. 'Toothless' forms in nematodes are referred to

a. Predentate b. Neodentate c. Edentate d. Adentate

49. Guiding ring is present at the tip of the stylet in

a. *Longidorus* b. *Xiphinema* c. *Meloidogyne* d. *Anguina*

50. 'Falciform' means

a. Rod shaped b. Circular shaped c. Sickle shaped d. Square shaped

ANSWERS KEY FOR XXI

Sl.No.	Ans.	Sl.No.	Ans.	Sl.No.	Ans.	Sl.No.	Ans.	Sl.No.	Ans.
1	a	11	c	21	b	31	b	41	c
2	d	12	d	22	a	32	d	42	d
3	c	13	a	23	c	33	a	43	a
4	b	14	d	24	d	34	d	44	b
5	a	15	c	25	c	35	c	45	d
6	d	16	b	26	b	36	b	46	c
7	b	17	a	27	d	37	d	47	d
8	b	18	c	28	a	38	b	48	c
9	d	19	b	29	c	39	a	49	a
10	a	20	d	30	a	40	c	50	c

XXII.

1. **Plant parasitic nematodes are usually confined to top cm of soil**
   a. 5 cm b. 10 cm c. 25 cm c. 35 cm
2. The most common tool used for taking soil sample in annual crop is
   a. Soil auger b. Hand shovel c. Spade d. Hand hoe
3. The most common tool used for taking soil sample in perennial crop is
   a. Hand hoe b. Hand shovel c. Crow bar d. Soil auger
4. Soil sample are collected from a depth varying between
   a. 4"-6" b. 6"-3" c. 6"-4" d. 6"-1.5"
5. Sampling depth for annual crop is
   a. 15-20cm b. 20-30 cm c. 20-25cm d. 25-30 cm
6. Sampling depth for deep rooted crop is
   a. 15-20cm b. 20-30 cm c. 30-40 cm d. 40-50 cm
7. The depth of soil samples during survey should be
   a. 5 cm b. 12 cm c. 18 cm d. 20 cm
8. Ideal soil sample quantity for extraction of potato cyst nematode
   a. 100 g b. 150 g c. 200 g d. 300 g
9. A tactic response in which gravitational attraction is the stimulating factor is known as
   a. Geotactis b. Geomorphic c. Geotatile d. Geotaxis
10. For nematicides treatments the samples to be taken for evaluation is
    a. Pre-treatment sample b. Intermediate post-treatment sample
    c. Mid season sample d. a + b + c
11. Sample size may be determined on the basis of
    a. Relative population densities b. Spatial pattern of nematode
    c. a + b d. Vertical distribution
12. Sampling units required for estimating large population
    a. 10 b. 30 c. 70 d. 100
13. For estimating low population densities sampling units required is
    a. 25 b. 60 c. 90 d. 100
14. Sampling patterns used for nematode assay include
    a. Random sampling b. Stratified random sampling
    c. Two stage d. a + b + c
15. The critical factor that affects nematode survival in storage is
    a. Temperature b. Soil moisture c. a + b d. Relative humidity
16. Soil and root samples should be stored at this temperature until further processing
    a. 5° b. 10° c. 15° d. 20°

17. **The natural environment of nematodes, specifically the locale in which they grow and live is termed**
    a. Ecoshell b. Localized area c. Niche d. Habitat
18. **Cobb's decanting and sieving method is based on the principle of**
    a. Hydrophilic nature of nematodes b. Gravitational force
    c. Nematode mobility **d. Pore size of filters**
19. **The genera Trichodorus and *Helicotylenchus* can be extracted using**
    a. Baermann funnel method **b. Centrifugal floatation technique**
    c. Cobb's decanting and sieving method d. Elutriation technique
20. **Nematodes from soil and organic debris can be extracted using**
    a. Cobb's decanting and sieving method b. Baermann funnel method
    c. **Mistifier** **d. Centrifugal floatation technique**
21. **Soil nematodes are rapidly extracted by**
    a. Sugar floatation b. Washing c. Incubation d. Elutriation
22. 'Helminth' simply means
    a. Nematode b. Animal c. Living organism d. Worm
23. **The specific gravity of nematodes is**
    a. 1.02 b. 1.04 c. 1.05 d. 1.08
24. **The specific gravity of the sugar solution used for nematode extraction is**
    a. 1.81 b. 1.18 c. 1.10 d. 1.08
25. **In centrifugal floatation technique sugar solution is used for**
    a. Lower specific gravity b. Easily soluble in water
    c. Easily available **d. High specific gravity**
26. *Circonema, Circonemoides* and *Hemicycliophora* nematodes can be extracted using
    a. Centrifugal floatation b. Elutriation technique
    c. Modified fenwick can method d. Root-incubation
27. **This technique is most suitable for the extraction of sluggish nematode from soil**
    a. Elutriation + modified baermann funnel method
    b. Sieving + floatation method
    c. Centrifugal sugar floatation method
    d. Modified baermann funnel method
28. **The two flask techniques used in the extraction of nematode is based on the principle of**
    a. Sedimentation **b. Counter flow of water**
    c. Centrifugal force **d. a + b**
29. **Oostenbrink elutriator apparatus is used for the separation of nematodes from**
    a. Soil b. Roots c. Leaf d. Stem
30. **The most useful method for collecting small nematodes from soil or endoparasitic nematodes is**
    a. Modified fenwick can method b. Baermann funnel method
    c. Elutriation technique d. Mist chamber

31. **The major principle involved in extraction of nematode by Baermann's filters is**
    a. Gravitational force b. Hydrophilic nature of nematodes
    c. Nematode mobility d. **Pore size of filters**
32. **For efficient extraction of nematodes from larger number of samples in a minimum period the suitable** method employed is
    a. Oostenbrink elutriator b. Cobb's sieving
    c. Centrifugation d. Mist chamber
33. Fenwick can method was developed in 1940 by
    a. N.A.Cobb b. Fenwick c. Oostenbrink d. Seinhorst
34. The best technique to extract cysts from soil
    a. Fenwick can b. Incubation method
    c. Seinhorst method d. **Centrifugal floatation method**
35. The most suitable technique for the extraction of 'full' cysts
    a. Fenwick can b. Schuelling apparatus c. 60-mesh sieve d. Fluidising column
36. **Modified Fenwick can method involves the use of**
    a. Carbon tetrachloride b. Acetone c. Chloroform d. a and b
37. The best method involved in extraction of nematodes from plant material
    a. Mechanical maceration technique b. **Mistifier technique**
    c. Root Incubation technique d. a + b + c
38. The root incubation method is excellent for experimental work involving
    a. Ectoparasites b. Endoparasites
    c. Sedentary ectoparasites d. a + b
39. The 'fenestra' refers to the
    a. Hatching pore b. Excretory pore c. Anal pore d. Vulvla cone
40. The immersion oil used to view the nematodes under 100x has this refracting index
    a. 1.350 b. 1.500 c. 1.505 d. 1.515
41. While using 100 X objective the immersion oil is also used for
    a. For better colour b. For sharp image
    c. Preventing the bleaching of light rays d. **Proper magnification**
42. To draw the nematodes accurately the equipment used is
    a. Occular micrometer b. Camera lucida
    c. Stage Micrometer d. Measuring scale
43. Histopathological technique was described by
    a. Barker b. N.A. Cobb c. Sasser d. Sass
44. Fixatives are mainly used for
    a. Cleaning of nematode b. Preserving of nematode
    c. Killing of nematode d. a + c

45. The common fixative used for fixing the nematode is
    a. Formaldehyde b. Glycerine c. Triethanol amine d. DPX mounts
46. This is the best general fixative for nematodes
    a. TAF b. Caroy's c. FAA d. Formalin 40%
47. Formalin acetolalcohol, a fixative consisits
    a. 90 ml of 50% ethanol b. 5 ml of glacial acetic acid
    c. 5 ml of 37% formaldehyde d. a + b + c
48. Fromalin propiono propanol (FPP) fixative mainly comprises
    a. 90 ml of 50% isopropyl alcohol b. 5 ml of propionic acid
    c. 5 ml of 37% formaldehyde d. a + b + c
49. This is the common anaesthetizer used for nematodes
    a. Dichloro- triethyl ether b. Dichloro- diethyl ether
    c. Diphemyl propyl d. Propylene dister
50. The mounting media used for the preparation of permanent nematode mount
    a. Glycerine jelly b. Lactophenol c. Glycerine d. DPX mountant

ANSWERS KEY FOR XXII

Sl. No.	Ans	Sl. No.	Ans	Sl. No.	Ans	Sl. No.	Ans
1	c	14	d	27	c	40	d
2	b	15	c	28	b	41	c
3	d	16	a	29	a	42	b
4	a	17	d	30	b	43	d
5	a	18	b	31	c	44	b
6	b	19	c	32	d	45	a
7	d	20	d	33	b	46	c
8	c	21	a	34	a	47	d
9	d	22	d	35	c	48	d
10	d	23	c	36	d	49	a
11	c	24	b	37	d	50	b
12	a	25	d	38	b		
13	d	26	a	39	a		

XXIII.

1. Homeostasis means
   a. An undisturbed ecosystem
   b. Ecosystem change induced by humans
   c. Ecosystem that resists change
   d. Ability to maintain constant body temperature
2. Nematode ecology related to individual species is termed as
   a. Auctecology
   b. Synecology
   c. Animal ecology
   d. Anchiology
3. Auctecology refers to
   a. Study of single individual in relation to environment
   b. Study of single individual in relation to other biota
   c. Study of whole community in relation to environment
   d. Study of whole community in relation to plants
4. The differently adapted nematode populations are termed as
   a. Pathovar
   b. Pathotype
   c. Ecotypes
   d. Biotype
5. Surveillance refers to
   a. Population density
   b. Population survey
   c. Sequential sampling
   d. b + c
6. The resemblance among nematode species which is not due to inheritance from common ancestery is termed as
   a. Homology
   b. Homotony
   c. Homoplasy
   d. Homoplane
7. **Nematode diversity is defined as a number of species of given taxon or the number of taxa in**
   a. Community
   b. Natality
   c. Mortality
   d. Fecundity
8. Density Frequency stands for
   a. Importance value
   b. Relative frequency
   c. Prominence value
   d. Relative density
9. This is the most accurate way of expressing nematode population densities
   a. No. of nematodes per g dried soil
   b. No. of nematodes per ml soil
   c. No. of nematodes per g soil
   d. No. of nematodes per cc soil
10. Prominence value in nematode distribution means
    a. Density. v/frequency
    b. Density/ frequency
    c. Frequency/Density
    d. Frequency/ Density
11. Relative frequency + Relative density + Relative dominance stand for
    a. Index of Similarity
    b. Importance value
    c. Absolute Biomass
    d. Prominence value
12. Index of similarity is used to compare
    a. Nematode densities
    b. Nematode population
    c. Nematode communities
    d. a + b
13. **It is difficult to frame any uniform model for nematodes mainly because of**
    a. Polyspecific community
    b. Small size
    c. Rapid multiplication
    d. b + c

14. According to de Man formula 'P' refers to
    a. Anterior position
    b. Position of phasmid
    c. Posterior postion of phasmid
    d. Position of pigment
15. Seinhorst's model is categorized under thish group of models
    a. Multiple point model
    b. Critical point model
    c. Regression model
    d. Community model
16. $Y = CZ^{P.T}$ refers to this model of nematode population density
    a. Regression equation model
    b. Multiple point model
    c. Critical point model
    d. Seinhorst model
17. The relationship between nematode population density and crop yield as expressed by the model $Y = CZ^{P.T}$ was propsed by
    a. Wallace
    b. Seinhorst
    c. Oostenbrick
    d. Nicholoson
18. Nematode species which have high rates of reproduction but have shortest life spans are termed
    a. m- selection
    b. r–selection
    c. t-selection
    d. g-selection
19. Taylor's power law pertains to
    a. Crop performance and nematode population densities
    b. Community analysis of nematode population
    c. Sampling scheme for nematode population estimation
    d. Nematode management optimization
20. The concept of growth stimulation at low densities is designated by
    a. y = (c-(c-1)e)
    b. y = (c-(c-1)e MP)
    c. y = (c-(c-1)e KP) ZP
    d. y = (c-(c-1e KP)
21. The most common type of locomotion found in nematodes
    a. Active movement
    b. Undulatory
    c. Passive movement
    d. Symmetrical
22. When nematode reverses its direction, the waves of bending pass from
    a. Anterior to posterior
    b. Posterior to anterior
    c. Lateral to posterior
    d. Anterior to lateral
23. In general, soil-inhabiting nematodes require ample supply of
    a. CO_2
    b. O_2
    c. H_2O
    d. b + c
24. During aerobic starvation usually nematodes utilize these stored nutrients
    a. Carbohydrates
    b. Proteins
    c. Lipids
    d. a + b
25. **Some nematode species that occur in soil obtain sufficient CO_2 from roots of plants. This statement is**
    a. False
    b. Partly true
    c. Controversial
    c. True
26. **The most obvious factor influencing development of a nematode within the host is**
    a. Soil
    b. Temperature
    c. Host specificity
    d. Moisture
27. The pre-hatch moult may also be absent in this form of nematodes
    a. Parasitic
    b. Free living
    c. Marine
    d. Saprophytic
28. The passive movement of nematode has been noticed in
    a. *Globodera rostochiensis*
    b. *Criconemoides curvatum*
    c. *Radopholus similis*
    d. a + b + c

29. Example for short range spread of nematodes through cultural operations
    a. Cyst nematodes b. Stem and bulb nematode
    c. Root knot nematode d. Both a and b
30. **The common dispersal agents for nematodes in field**
    a. Irrigation water b. Strong winds
    c. **Drainage of flood water** d. a + b + c
31. **The most efficient mounting media used for the preparation of permanent mount is**
    a. Phenol b. Lactophenol c. DPX mountant d. Glycerine
32. The material used for sealing the cover slip while mounting nematodes is
    a. Glyceel b. Zut c. Glycerine d. a and b
33. For mounting the ribbons of sections in histopathological studies, glass slides are smeared with
    a. Keyer's albumen b. Meyer's cartinin c. Meyer's albumen d. Meyer's alanine
34. During preparation of glycerin mounts of nematodes, this is needed
    a. Triethanol amine b. Trimethanol amine c. Diethanol amine d. Dimethanol amine
35. **For the identification of cyst nematode, the mount prepared is**
    a. Perineal pattern b. Glycerol mount c. Vulval cone top d. Permanent mount
36. **In histopathological techniques, mounted slides are finally cleared in a mixture of alcohol, xylol and**
    a. Olive oil b. Canada blasm oil c. Balsam oil d. Clove oil
37. The aluminium slide normally has a central hole of diameter
    a. 10 mm b. 21 mm c. 10 cm d. 21 cm
38. The thickness of cover slips used is
    a. 130 to 200 μ b. 50 to 100 μ c. 250 to 300 μ d. 275 to 400 μ
39. For cutting enface view sections of nematodes, this is used
    a. Gelatin gel b. Gelatin + Phenol gelly c. Glycerin gelly d. Phenolic gelly
40. Most common stain used for nematode-infested root materials
    a. Johnsen's Quadruple stain b. Sass Safranin
    c. Fast green stain d. a + b + c
41. The most widely used procedure for staining nematodes from plant tissues
    a. Lactophenol-Acid fuchsin method b. Maceration technique
    c. Both a and b d. Cotton blue technique
42. The excellent method for staining endoparasitic nematode is
    a. Lactophenol method b. Maceration technique
    c. Sodium hypochloride Acid fuchsin method d. Glycerine technique
43. During staining of nematodes lactophenol solution is cleared with this solution
    a. Acid fuchsin b. Glycerin c. Lactic acid d. Phenol
44. Double staining in histology consists of these stains
    a. Cotton blue and fast green b. Lactophenol and cotton blue
    c. Safranin and fast green d. Safranin and cotton blue

45. *Pratylenchus* species can be cultured using

a. Agrobacterium transformed roots   b. Carrot disc
c. Callus tissue   d. Potato disc.

46. The genus *Longidorus* was created by

a. M.R.Siddiqi   b. D.J.Hopper   c. E.Khan   d. a + b + c

47. This genus retains virus particle for shorter time compared to other genera

a. *Trichodorus*   b. *Paralongidorus*   c. *Xiphinema*   d. *Longidorus*

48. Amphids are funnel or stirrup shaped in this genus

a. *Trichodorus*   b. *Paralongidorus*   c. *Heterodera*   d. *Pratylenchus*

49. Longest nematode is

a. *Belonolaimus longicadatus*   b. *Longidorus elongates*
c. *Radopholus similis*   d. *Paralongidorus maximus*

50. Xiphinema and Trichodorus can be differentiated by

a. Amphid around the lip   b. Three parted solid stylet
c. Body length   d. Relaxation posture

ANSWERS KEY FOR XXIII

Sl. No.	Ans	Sl. No.	Ans	Sl. No.	Ans	Sl. No.	Ans	Sl. No.	Ans
1	d	11	b	21	b	31	d	41	a
2	b	12	c	22	a	32	d	42	c
3	a	13	a	23	b	33	d	43	b
4	c	14	b	24	c	34	a	44	d
5	b	15	c	25	a	35	c	45	b
6	d	16	d	26	c	36	b	46	d
7	a	17	b	27	b	37	a	47	b
8	c	18	a	28	d	38	b	48	b
9	d	19	c	29	d	39	c	49	d
10	a	20	d	30	d	40	d	50	c

XXIV.

1. Nematodes are strictly

   a. Obligate parasites   b. Facultative saprophytes

   c. Facultative parasites   d. Obligate saprophytes

2. The following nematode genera can live in a wide variety of habitats

   a. *Helicotylenchus*   b. *Tylenchorhynchus*   c. *Hemicycliophora*   d. a + b + c

3. Plant parasitic nematodes are mostly concentrated up to thisdepth in rhizospheric zone

   a. 5 cm   b. 10cm   c. 25 cm   d. 35 cm

4. Generally, the distribution of nematodes in top surface layer of sandy soil will be

   a. Maximum   b. Minimum   c. Scattered   d. Varies

5. Peak distribution of *Globodera rostochiensis* is noticed in the surface layer upto

   a. 5 cm   b. 10cm   c. 13 cm   d. 19 cm

6. *Tylenchulus semipentrans* is distributed in the soil upto a depth of

   a. 1.2 m   b. 2.4 m   c. 3.6 m   d. 4.8 m

7. *Radopholus similis* is found in citrus orchard from surface to feeder roots up to a depth of

   a. 0.3 to 1.5 m   b. 3.0 to 3.16 m   c. 1.5 to 3.0 m   d. a and b

8. Most phyto nematodes spend their entire life cycle in

   a. Water   b. Soil   c. Plant tissue   d. a + b + c

9. In soil, phytonematodes can navigate in pores of following diameter

   a. 5 – 10 µm   b. 20 – 30 mµ   c. 20 – 30 µm   d. 50 µm

10. Most phyto nematodes become inactive at a temperature of

    a. 5 – 15° C   b. 30 – 40° C   c. 45 – 55° C   d. a and b

11. The favourable temperature for most plant parasitic nematodes is

    a. 10 – 15°C   b. 15 – 30°C   c. 30 – 40°C   d. <15°C

12. The nativity of *Tylenchulus semipentrans* was reported by

    a. **Chitwood and Birhcfield** b.   Raski

    c. Jones   d. Sass

13. The California nativity of *Meloidogyne hapla* was mentioned by

    a. Jones   b. Saaaer   c. N.A.Cobb   d. Raski

14. The optimum temperature required for activity of potato cyst nematode including invasion of the host is

    a. 5 – 8° C   b. 10 – 20° C   c. 20 – 30° C   d. a + b

15. The optimum temperature required for root invasion in *Heterodera schachtii* is

    a. 5 – 10° C   b. 10 – 20° C   c. 25 – 30°C   d. 35 – 38°C

16. *Heterodera avenae* is essentially prefers an habitant of

    a. Warm tropical climate   b. Cool temperature climate

    c. Both a and b   d. Warm temperate climate

17. In the tropical zone the most common species of root knot nematodes are
    a. *Meloidogyne incognita* b. *Meloidogyne javanica*
    c. *Meloidogyne exigua* d. a and b
18. Very high temperature normally inhibit this activity of nematodes
    a. Movement b. Invasion c. Hatching d. Reproduction
19. This factor affects the life cycle duration, the most
    a. Soil texture b. Soil moisture c. Plant host d. Soil temperature
20. Most important and available soil water for nematode activity is
    a. Gravitational water b. Capillary water
    c. Hygroscopic water d. Sore leach water
21. The soils which are not generally favorable for nematodes are
    a. Dry soils b. Fallow soils
    c. Highly saturated soils d. a + b + c
22. Nematodes are generally active at a field capacity of
    a. 30 – 40 % b. 40 – 60 % c. 60 – 80 % d. Above 80%
23. The following nematodes are chiefly found in wet locations
    a. *Ditylenchus angustus* b. *Radopholus similis*
    c. *Hirschmanniella oryzae* d. a + b + c
24. This factor is most important for the biological activities of nematodes
    a. Cation exchange capacity b. Soil texture
    c. Soil moisture d. Soil pH
25. Excess moisture tends to inhibit this activity of nematodes
    a. Reproduction b. Locomotion c. Growth d. a + c
26. Nematodes undergo anoxybiosis in
    a. Wet soils due to less O_2 b.Wet soils due to more CO_2
    c. Dry soils due to more O_2 d.Dry soils due to more CO_2
27. High soil moisture content may favour population increase of
    a. *Meloidogyne incognita* b. *Ditylenchus dipsaci*
    c. *Tylenchulus semipenetrans* d. *Heterodera glycine*
28. The soils which are better suited for most nematode species
    a. Heavy sandy b. Sandy loam c. Clayey d. a + c
29. Some species of lesion and stunt nematodes are numerous in
    a. Light sandy soils b. Alkaline soils c. Clayey soils d. b + c
30. Citrus nematode occurs frequently in
    a. Sandy soils b. Clay soils c. Red soils d. a and b
31. This nematode usually prefers porous soils
    a. *Tylenchorhynchus* b. *Pratylenchus* c. *Longidorus* d. *Heterodera*

32. These nematodes usually cause more damage in soils with high organic contents
    a. Cyst nematode b. Root-knot nematode c. Lesion nematode d. Burrowing nematode
33. Though sandy soils favour nematode problems, this nematode invades heavy soils better than sandy soils
    a. Root-knot nematode b. Spiral nematode
    c. Cyst nematode d. Stem and bulb nematode
34. Highest hatching rate of nematode eggs occur at this pH
    a. 4-6 b. 7 c. 7 and above d. Not specific
35. In general, pH range of 3.0 – 10.6 has this effect on the orientation of root knot nematodes
    a. Varied b. Favorable c. Negative d. a or b
36. Dyuggan found that the population of *Heterodera avenae* increase at a pH of
    a. 4.0 – 6.0 b. 4.0 – 8.0 c. 5.0 – 7.5 d. 8.0 – 9.5
37. The positive correlation in *Helicotylenchus platyurus* at 4.5 – 6.9 pH was reported by
    a. Norton and Hoffman, 1974 b. Sasser and Jenkins, 1959
    c. Ellenby, 1946 d. Breakley et *al.* 1976
38. Soil and plant nematodes can withstand pressure close to
    a. 2 atmospheres b. 10 atmospheres c. 5 atmospheres d. 7 atmospheres
39. At a constant concentration of nitrogen the ratio of O_2 to CO_2 increases, hence emergence of nematodes larvae
    a. Decreases b. May increase or decrease
    c. Increases d. Remains constant
40. Gas known to attract phytonematodes to host roots
    a. O_2 b. CO_2 c. NH_3 d. a and b
41. Distribution of this nematode down to 100 cm soil depth is very common
    a. *Rotylenchus* b. *Trichodorus* c. *Heterodera* d. *Helicotylenchus*
42. *Radopholus citrophilus* occurs in maximum number at a soil depth of
    a. 0.5 – 1.0 m b. 2.0 – 3.0 m c. Up to 4m d. a and b
43. Surface tension of the following controls its availability to nematodes in soil
    a. Soil b. Water c. Organic matter d. Root chemicals
44. The nematodes are attached to roots in gradients of electrical potential due to
    a. Lower redox b. Sodium thionite c. Higher redox d. Sodium dithionide
45. High frequency of plasmodesmata is found in feeding sites of this nematode
    a. *Heterodera* b. *Meloidodera* c. *Meloidogyne* d. *Nacobbus*
46. Anhydrobiotic nematode being used in US space programmes to study the effect of high energy particles in space
    a. *Caenorhabditis elegans* b. *Paragrellus redivivus*
    c. *Turbatrix aceti* d. *Ditylenchus dipsaci*

47. In soil fauna, phytonematodes contribute to
    a. High number
    b. Moderate number
    c. Low number
    d. Equal to bacterial feeders
48. At high densities, the egg production per cyst
    a. Varies
    b. Raises
    c. Decreases
    d. Remains unchanged
49. **The nematode population after reaching a high level fluctuates around equilibrium. This behavior is termed**
    a. Static effect
    b. Variation effect
    c. Equilibrium effect
    d. Ceiling effect
50. The factor preventing a nematode population from growing at its full potential rate comes under
    a. Environment potential
    b. Environmental resistance
    c. Biotic potential
    d. Biotic resistance

ANSWERS KEY FOR XXIV

Sl. No.	Ans	Sl. No.	Ans	Sl. No.	Ans	Sl. No.	Ans
1	a	17	d	33	d	49	d
2	d	18	a	34	a	50	a
3	c	19	d	35	b		
4	b	20	b	36	c		
5	a	21	d	37	a		
6	b	22	b	38	b		
7	d	23	d	39	c		
8	b	24	c	40	d		
9	c	25	b	41	a		
10	d	26	a	42	c		
11	b	27	b	43	b		
12	a	28	b	44	d		
13	d	29	a	45	a		
14	b	30	d	46	b		
15	c	31	b	47	c		
16	b	32	d	48	c		

XXV.

1. Based on feeding habit nematodes are divided into
   a. Ecto and endo parasitic nematodes
   b. Endo migratory and sedentary parasitic nematodes
   c. Semi-endoparasitic nematodes
   d. a + b + c
2. Nematode which live freely in rhizosphere and feed intermittently on the epidermis and root hairs without entering inside the plant tissues is known as
   a. Ectoparasitic nematode b. Semi-endoparasitic nematode
   c. Endoparasitic nematode d. Migratory endoparasite
3. Ectoparasitic nematodes are subdivided into
   a. Migratory ectoparasitic nematodes b. sedentary ectoparasitic nematodes
   c. a and b d. Sluggish ectoparasitic nematodes
4. The nematodes which feed at a particular site for a brief period and then move the next is called as
   a. Sedentary ectoparasites b. Migratory ectoparasites
   c. Semi-endoparasites d. Sluggish ectoparasites
5. This nematode feeds on subsurface tissues.
   a. *Hoplolaimus* b. *Trichodorus* c. *Pratylenchus* d. *Tylenchorhynchus*
6. The nematode belonging to migratory ectoparasites group
   a. *Paratylenchus* b. *Criconemoides* c. *Belonolaimus* d. a + b + c
7. This is an example of migratory ectoparasites
   a. *Xiphinema* b. *Trichodorus* c. *Longidorus* d. a + b + c
8. *Trichodorus* is a
   a. Migratory ectoparasites b. Semi-endoparasites
   c. Sedentary endoparasites d. Migratory endoparasites
9. The nematodes which feed at a particular site for a long time
   a. Migratory ectoparasites b. Semi-endoparasites.
   c. Sedentary endoparasites d. Migratory endoparasites
10. The nematodes belonging to sedentary ectoparasites group
    a. *Hemicycliophora arenaria* b. *Cacopaurus pestis*
    c. *Tylenchulus semipenetrans* d. a and b
11. Nematodes which spend their entire life cycle inside the root system and major portion of the body is found inside the plant tissue are termed as
    a. Ectoparasites b. Endoparasites c. Semi-endoparasites d. a + b + c
12. The most destructive forms of plant parasitic nematodes are
    a. Endoparasites b. Ectoparasites c. Semi-endoparasites d. b + c

13. Endoparasitic nematodes are categorized into
    a. Migratory endoparasites
    b. Sedentary endoparasites
    c. a and b
    d. Sluggish ectoparasites
14. Nematodes which move in the cortical parenchyma of host root and feed on cells, multiply and cause necrotic lesions are termed as
    a. Migratory endoparasites
    b. Sedentary endoparasites
    c. Semi-endoparasites
    d. Sluggish ectoparasites
15. *Pratylenchus* is
    a. Migratory ectoparasite
    b. Sedentary endoparasite
    c. Sedentary ectoparasite
    d. Migratory endoparasite
16. This is a migratory endoparasitic nematode
    a. *Xiphinema americanum*
    b. *Hoplolaimus sienhortii*
    c. *Radopholus similis*
    d. *Longidorus elongates*
17. *Hirschmanniella* is
    a. Migratory ectoparasite
    b. Migratory endoparasite
    c. Sedentary ectoparasite
    d. Sedentary endoparasite
18. Nematodes in which second stage juveniles penetrate root lets and become sedentary throughout the life cycle inside the root cortex are referred as
    a. Migratory ectoparasite
    b. Semi-endoparasite
    c. Sedentary ectoparasite
    d. Sedentary endoparasite
19. The most advanced form of plant parasitic nematodes are
    a. Sedentary endoparasites
    b. Semi-endoparasites
    c. Migratory ectoparasites
    c. b + c
20. *Meloidogyne* sp and *Heterodera* sp are
    a. Migratory endoparasites
    b. Endoparasites
    c. Sedentary endoparasites
    d. Sedentary ectoparasites
21. Nematodes which partially penetrate the root and become immobile is known as
    a. Migratory endoparasites
    b. Semi-endoparasites
    c. Sedentary ectoparasites
    d. a or c
22. Examples of semi- endoparasites
    a. *Tylenchulus semipenetrans*
    b. *Rotylenchulus reniformis*
    c. *Nacobbus batatiformis*
    d. a + b + c
23. The most primitive forms of plant parasitic nematodes are
    a. *Trichodorus*
    b. *Paratrichodorus*
    c. *Longidorus*
    d. a + b + c
24. Highly evolved phytonematode is
    a. *Anguina tritici*
    b. *Meloidogyne incognita*
    c. *Rotylenchulus reniformis*
    d. *Tylenchulus semipenetrans*
25. Most common enzymes found in nematodes
    a. Cellulase
    b. Protease
    c. Protease
    d. a + b + c

26. Lesions produced by *Pratylenchus penetrans* on peach roots are due to
    a. Amygdalin　　b. à-Glucosidase
    c. Benzaldehyde and HCN　　d. Phenolic substances
27. The only ectoparasite which induces galls on root tips is
    a. *Xiphinema diversicaudatum*　　b. *Trichodorus christei*
    c. *Tylenchorhynchus brevilineatus*　　d. *Nacobbus spp.*
28. This species usually produces small galls on roots
    a. *Meloidogyne hapla*　　b. *M. incognita*　　c. *M. arenaria*　　d. *M. javanica*
29. Indole-3-aectic acid(IAA), Indole-3-acetic ethylester(IAE), Indole- 3- acetonitrile (IAN) and Indole-3-butyric acid (IBA) are detected in
    a. *Meloidogyne incognita*　　b. *M. hapla*
    c. *M. javanica*　　d. a + b + c
30. In syncytia there is intense synthesis of
    a. RNA　　b. DNA　　c. Proteins　　d. a and b
31. The galled tissues contain increased amount of
    a. Cellulose　　b. Free amino acids　　c. Phosphorus　　d. a + b + c
32. Indole-3-aectic acid(IAA) is detected in larvae of
    a. *Heterodera schachtii*　　b. *Tylenchulus semipenetrans*
    c. *Rotylenchulus reniformis*　　d. *Anguina tritici*
33. Indole-3-aectic acid methylester (IAM) is detected in larvae of
    a. *Nacobbus batatiformis*　　b. *Ditylenchus dipsaci*
    c. *Rotylenchulus reniformis*　　d. *Tylenchulus semipenetrans*
34. **The anti-nematode compounds produced by Asparagus officinalis var Mary Washington is toxic to**
    a. *Trichodorus christei*　　b. *Pratylenchus penetrans*
    c. *Tylenchorhynchus claytoni*　　d. a + b + c
35. Root knot nematodes fail to induce gall formation in the roots of
    a. *Tagetes patula*　　b. *T. erecta*　　c. a and b　　d. *Crotolaria*
36. The nematicidal principle in *Tagetes* is
    a. à-terthienyl　　b. Bithienyl　　c. a + b　　d. Lipids
37. *Longidorous* induced galls of marigold have more of these compared to healthy plants
    a. Free aminoacids　　b. Enzyme activity　　c. Lipids　　d. a + b + c
38. The resistance and nematicidal effect of *Eragrostris curvula* against root knot nematodes is due to the presence of high level of
    a. Catechol　　b. Catechuic acid　　c. Pyrocatechol　　d. a + b + c
39. **Race identification is based on pathogenecity of populations on**
    a. Identical hosts　　b. Different hosts　　c. Indicator hosts　　d. a + c
40. **Biotypes identified in golden nematode of potato are**
    a. Biotype A (Original trace)　　b. Biotype B (resistance breaking race)
    c. Biotype C　　d. a and b

41. Occurrence of four races of *Meloidogyne incognita* was reported by
    a. Taylor b. Sasser c. Stokes d. a and b
42. Four races of *Meloidogyne arenaria* was reported by
    a. Taylor and Sasser b. Paulson and Webster c. Epps and Golden d. Johnson and Thorne
43. Following distinguished 11 biological races in Ditylenchus dipsaci on the basis of their pathogenecity on nine plant species
    a. Paulson b. Webster c. Seinhorst d. a + b
44. Radopholus *similis* races identified in Florida are
    a. Banana race b. Citrus race c. Pepper race d. a and b
45. Banana race Radopholus *similis* parasitizes
    a. Citrus only b. Banana only c. Pepper only d. Citrus and banana
46. Citrus race of Radopholus *similis* attacks
    a. Banana b. Citrus c. Both banana and citrus d. Pepper
47. This biotype of potato cyst nematode is not present in India
    a. Ro1 b. Ro5 c. Pa1 d. Pa3
48. This species/ race of root-knot nematode attacks cotton
    a. *Meloidogyne incognita race 3* b. *M. incognita race 1*
    c. *M. javanica race 1* d. *M. arenaria race 2*
49. Cluster bean can be used to distinguish host races of
    a. *Heterodera glycines* b. *H. cajani*
    c. *H. trifolii* d. *H. ciceri*
50. Multinucleate giant cells induced by root-knot nematodes are formed due to
    a. Modification of transfer cells b. Cell wall dissolution of adjacent cells
    c. Hypertrophy and hyperplasia of cells d. Karyokinesis without cytokinesis

ANSWERS KEY FOR XXV

Sl. No.	Ans	Sl. No.	Ans	Sl. No.	Ans	Sl. No.	Ans	Sl. No.	Ans
1	d	11	b	21	b	31	d	41	d
2	a	12	a	22	d	32	a	42	a
3	c	13	c	23	d	33	b	43	c
4	b	14	a	24	a	34	d	44	d
5	b	15	d	25	d	35	c	45	b
6	d	16	c	26	b	36	a	46	c
7	d	17	b	27	a	37	b	47	b
8	a	18	d	28	c	38	c	48	a
9	c	19	a	29	d	39	b	49	b
10	d	20	c	30	d	40	d	50	c

XXVI

1. The tissues involved in the induction of nurse cells by citrus nematode are
   a. Xylem and phloem b. Pericycle
   c. Endodermis d. Cortical parenchyma cells
2. The leaf tip of rice turns to white colour due to the attck of
   a. *Aphelenchoides ritzemabosi* b. *A.fragariae*
   c. *A. besseyi* d. a + b
3. 'Stylet aperture' means
   a. The anterior opening into the stylet b. the posterior opening into the stylet
   c. aperture on the host tissue caused by the stylet d. b + c
4. Greatly enlarged phasmid in some species of Hoplolaiminae are termed as
   a. Epinum b. Scutellum c. Scrotum d. Sintellum.
5. Distortion of aerial parts is caused by stem and leaf feeding nematodes such as
   a. *Ditylenchus* b. *Aphelenchoides* c. *Anguina* d. a + b + c
6. In onion, the basal leaves become twisted when infested with
   a. *Ditylenchus destructor* b. *D. angustus*
   c. *D. dipsaci* d. *D. myceliophages*
7. In rice the top leaves become twisted due to feeding of
   a. *Ditylenchus destructor* b. *D. angustus*
   c. *D.myceliophages* d. *Aphelenchoides besseyii*
8. Crinkling, curling, twisting and distortion of stems and leaves is noticed in oat plants infested with
   a. *Anguina tritici* b. *A. agrostis* c. *A. graminis* d. *A. tumifaciens*
9. Coconut trees exhibit a ring of reddish discoloured tissue in the trunk due to the attck of
   a. *Bursaphalenchus xylophilus* b. *Nacobbus batatiformis*
   c. *Rotylenchulus reniformis* d. *Radinophilenchus cocophilus*
10. Cell necrosis of localized areas is induced by
    a. *Helicotylenchus* b. *Ditylenchus* c. *Pratylenchus* d. *Scutellonima*
11. Interveinal necrosis in plants is caused by
    a. *Criconimoides* b. *Hirshmaniella* c. *Aphelenchoides* d. *Meloidogyne*
12. Necrosis and discolouration of tissues are mainly due to accumulation of
    a. Starch b. Pigment granules c. Phenolic substances d. a + b
13. Typical lesions on leaf caused by *Aphelenchoides ritzemabosi* is exhibited on
    a. Citrus b. Strawberry c. Rice d. Chrysanthimum
14. Seed galls on grasses and cereals are caused by the species of the genus
    a. *Ditylenchus* b. *Pratylenchus* c. *Anguina* d. *Radopholous*

15. Typical leaf gall is produced by
    a. *Anguina balsamophila*
    b. *A. millefolii*
    c. *A. agrostis*
    d. a and b
16. Decline and die back in citrus is caused by
    a. *Tylenchulus semipenetrans*
    b. *Trichodorus christei*
    c. *Globodera rostochiensis*
    d. *Tylenchorhynchus claytoni*
17. Patchy yellow appearance in coffee is due to
    a. *Paratylenchus coffeae*
    b. *Pratylenchus coffeae*
    c. *Pratylenchus vulnus*
    d. *Pratylenchus pratensis*
18. Potato plants exhibit light green foliage due to infestation with
    a. *Meloidogyne hapla*
    b. *M. incognita*
    c. *Globodera rostochiensis*
    d. *Nacobbus batatus*
19. This nematode is intermediate which can live entirely above ground and subterranean form also
    a. *Ditylenchus destructor*
    b. *Ditylenchus dipsaci*
    c. *Trichodorus christei*
    d. *Tylenchulus semipenetrans*
20. This genus usually remains attached to the host tissue for long time
    a. *Tylenchorhynchus*
    b. *Tylenchulus*
    c. *Trichodorus*
    d. *Paratylenchus*
21. If two spicules are fused together, it is referred to as
    a. Spiculum
    b. Spicullane
    c. Spiculonum
    d. Spiculin
22. 'Segging in oats' is a disease caused by this nematode
    a. *Anguina tritici*
    b. *Ditylenchus dipsaci*
    c. *Heterodera avenae*
    d. *Anguina balsamophila*
23. False root knots are produced on sugar beet and tomato by
    a. *Nacobbus batatiformis*
    b. *Nacobbus serendipiticus*
    c. *Ditylenchus radicicola*
    d. a and b
24. *Ditylenchus radicicola* causes small galls on the roots on
    a. Wheat
    b. Rye
    c. Oats
    d. a + b + c
25. *Hemicycliophora arenaria* induces galls in the roots of this crop
    a. Caster
    b. Lemon
    c. Pepper
    d. Tobacco
26. Knot like galls on rose are induced by
    a. *Xiphinema diversicaudatum*
    b. *Hemicycliophora arenaria*
    c. *Criconema*
    d. b + c
27. Short stubby and malformed roots are symptoms of attack of
    a. *Radopholus similis*
    b. *Helicotylenchus multicinctus*
    c. *Trichodorus christei*
    d. *Scutellonema bradys*
28. 'The root system appears devoid of branch rootlets and made up of mostly main roots' is the characteristic symptom observed by the attack of
    a. *Meloidogyne hapla*
    b. *Belonolaimus longicaudatus*
    c. *Ditylenchus destructor*
    d. a + b

29. **Elliptical necrotic lesions are chiefly produced by**
   a. *Pratylenchus spp* b. *Radopholus similis*
   c. *Helicotylenchus multicinctus* d. a + b + c
30. Examples of root rot symptoms caused by nematode attack
   a. *Ditylenchus destructor* b. *Scutellonema bradys*
   c. *Meloidogyne hapla* d. a and b
31. **Koch's postulates were proved for the first time with this nematode involved in peach replant problem**
   a. *Pratylenchus pratensis* b. *P. penetrans*
   c. *P. vulnus* d. *P. scribneri*
32. Gall forming nematodes are
   a. *Angulnu trltlcl* b. *A. balsomophila* c. *Ditylenchus dipsaci* d. a and b
33. Non- gall forming nematodes are
   a. *Ditylenchus dipsaci* b. D. *angustus*
   c. *Aphelenchoides fageriae* d. a + b + c
34. Insect transmitted phytonematodes are
   a. *Bursaphelenchus xylophilus* b. *Radinaphelenchus cocophilus*
   c. *Tylenchulus semipenetrans* d. a and b
35. The genus Xiphinema was created by
   a. N.A.Cobb b. Clark c. Thorne d. Hooper
36. *Xiphinema cobbi* was described and reported in 1980 from
   a. Brazil b. India c. South Africa d. Italy
37. Guiding ring is present at the base of the stylet in
   a. *Longidorus* b. *Criconema* c. *Hoploaimus* d. *Xiphinema*
38. 'Z' organ is of taxonomic importance in
   a. *Xiphinema* b. *Longidorus* c. *Criconema* d. *Hoplolaimus*
39. **Auto fluorescent guilding sheath is found in this species**
   a. *Xiphinema index* b. *X.americanum* c. *X. diversicadatum* d. *X. coxi*
40. The number of rectal glands in male nematodes is usually
   a. 3 b. 6 c. 4 d. Varies
41. Super family Trichodoridea consists the following nematode
   a. *Trichodorus* b. *Paratrichodorus* c. *Longidorus* d. a and b
42. The superficial region about the anus is termed as
   a. Perineum b. Perenna c. Pyrineum d. Epineum
43. The genus *Trichodorus* was created by
   a. Filipjev b. Siddiqi c. N.A.Cobb d. Sasser
44. The term 'pisiform' means
   a. Pear shaped b. Tubular c. Pea shaped d. Irregular shaped

45. Tooth-like stylet is present in

a. *Xiphinema* b. *Trichodorus* c. *Meloidogyne* d. *Heterodera*

46. Among the following, this nematode is smaller in size

a. *Trichodorus* b. *Xiphinema* c. *Radopholus* d. *Tylenchorhynchus*

47. 'Mesenteron' also means the

a. Pharynx b. Rectum c. Cloaca d. Intestine

48. The genus Paratrichodorus was discovered by

a. N.A.Cobb b. Siddiqi c. Hooper d. Filipjev

49 The existence of physiological races in phytonematodes was reported as early as 1888 by

a. Carter b. Mountain c. Sturhan d. J. Ritzemaboss

50. **Goodey was known for this contribution in the field of Nematology**

a. Ecpology b. Morphology c. Taxonomy d. Anatomy

ANSWERS KEY FOR XXVI

Sl. No.	Ans	Sl. No.	Ans	Sl. No.	Ans	Sl. No.	Ans	Sl. No.	Ans
1	d	11	b	21	a	31	b	41	d
2	c	12	c	22	b	32	d	42	a
3	a	13	d	23	d	33	d	43	c
4	b	14	c	24	d	34	d	44	c
5	d	15	d	25	b	35	a	45	b
6	c	16	a	26	a	36	b	46	a
7	b	17	b	27	c	37	d	47	d
8	a	18	c	28	b	38	a	48	b
9	d	19	d	29	d	39	c	49	d
10	a	20	b	30	d	40	b	50	c

XXVII.

1. **The first record of involvement of root knot nematode and *Fusarium* complex in vascular wilt of cotton was in 1892 by**
   a. Atkinson b. Carter c. Oostenbrink d. Christie
2. Augmentation of *Fusarium* wilt of cotton by *Belanolaimus longicaudatus* **was first reported by**
   a. Christie and Perry b. Holdeman and Graham
   c. Barron and Duddington d. Jenson and Hawkins
3. The root knot- *Fusarium* interaction in tobacco was reported by
   a. Goodey and McGuire b. Minton and Minton
   c. Melendez and Powell d. Buzan de Segura and Aguilar
4. Interaction of *Fusarium* wilt with *Belanolaimus* is usually found in this crop
   a. Cotton b. Onion c. Peas d. Celery
5. Black shank disease of tobacco, Phytophthora nicotianae is greatly enhanced by early invasion of
   a. *Meloidogyne incognita* b. *Globodera rostochiensis*
   c. *Tylenchulus semipenetrans* d. *Radopholus similis*
6. Panama wilt is induced by
   a. *Meloidogyne incognita and Fusarium*
   b. *M. incognita and Verticillium*
   c. *Radopholus similis and Fusarium oxysporum f.sp. cubense*
   d. *Hoplolaimus uniformis and Fusarium oxysporum f.sp. pisi*
7. Stunt nematodes form disease complex frequently with these fungi
   a. *Pythium ultimum* b. *P. agoni* c. *P. claytoni* d. *P.debarynum*
8. 'Fusiform' means
   a. Bottle shaped b. Tapering toward posterior end
   c. Tapering toward anterior end d. Tapering toward each end
9. **Breakdown of resistance to fungal pathogens due to root-knot nematodes was first reported by**
   a. Mountain b. Atkinson c. Powell d. Patrick Pitcher
10. *Rhizoctonia solani* is able to attack root-knot infected plants because
    a. Galled tissues exude more carbohydrates
    b. Nematodes rupture galled tissues for egg laying
    c. J2 leave micropunctures on roots while invading
    d. Nematodes carry fungus spores on their surface
11. In most cases involving nematode and pathogenic fungal interactions, this is most possible role of nematodes
    a. Mechanical wounding agents b. Vector
    c. Physiological modification of host plant d. Resistant breakers
12. **Association of root knot nematode with bacterial wilt of tomato was first reported by**
    a. Atkinson b. Goheen c. Hewitt d. Hunger

13. First person to demonstrate that bacteria are carried through digestive tract of nematodes
 a. Stewart b. Jensen c. Kalinenko d. Schindler

14. 'Wounding of roots by Meloidogyne facilitates infection of tobacco roots by '*Ralstonia solanacearum*' was reported by
 a. Hewitt *et al.* b. Lucas *et al.* c. Pitcher *et al.* d. Johnson *et.al.*

15. Association of *Aphelenchoides ritzemabosi* with the bacterium *Clavibacter fascians* was demonstrated by
 a. Hewitt *et al* b. Lucas *et al* c. Pitcher and Crosse d. Stewart

16. The cauliflower disease is due to the interaction of *Clavibacter fascians* with this nematode
 a. *Meloidogyne incognita* b. *Anguina tritici*
 c. *Rotylenchulus reniformis* d. *Aphelenchoides ritzemabosi*

17. External genital armature usually means the
 a. Spicule b. Gubernaculum c. Bursa d. a + b

18. Bacteria are transmitted through this organ of female nematode
 a. Ovule b. Oocytes c. Ovary d. Vulva

19. The root-knot generally has interaction with
 a. *Xanthomonas citri* b. *Ralstonia solanacearum*
 c. *Bacillus subtilis* d. *Streptomyces scabies*

20. The tundu disease of wheat is usually associated with this nematode
 a. *Anguina avenae* b. *A. millifoli* c. *A. agrostis* d. *A. tritici*

21. In the 'tundu' disease of wheat, the role of nematode is
 a. Vector b. Pathogen c. Incitant d. Aggravator

22. Bacterial wilt of Lucerne caused by Corynebacterium insidiosum is increased in the presence of
 a. *Meloidogyne incognita* b. *Heterodera glycines*
 c. *Ditylenchus dipsaci* d. *Pratylenchus vulnus*

23. Crown gall of peach caused by Agrobacterium tumifaciens is increased in the presence of high population of
 a. *Meloidogyne incognita* b. *M. javanica* c. *M. arenaria* d. *M. hapla*

24. This pathogen causing scab of gladiolus is involved with root knot complex
 a. *Pythium gladioli* b. *Fusarium gladioli* c. *Bacillus gladioli* d. *Pseudomonas gladioli*

25. The first report of virus transmission by a phytonematode was by
 a. Johnson, 1945 b. Hewitt, Raski and Goheen, 1958
 c. Harrison & Cadman, 1959 d. Christie, 1951

26. Hewitt and associates experimentally demonstrated the virus transmission by a plant parasitic nematode in
 a. Grapevine fan leaf virus transmitted by Xiphinema index
 b. Tomato ring spot virus transmitted by Xiphinema americanum
 c. Tobacco rattle virus transmitted by Paratrichodorus christei
 d. Grapevine rosette virus transmitted by Xiphinema index

27. Nematodes of this order are involeved in the virus transmission
    a. Tylenchida b. Aphelenchida c. Dorylaimida d. a and b
28. All virus transmitting nematodes possess this character
    a. Spear without a dorsal groove b. Spear with a dorsal groove
    c. Spear without a sucking tube d. Spear with a ventral groove
29. In connection to virus nematode relationship following statement is true
    a. Virus can invade nematode tissue b. Virus is lost after a molt
    c. Virus is retained after molt d. Virus can multiply inside nematode body
30. **Specificity of transmission of a virus by nematode is determined by**
    a. Nucleic acid of virus b. Shape of virus c. Protein coat of virus d. Size of virus
31. **Nematodes that do not kill cells rapidly during feeding are efficient transmitters of**
    a. Bacteria b. Fungi c. Viruses d. Mycorrhizae
32. Virus is transmitted by this nematode genus
    a. *Globodera* b. *Longidorus* c. *Anguina* d. *Tylenchorhynchus*
33. NEPO viruses are transmitted by
    a. *Xiphinema* b. *Longidorus* c. *Paralongidorus* d. *a + b + c*
34. NEPO viruses are transmitted by
    a. *Xiphinema index* b. *Trichodorus christei* c. *Pythium ultimum* d. *Meloidogyne incognita*
35. The diseases caused by NEPO viruses
    a. Arabis mosaic b. Grapevine fan leaf virus
    c. Tobacco ring spot virus d. a + b + c
36. Grapevine fan leaf virus is transmitted by
    a. *Xiphinema americanum* b. *X. index*
    c. *X. diversicaudatum* d. *X.brevicolle*
37. Tobacco ring spot virus is transmitted by
    a. *Xiphinema brevicolle* b. *X. index*
    c. *X. americanum* d. *X. diversicaudatum*
38. Shock and recovery phases are the characteristic symptoms exhibited by
    a. Arabis mosaic virus b. Tomato ring spot virus
    c. Strawberry latent ring spot d. Tobacco ring spot
39. *Xiphinema diversicaudatum* transmits this virus
    a. Grapevine fan leaf b. Arabis mosaic c. Tomato ring spot d. Pea early browning
40. This nematode can remain for hours or days at one feeding site
    a. *Xiphinema* b. *Trichodorus* c. *Tylenchorhynchus* d. *Longidorus*
41. The Scottish strain of Raspberry ring spot virus is transmitted by
    a. *Longidorus macrosoma* b. *L. elongatus*
    c. *L. attenuatus* d. *Paralongidorus maximus*
42. Tomato black ring virus is transmitted by
    a. *Longidorus macrosoma* b. *Paralongidorus maximus*
    c. *L. attenuatus* d. *L. elongatus*

43. Mulberry ring spot virus is transmitted by this nematode species
   a. *Longidorus apulus* b. *L. martini* c. *L. elongates* d. *L. enonymus*
44. Strawberry latent ring spot is transmitted by
   a. *Xiphinema diversicaudatum* b. *Longidorus macrosoma*
   c. *Paralongidorus maximus* d.*Paratrichodorus christei*
45. **Specificity of virus transmission is comparatively more pronounced in this genus**
   a. *Xiphinema* b. *Trichodorus* c. *Paratrichodorus* d. *Longidorus*
46. **Following is the efficiency order of nematodes in transmitting viruses**
   a. *Longidorus, Xiphinema, Paralongidorus*
   b. *Paralongidorus, Xiphinema, Longidorus*
   c. *Xiphinema, Longidorus, Paralongidorus*
   d. *Xiphinema, Paralongidorus, Longidorus*
47. NETU viruses are transmitted by
   a. *Trichodorus* b. *Paratrichodorus* c. *Longidorus* d. *a and b*
48. This is an example for a NETU virus
   a. Tomato black virus b. Pea early browning
   c. Arabis mosaic d. Strawberry latent ring spot
49. Tobacco rattle virus is transmitted by
   a. *Longidorus elongates* b. *Paratrichodorus teres*
   c. *Trichodorus christei* d. *Xiphinema index*
50. 'Dorylaimoid esophagus' has this shape
   a. Kidney shaped b. Short necked bottle shape
   c. Bulbous shaped d. Long necked bottle shape

ANSWERS KEY FOR XXVII

Sl. No.	Ans	Sl. No.	Ans	Sl. No.	Ans	Sl. No.	Ans	Sl. No.	Ans
1	a	11	c	21	a	31	c	41	b
2	b	12	d	22	c	32	b	42	d
3	c	13	a	23	b	33	d	43	b
4	d	14	b	24	d	34	a	44	c
5	a	15	c	25	b	35	d	45	a
6	c	16	d	26	a	36	b	46	c
7	b	17	c	27	c	37	c	47	d
8	d	18	a	28	b	38	d	48	b
9	b	19	b	29	d	39	c	49	c
10	a	20	d	30	a	40	a	50	d

XXVIII.

1. In USA the following nematode has been subjected to the Federal Quarantine since 1954
   a. *Heterodera glycines*
   b. *Globodera pallid*
   c. *Meloidogyne incognita*
   d. *Tylenchulus semipenetrans*
2. State quarantine was enforced in USA against this nematode
   a. *Heterodera schachtii*
   b. *Globodera rostochiensis*
   c. *Radopholus similis*
   d. *Pratylenchus coffeae*
3. *Radopholus citrophilus* was brought under the State quarantine act in Florida in
   a. 1941 b. 1950 c. 1953 d. 1958
4. In India Quarantine act has been imposed against
   a. *Pratylenchus coffeae*
   b. *Radopholus similis*
   c. *Meloidogyne incognita*
   d. *Globodera rostochiensis*
5. Most phytonematodes spread mainly through
   a. Air b. Soil c. Water d. b + c
6. Most nematodes spread by
   a. Active means b. Passive means c. Lathargic d. a + c
7. Movement of irrigation water and floods due to severe rain aids in the
   a. Passive dispersal of nematodes
   b. Active dispersal of nematodes
   c. a and b
   d. None of the above
8. These are seed borne nematodes
   a. *Ditylenchus dipsaci*
   b. *Aphelenchoides besseyi*
   c. *Anguina tritici*
   d. a + b + c
9. Seed borne nematodes can be controlled by fumigation with
   a. Chloropicrin b. Formalin c. Methyl bromide d. Aldicarb
10. Hot water treatment is the effective method for the management of
    a. *Pratylenchus coffeae*
    b. *Globodera rostochiensis*
    c. *Hirschmanniella oryza*
    d. *Aphelenchoides besseyi*
11. 'Nematode mortality when sucrose or dextrose is added to nematode infested soil' was reported by
    a. N.A. Cobb b. Feder c. Norton d. Gubanov
12. The most important nematode pests controlled by greenhouse steaming
    a. Golden nematode of potato
    b. Root knot nematode
    c. Burrowing nematode
    d. a and b
13. The most effective means of control of Molya disease of wheat is
    a. Chemical spraying
    b. Flooding
    c. Crop rotation
    d. Seed treatment

14. *Meloidogyne hapla*, a serious pest of groundnut and soybean can be effectively controlled by crop rotation with
   a. Sesamum b. Linseed c. Sunflower d. Cotton
15. Thisis the best method of controlling potato cyst nematode
   a. Seed cleaning b. Crop rotation
   c. Soil application with systemic granules d. Nematicide folair spray
16. Fallowing is more effective particularly, against
   a. Root knot nematode b. Burrowing nematode
   c. Cirtus nematode d. Red ring nematode
17. Anaerobic conditions of flooded fields kill nematodes due to production of
   a. Hydrogen sulphide b. Butyric acid c. Ammonia d. a + b + c
18. Flooding is very effective against
   a. Root knot nematode b. Stunt nematode c. Reniform nematode d. a and b
19. The following nematodes can be eliminated by selecting nematode free planting material
   a. *Tylenchulus semipenetrans* b. *Globodera rostochiensis*
   c. *Radopholus similis* d. a + b + c
20. This nematode can be eliminated by planting nematode free seeds
   a. *Anguina tritici* b. *Aphelenchoides besseyi*
   c. *Trichidorus christei* d. a and b
21. The plant that can be used as trap crop for nematodes is
   a. *Tagetes* b. Sweet potato c. Sasamum d. Neem
22. The following are responsible to suggest that *Crotalaria* root system exerts a toxic effect on *Meloidogyne* populations
   a. Meijneke and Oosternbrink b. Atwal and Mangar
   c. Rhode and Jenkins d. Ochse and Brewton
23. The toxic material produced by ***Asperagus officinalis***, which is effective against *Trichodorus christei* and Pratylenchus curvatus
   a. Asparaguisic acid b. Maleic acid c. Aspartic acid d. Chlorogenic acid
24. The toxic material produced by Asperagus officinalis effective against Trichodorus christei was first demonstrated by
   a. Meijneke and Oosternbrink b. Ochse and Brewton
   c. Rhode and Jenkins d. Atwal and Mangar
25. Root exudates of *Brassica nigra* and *Brassica hirta* neutralizes the effect of
   a. Root knot nematode b. Potato cyst nematode
   c. Sheath nematode d. Ring nematode
26. The compound produced by cucumber which acts as a repellant for nematodes
   a. Cucurbitacin b. Cucubirmacin c. Cucurbicin d. Cucubirneosin

27. The leaves of *Pongamia labra* and *Azadirachta indica* reduces the population of
    a. Buib nematode
    b. Cyst nematode
    c. Burrowing nematode
    d. Root-knot nematode
28. The follwoing showed that the populations of *Pratylenchus, Pratylenchus, Tylenchorhynchus* and *Rotylenchulus* could be considerably reduced by growing *Tagetes erecta* and *Tagetes patula*
    a. Meijneke and Oostenbrink
    b. Ochse and Brewton
    c. Rohde and Jenkins
    d. Atwal and Mangar
29. 'Corpus' refers to the
    a. Most posterior segment of the oesophagus
    b. Mid part of the oesophagus
    c. Extension of the intestine
    d. Most anterior segment of the oesophagus
30. This marigold species gives better management of nematodes
    a. *Tagetes erecta*
    b. *T. patula*
    c. *T. minuta*
    d. *a and b*
31. This is a highly effective product in reducing root-knot nematode population
    a. *Azardirachta arabica*
    b. *Azardirachta indica*
    c. *Azardirachta indicus*
    d. *Azardirachta glabra*
32. The Following showed that root exudates from *Esamum orientale* have nematicidal properties against *Meloidogyne incognita*
    a. Meijneke and Oostenbrink
    b. Ochse and Brewton
    c. Rohde and Jenkins
    d. Atwal and Mangar
33. The active principle involved in mustard which is toxic to nematodes
    a. Allyl isothiocyanate
    b. Protocatechuic acid
    c. Catechol
    d. à-terthienyl
34. Early planting of this crop can escape from nematode attack
    a. Sugar cane
    b. Potato
    c. Tomato
    d. Sugar beet
35. **The signifincance of decomposition of organic matter in soil in the reduction of nematode damage was first demonstrated by**
    a. Johnston
    b. Rhoades
    c. Hollis
    d. Linford et al
36. Root knot nematode can be effectively minimized by incorporation of these into the soil
    a. Chopped sapota leaves
    b. Chopped mango leaves
    c. Chopped pineapple leaves
    d. Chopped guava leaves
37. The phenomenon where 'an individual which exhibits a blending of male and female characters and which functions as one or the other or neither sex, but never as both' is termed as
    a. Bisexual
    b. Amphibial sex
    c. Dioecious
    d. Intersex
38. Sawdust application to reduce nematode population is not very effective in this soil type
    a. Clayey
    b. Heavy sandy
    c. Light sandy
    d. Porous
39. Earcockle disease of wheat can be easily eradicated by
    a. **Use of certified seed**
    b. Crop rotation
    c. **Removal of galls by water floatation**
    d. Winnowing
40. The scientist who used CS_2 as soil fumigant for the control of sugar beet cyst nematode
    a. Schmidt
    b. Julius Kuhn
    c. Schacht
    d. Hue

41. The effect of chloropicrin against phytonematodes was discovered in 1919 by
    a. Kagy b. Youngson c. Mathews d. Carter
42. The nematicidal value of DD was discovered in 1943 by
    a. Perry b. Christie c. Dastur d. Carter
43. The nematicidal value of EDB was discovered in 1945 by
    a. Christie b. Perry c. Chitwood d. N.A. Cobb
44. This belongs to 'Halogenated hydrocarbons'
    a. Chloropicrin b. Methyl bromide c. Ethylene dibromide d. a + b + c
45. Metham sodium and Dibromochloropropane belong to this group of nemaicide
    a. Organo phosphate b. Oxime carbamate
    c. Halogenated hydrocarbon d. Inorganic
46. The nematicide which is effective against nematodes, soil insects, fungi, rodents and weeds
    a. Methyl bromide b. Chloropicrin c. Carbofuran d. a and b
47. This is a chemical with an outstanding solubility in water
    a. DD b. MB c. EDB d. DBCP
48. DD contains
    a. 1,3 dichloropropene b. 1,2 dichloropropane
    c. 1,4 dichloropropane d. a and b
49. This chemical does not hydrolyse readily and persists longer
    a. EDB b. MB c. DD d. DBCP
50. This fumigant dissolves slowly compared to others
    a. DD b. EDB c. MB d. DBCP

ANSWERS KEY FOR XXVIII

Sl. No.	Ans	Sl. No.	Ans	Sl. No.	Ans	Sl. No.	Ans	Sl. No.	Ans
1	a	11	b	21	b	31	b	41	c
2	b	12	d	22	d	32	d	42	d
3	c	13	c	23	a	33	a	43	a
4	d	14	d	24	c	34	b	44	d
5	d	15	b	25	b	35	d	45	c
6	b	16	a	26	a	36	c	46	d
7	c	17	d	27	d	37	d	47	b
8	d	18	d	28	a	38	a	48	d
9	c	19	d	29	d	39	c	49	c
10	d	20	d	30	a	40	b	50	a

XXIX.

1. Trade name of EDB is
   a. Dow fume b. Bromofume 40 c. Nemex 42 d. a,b & c
2. This compound has a very high vapour pressure
   a. DD b. MB c. DBCP d. EDB
3. The crop less sensitive to DBCP
   a. Onion b. Potato c. Tobacco d. Coconut
4. Trade name of DBCP
   a. Nemagon b. Telone c. Dowfume d. Kelone
5. Nematicidal activity of halogenated hydrocarbons is
   a. Alkylation b. Reduction c. Hydrolysis d. Carbonation
6. The organphorus compounds which are employed to kill nematodes are
   a. Fensulphothion b. Phenamiphos c. Thionazin d. a,b & c
7. Dichlorofenthion, Ethoprop, Phorate and Parathion belong to this group of nematicides
   a. Organo phosphate b. Oxime carbamate
   c. Halogenated hydrocarbons d. Mercuric
8. Trade name of Thionazin
   a. Dasanit b. Nemaphos c. Nemacur d. Thimet
9. Trade name of Fensulphothion
   a. Nemacur b. Nemaphos c. Dasanit d. Thimet
10. Trade name of Ethoprop
    a. Thimet b. Thiophos c. Hexanema d. Mocap
11. Trade name of Phenamiphos
    a. Thimet b. Nemacur c. Hexanema d. Furadon
12. Trade name of Phorate
    a. Thimet b. Furadon c. Blitox d. Terracur
13. Persistence of this nematicide is longer in soils having low pH
    a. Aldicarb b. Carbofuran c. Phorate d. Oxamyl
14. The following nematicide exhibit irreversible bibding of acetyl cholinesterase
    a. Oxamyl b. Fensulfothion c. Phorate d. Phenamiphos
15. After hydrolysis of acetyl choline, acetic acid can resythesize acetyl choline in the presence of co-enzyme
    a. Choline esterase b. Choline acetylase c. Choline acetyliase d. Choline synthetase
16. This is a carbomate group of compound
    a. Aldicarb b. Carbofuran c. Metham sodium d. a,b &c

17. The metabolic product of carbofuran is

a. Carbofuran anide
b. Carbofuran phenol
c. Carbofuran sulfone
d. Carbofuran phosphate

18. Most common brand name of Carbofuran

a. Thiodon
b. Furadon
c. Celdon
d. Paldon

19. Trade name of Aldicarb

a. Furadan
b. Vapam
c. Vydate
d. Temik

20. The following nematicide liberates methyl isothiocyanate

a. Oxamyl
b. Aldicarb
c. Dazomet
d. Ethoprop

21. Trade name of Metham sodium

a. Temik
b. Vapam
c. Vydate
d. Furadan

22. Most nematicides kill nematodes by affecting respiration through

a. Affecting enzymes
b. Denaturing organs
c. Blocking cytochromes
d. Imbalancing CO_2

23. Narcotic action refers to

a. Mode of predation
b. Parasitization
c. Nematicidal action
d. Hypersensitivity

24. Penetration of nematicides to the receptor site follows this sequence

a. Detoxification- Excretion -Toxification
b. Toxification –Excretion- Detoxification
c. Excretion –Toxification- Detoxification
d. Toxification- Excretion -Detoxification

25. Mode of action of most nematicides occurs in this route of inhibition

a. Ach E Lipoprotein Cytochrome
b. Haemoprotein Ach E Lipoprotein
c. Ach E Haemoprotein Cytochrome
d. Lipoprotein Ach E Haemoprotein

26. With respect to toxicity of nematicides, following is the ascending order

a. Endosulfan, Diazinon, and Fenvalerate
b. Carbofuran, Demethon and Chloropicrin
c. Aldicarb, Fenvalerate and DD
d. Endosulfan,Carbofuran and Fenvalerate

27. Addition of this compound to soil can protect plants from the combined pathological action of nematodes and other organisms

a. Chitin
b. Pectin
c. Lignin
d. Cellulose

28. Following concentration of dextrose may be detrimental to nematodes

a. 100 ppm
b. 1000 ppm
c. 10 ppm
d. 10000 ppm

29. For killing of nematodes chemicals should pass through this intracellular barrier

a. Cell wall
b. Plasma membrane
c. Cytoplasmic membrane
d. Mitochondrial membrane

30. This is the chemical transmitter, which contracts muscles in nematodes

a. Chlorogenic acid
b. Acetylcholine
c. Pentose monophosphate
d. Lactose dinine

31. Most of the nematophagous fungi belong to this order
    a. Moniliales b. Zoophagales c. Saproleginales d. a and b
32. Nematophagous fungi have been mainly divided into
    a. Trap forming fungi b. Non trap forming endozoic parasites
    c. Spore forming fungi d. a and b
33. The active morphogenic substance believed to induce trap formation in Arthrobotyrsn is
    a. Nimbin b. Nemin c. Epinimbin d. all the above
34. The sticky knobs formed by the predacious fungi to trap the nematodes
    a. *Monacrosporium ellipspora* b. *Dactyella cinophaga*
    c. *Athrobotrys oligospora* d. a + b + c
35. Following are the predaceous fungi which form constricting ring to trap the nematodes
    a. *Dactylella bemicoides* b. *Monosporium bembicoides*
    c. *Dactylaria brochopaga* d. a + b + c
36. The predaceous fungi which form non- constricting ring to trap the nematodes
    a. *Dactylaria candida* b. *Catenaria vermicola*
    c. *Phialophora heteroderae* d. *Dactylella primus*
37. The nematode trapping fungus is
    a. *Dactylella bemicoides* b. *Dactylaria candida*
    c. *Dactylella cinophaga* d. a + b + c
38. The endoparasitic fungi which frequently attack sugarcane nematode
    a. *Catenaria anguillulae* b. *Dactylaria candida*
    c. *Nematoctonus heptocladus* d. *Dactylella cinophaga*
39. The fungus which penetrates the egg shell of *Globodera rostochiensis* and decrease the number of viable larvae
    a. *Harposporium anguillulae* b. *Catenaria anguillulae*
    c. *Phialophora heteroderae* d. a + b
40. The endozoic parasites which kill the nematodes on contact before penetrating the cuticle
    a. *Nematoctonus heptocladus* b. *Nematoctonus concurrence*
    c. *Curvularia pallescence* d. a and b
41. The endozoic fungi commonly used for the control of nematode is
    a. *Dactylella cinophaga* b. *Dactylaria spp.*
    c. *Paecilomyces lilacinus* d. *Pasteuria penetrans*
42. This predaceous nematode possess nerve poison to kill plant parasitic nematodes
    a. Mononchus b. Seinura c. Chambersiella d. Strongylus
43. Following demonstrated that the tardigrade, *Hypsibius myrops* can reduce the population of phytonematodes
    a. Sayre b. Doncaster c. Murphy d. Schaeffenberg

44. These bio agents kill nematodes by secreting cuticle dissolving enzymes

a. Arthropods b. Protozoa c. Fungi d. Bacteria

45. The endospore forming bacteria effective in nematode management is

a. *Pasteuria penetrans* b. *Paecilomyces lilacinus*

c. *Achromobacter nematophilus* d. *Bacillus subtilis*

46. 'Coelomate' refers to

a. Having a false body cavity b. Having a incomplete body cavity

c. Having a true body cavity d. Having a non-functional body cavity

47. The nerve ring is also called as the

a. Circompassing ring b. Circomerian ring

c. Cirraceous ring d. Circumenteric ring

48. 'DEGO' refers to the

a. Dominant Enlarged Gene Organ b. Dorsal Esophageal Gland Orifice

c. Dendro Esophageal Gland Orifice d. Delta Enlarged Gut Opening

49. A mixture of chemicals that harden and preserve the bodies of relaxed nematode specimen are referred to

a. Fixative b. Preservative c. Hardener d. Mounting medium

50. A nematode that can directly kill another nematode

a. Mononchoid b. Discolaimus c. Mononchus d. a + b + c

ANSWERS KEY FOR XXIX.

Sl. No.	Ans	Sl. No.	Ans	Sl. No.	Ans	Sl. No.	Ans	Sl. No.	Ans
1	d	11	b	21	b	31	b	41	c
2	c	12	a	22	c	32	d	42	b
3	d	13	c	23	d	33	b	43	a
4	a	14	d	24	c	34	d	44	d
5	b	15	a	25	a	35	d	45	a
6	d	16	d	26	b	36	a	46	c
7	a	17	c	27	d	37	d	47	d
8	b	18	b	28	b	38	b	48	b
9	c	19	d	29	c	39	c	49	a
10	d	20	b	30	a	40	d	50	d

XXX.

1. Occurrence of root knot nematode in cucumber was first reported in 1855 by
   a. M.J. Berkeley b. C. Liebscher c. Barber d. A. Strubell
2. The first specific mention of root knot nematode, *Anguillula marioni* was described in 1879 by
   a. E.A. Goeldi b. M. Cornu c. Schacht d. Needham
3. Root knots are produced due to the attack of *Meloidogyne* that involves
   a. Hypertrophy b. Hyperplasia
   c. Hypertrophy and hyperplasia d. Hyptrophy and hypoplassia
4. 'Germarium' refers to the
   a. Germinal organ b. Germinal cells c. Germinal oocytes d. Germinal zone
5. Root knot infected plants contain more of
   a. Adenine b. Guanine c. Tryptophan d. Thiamine
6. A natural pening in the egg shell is tered as
   a. Hiatus b. Gelatus c. Geltifice d. Hiatifice
7. *Meloidogyne* hapla may be restricted to
   a. Temperate regions b. High altitudes of tropics
   c. High altitudes of sub-tropics d. a, b & c
8. *Meloidogyne incognita* and *Meloidogyne javanica* are found in
   a. Tropical regions b. High altitudes c. Temperate regions d. a,b & c
9. This species usually produces small galls on roots
   a. *Meloidogyne incognita* b. *M. hapla*
   c. *M. arenaria* d. *M. javanica*
10. The Meloidogyne species which produces large galls on roots
    a. *Meloidogyne javanica* b. *M. incognita*
    c. *M. arenaria* d. *a and b*
11. The number of races identified by International *Meloidogyne* Project in *Meloidogyne incognita* and *Meloidogyne arenaria* are
    a. 2 & 3 b. 3 & 3 c. 4 & 2 d. 2 & 5
12. *Meloidogyne graminacola* is generally a serious problem in
    a. Low land rice b. Upland rice c. Wheat d. a and b
13. Rice root-knot nematode normally is favoured by
    a. Acidic pH b. Alkaline pH c. Neutral pH d. High salinity
14. The *Meloidogyne* species most commonly recorded on tobacco is
    a. *Meloidogyne incognita* b. *M. javanica*
    c. *M. hapla* d. a and b
15. The alternate hosts of *Meloidogyne graminacola* are
    a. *Amaranthus viridis* b. *Desmodium triflorum*
    c. *Portulaca oleracea* d. a, b & c

16. This root-knot nematode species is more serous in citrus orchards in India
    a. *Meloidogyne javanica* b. *Meloidogyne africana*
    c. *Meloidogyne incognita* d. *M. indica*
17. The most important nematode infesting the grape is
    a. *Trichodorus christei* b. *Tylenchulus semipenetrans*
    c. *Meloidogyne incognita* d. a & b
18. The root-knot nematode species causing more damage to the papaya crop is
    a. *Meloidogyne javanica* b. *M. incognita*
    c. *M. arenaria* d. *M. hapla*
19. Occurrence of root-knot nematode on vegetables in India was first reported by
    a. Ayyer b. Thirumala Rao c. Butler d. Gupta
20. Sensory papillae situated toward the apex of the lip segment forming a circlet of six in primitive forms are known as
    a. Outer labial papillae b. Inner labial papippae
    c. Paired labial papillae d. Complex labial papillae
21. *Meloidogyne* infection leads to high accumulation of potassium and reduction in calcium content in roots of this crop
    a. Pepper b. Papaya c. Tomato d. Cotton
22. The root-knot nematode commonly found associated with crossandra
    a. *Meloidogyne incognita* b. *M. javanica*
    c. *M. arenaria* d. a and b
23. *Meloidogyne* species commonly found associated with tuberose and jasmine
    a. *Meloidogyne incognita* b. *M. hapla*
    c. *M. arenaria* d. *M. brevicauda*
24. Root knot nematode on black pepper in Kerala was first reported by
    a. Menon b. L.C. Coleman c. Nair d. Butler
25. *Meloidogyne arenaria* is serious problem on groundnut in
    a. Karnataka b. Gujarat c. Andhra Pradesh d. Maharashtra
26. Root knot nematode on tea in South India was first reported by
    a. Siddiqi b. Butler c. Barber d. Ayyer
27. The root knot nematode infecting tea is
    a. *Meloidogyne brevicauda* b. *M. africana*
    c. *M. arenaria* d. *M. incognita*
28. Detailed description of *Meloidogyne exigua* was published in 1887 by
    a. M. Cornu b. E.S. Bessey c. G.F. Atkinson d. E.A. Goeldi
29. The *Meloidogyne* species infecting coffee in Karnataka
    a. *Meloidogyne arenaria* b. *M. exigua* c. *M. javanica* d. *M.hapla*

30. High levels of potassium favor the development of this nematode in certain hosts
    a. Foliar b. Cyst c. Root knot d. Dagger
31. The method to raise root-knot nematode free seedlings is
    a. Flat bed method b. Rabbing c. Raised bed method d. a and c
32. The promising biocontrol agent of root- knot nematodes are
    a. *Purpureocillium lilacinum* b. *Pochonia chlamydosporia*
    c. *Pseudomonas solanacearum* d. a and b
33. This is not a seed borne nematode
    a. *Meloidogyne arenaria* b. *Aphelenchoides besseyi*
    c. *Globodera rostochiensis* d. *Anguina tritici*
34. 'Linguliform' refers to the
    a. Cup shaped b. Rod shaped c. Tongue shaped d. Forcep shaped
35. 'Labium' means
    a. Lip b. Tail tip c. Stoma d. Anus
36. False root-knot nematode, *Nacobbus batatiformis* can infect
    a. Sugar beet b. Cucumber c. Turnip d. a, b & c
37. The major host of *Nacobbus batatiformis* are
    a. Cotton b. Pumpkin c. Black pepper d. Arecanut
38. Bearding or hairy root and witches broom symptoms are caused by
    a. *Meloidogyne* b. *Nacobbus* c. *Heterodera* d. *Rotylenchulus*
39. Molya disease of rye is caused by
    a. *Heterodera schachti* b. *H. radicicola* c. *H. avenae* d. *H. glycines*
40. 'Obtuse' refers to
    a. Blunt, not pointed b. Sharp, pointed c. Long, rounded tip d. Tubular, sharp tip
41. The number of papillary nerves in nematodes is
    a. 8 b. 4 c. 2 d. 6
42. Wheat variety resistant to *Heterodera avenae* that was developed in India
    a. Katyil b. C-164 c. Raj Kiran d. Raj MR 1
43. Barley varietie resistant to *Heterodera avenae*
    a. Raj Kiran b. C-162 c. Raj Kaushal d. Both a and b
44. *Heterodera avenae* is confined to the graminae family except this single legume
    a. *Senebiera pinnatifolia* b. *Cajanus cajan*
    c. *Vigna mungo* d. *Pisum sativum*
45. *Heterodera delvi* is parasitic on this crop
    a. Rice b. Finger millet c. Maize d. Cowpea
46. *Heterodera glycines* was first described in 1952 from
    a. USA b. China c. Japan d. India

47. The segment of the tail posterior to the bursa is named as
    a. Postbursal tail   b. Postanal tail   c. Postcloacal tail   d. Postalar tail
48. Cluster bean can be used to distinguish host races of
    a. *Heterodera glycines*   b. *H. cajani*   c. *H. trifolii*   d. *H. ciceri*
49. The only non-leguminous host of Heterodera cajani is
    a. Soybean   b. Castor   c. Niger   d. Sesamum
50. Tactile organs also referred to as
    a. Sense organs   b. Excretory organs   c. Locomotary organs   d. Molting organs

ANSWERS KEY FOR XXX

Sl. No.	Ans	Sl. No.	Ans	Sl. No.	Ans	Sl. No.	Ans	Sl. No.	Ans
1	a	11	c	21	c	31	b	41	d
2	b	12	a	22	d	32	d	42	a
3	c	13	c	23	a	33	a	43	a
4	d	14	d	24	d	34	c	44	a
5	c	15	d	25	b	35	a	45	b
6	a	16	d	26	c	36	d	46	c
7	d	17	c	27	a	37	b	47	d
8	a	18	b	28	d	38	b	48	b
9	b	19	a	29	b	39	c	49	d
10	d	20	b	30	c	40	a	50	a

XXXI

1. The first ever soil fumigation using carbon disulphide was carried out against
   a. *Globodera rostochiensis* b. *Heterodera schachtii*
   c. *Meloidogyne incognita* d. *Anguina tritici*
2. *Heterodera vignae* is parasitic on this crop
   a. Cowpea b. Mungbean c. Urd bean d. a, b & c
3. *Heterodera oryzicola* was first reported by
   a. Rao b. Jayaprakash c. Edward d. a and b
4. Maize cyst nematode is
   a. *Heterodera zenis* b. *H. zeae* c. *H. zadis* d. *H. zenis*
5. Three host races of Heterodera zeae have been recorded from India on the basis of their multiplication on
   a. Maize only b. Vetiver only c. Maize and Sorghumr d. a & b
6. This disaccharide sugar is stored in the nematode body in place of starch
   a. Dialose b. Trehalose c. Glucose d. Fructose
7. Presence of brown to dark brown cysts projecting on the root surface is the characteristic symptom of this nematode attack
   a. *Longidorus* b. *Globodera* c. *Meloidogyne* d. *Heterodera*
8. Cysts of golden nematode of potato can remain viable for
   a. Less than 1 year b. Up to 5 years c. Up to 15 years d. Up to 50 years
9. Approximate threshold density of Globodera rostochiensis for causing damage in potato
   a. 10 cysts/ 100 g of soil b. 4-10 cysts/ 100 g of soil
   c. 31 cysts/ 100 g of soil d. 50 cysts/ 100 g of soil
10. Potato cyst nematode spreads by
    a. Infested seed tubers b. Movement of soil c. Insect vectors d. a and b
11. 'Kufri Swarna' variety of potato is resistant to
    a. Root-knot nematode b. Golden nematode c. Potato rot nematode d. Reniform nematode
12. Nematode implicated in the Peach Replant problem in Canada
    a. *Pratylenchus penetrans* b. *Meloidogyne hapla*
    c. *Globodera pallida* d. *Radopholus similis*
13. *Rotylenchulus reniformis* was first identified as a parasitic of cotton by
    a. Balasubramanian b. Palaniswamy c. Nandwana d. Steiner
14. First to demonstrate the pathogenecity of a nematode to the satisfaction of Koch's postulates
    a. Mountain b. Patrick c. Filipjev d. a and b
15. 'Truncate' refers to
    a. Squared terminus b. Spherical terminus c. Pointed terminus d. Swollen terminus

16. *Pratylenchus indicus* was first described on tomato and brinjal by
    a. Goodey b. Mahajan c. Colbran d. Das
17. Lesion nematode infecting rice
    a. *Pratylenchus coffeae* b. *P. indicus* c. *P. oryzae* d. *P. pratensis*
18. The alternate hosts of *Pratylenchus indicus* are
    a. Finger millet b. Wheat c. Sorghum d. a + b + c
19. *Pratylenchus zeae* does not attack this crop
    a. Maize b. Sugarcane c. Banana d. Sorghum
20. The major nematode of fruit trees in the nursery stage
    a. *Pratylenchus coffeae* b. *P. pratensis* c. *P. penetrans* d. *Ditylenchus angustus*
21. This nematode is also called as 'Black currant nematode'
    a. *Aphelenchoides ritzemabosi* b. *Aphelenchoides besseyi*
    c. *Ditylenchus dipsaci* d. *Ditylenchus destructor*
22. The typical symptoms of Hirschmanniella oryzae on rice roots are
    a. Cracking up of roots b. Discrete lesions and reduction of root system
    c. General necrosis and reduction of root system d. Coarse and reduced root system
23. Interveinal necrosis in plants is caused by
    a. *Criconemaides* b. *Hirschmanniella* c. *Aphelenchoides* d. a & c
24. Infection of rice by *Rhizoctonia solani* increases in the presence of this nematode
    a. *Meloidogyne graminicola* b. *Aphenchoides besseyi*
    c. *Hoplolaimus indicus* d. *Hirschmanniella oryzae*
25. Miti-miti disease of corm taro (*Colocasia esculenta*) is caused by
    a. *Scutellonema mitium* b. *Hirschmanniella miticausa*
    c. *Aphasmatylenchus mitalis* d. *Scutellonema brachyurus*
26. Spreading decline of citrus is caused by
    a. *Radopholus citrophilus* b. *Tylenchulus semipenetrans*
    c. *Pratylenchus coffeae* d. *Rotylenchulus reniformis*
27. *Radopholus similis* is considered most serious pest of
    a. Brinjal b. Citrus c. Cardamom d. Pepper vine
28. Infection by *Radopholus similis* leads to decrease in organic acid contents in roots of this crop
    a. Banana b. Grape fruit c. Black Pepper d. Grapevine
29. Two distinct races of burrowing nematode identified by Du Charme and Birchfield are
    a. Banana race and Coconut race b. Citrus race and Orange race
    c. Coffee race and Citrus race d. Banana race and citrus race
30. Banana race is parasitic only on
    a. Banana b. Citrus c. Coconut d. a and b
31. Citrus race is parasitic on
    a. Banana b. Citrus c. Arecanut d. a and b

32. The disease caused by *Radopholus similis* on banana is known by
    a. Black head b. Banana decline c. Rhizome rot d. a + b + c
33. Paring and prolinage is a method followed before planting to control burrowing nematode infecting
    a. Rice b. Citrus c. Banana d. Any planting material
34. The causal agent of yellows disease in Black pepper is
    a. *Radopholus citrophilus* b. *Radopholus similis*
    c. *Rotylenchulus reniformis* d. *Pratylenchus penetrans*
35. Slow decline of citrus is caused by
    a. *Tylenchulus semipenetrans* b. *Radopholus similis*
    c. *Helicotylenchus multicinctus* d. *Dolichodorus indicus*
36. Spreading decline differs from slow decline by exhibiting this symptom
    a. Sparse foliage b. Die back c. Wilt d. Cupped leaves
37. 'Race A' of *Rotylenchulus reniformis* multiplies on
    a. Castor b. Cotton c. Cowpea d. a + b + c
38. 'Race B' of *Rotylenchulus reniformis* multiplies only on
    a. Cowpea b. Cotton c. Castor d. a + b + c
39. *Rotylenchulus robustus* is known to cause severe damage in
    a. Citrus b. Pea c. Cotton d. Castor
40. 'Tulip root nematode' is the common name of
    a. *Aphlenchoides besseyi* b. *Ditylenchus dipsaci*
    c. *Radopholus similis* d. *Heterodera vulnus*
41. *Pratylenchus vulnus* is also called as
    a. Vinegar nematode b. Green meadow nematode
    c. Awl nematode d. Walnut meadow nematode
42. The nematodes causing serious damage to papaya
    a. *Rotylenchulus reniformis* b. *Meloidogyne incognita*
    c. *Belanolaimus longicaudatus* d. a and b
43. This species of lance nematodes is most commonly associated with sugarcane
    a. *Hoplolaimus columbus* b. *H. indicus*
    c. *H. galeatus* d. *H. seinhorsti*
44. Hoplolaimus galeatus causes serious damage in
    a. *Soybean* b. *Sugarcane* c. *Cotton* d. *Caster*
45. Extensive discolouration of roots of rose which is often termed as 'rot' is due to
    a. *Pratylenchus zeae* b. *Tylenchorhyncgus claytoni*
    c. *Hoplolaimus denatus* d. *Helicotylenchus multicinctus*
46. *Helicotylenchus multicinctus* is an important parasite of
    a. Banana b. Citrus c. Coconut d. Arecanut

47. This nematode causes superficial lesions on banana roots
    a. *Pratylenchus coffeae*
    b. *Radopholus similis*
    c. *Helicotylenchus multicinctus*
    d. *Hoplolaimus seinhorsti*
48. The most important economically important nematode in betel vine
    a. *Meloidogyne incognita*
    b. *Rotylenchulus reniformis*
    c. *Helicotylenchus incisus*
    d. *Hoplolaimus deinhorsti*
49. 'Fern nematode' is referred to
    a. *Anguina millefoli*
    b. *Aphelenchoides fragariae*
    c. *Ditylenchus angustus*
    d. *Belonolaimus indicus*
50. This nematode is implicated in 'Kalahasti' melody of groundnut
    a. *Tylenchorhynchus brevilineatus*
    b. *Meloidogyne arenaria*
    c. *Scutellonema cavenessi*
    d. *Aphelenchoides arachidis*

ANSWERS KEY FOR XXXI

Sl. No.	Ans	Sl. No.	Ans	Sl. No.	Ans	Sl. No.	Ans	Sl. No.	Ans
1	b	11	b	21	a	31	d	41	d
2	d	12	a	22	c	32	d	42	d
3	d	13	d	23	b	33	c	43	b
4	b	14	d	24	d	34	b	44	c
5	d	15	a	25	b	35	a	45	d
6	c	16	d	26	a	36	c	46	a
7	d	17	b	27	d	37	d	47	c
8	c	18	d	28	c	38	a	48	a
9	c	19	c	29	d	39	b	49	b
10	d	20	c	30	a	40	b	50	a

XXXII.

1. *Tylenchorhynchus vulgaris* is associated with
   a. Pearl millet b. Sorghum c. Chickpea d. a + b + c
2. Hosts of *Tylenchorhynchus brassicae* are
   a. Cabbage b. Cauliflower c. Knol khol d. a + b + c
3. This nematode has been implicated as one of the causal agents for peach tree short life disease (PTSL) in USA
   a. *Mesocriconema xenoplax* b. *Hemicriconema peanus*
   c. *Criconema mutabile* d. *Tylenchus maydis*
4. Gradual die back of walnut is caused by
   a. *Criconema* b. *Hemicriconemoides* c. *Cacopaurus* d. *Bursaphelenchus*
5. *Hemicycliophora similis* on lemon root produces
   a. Excessive branching b. Galls c. Lesions d. Proliferation
6. The only ectoparasite which induces galls on root tips is
   a. *Trichodorus christei* b. *Xiphinema diversicaudatum*
   c. *Tylencorhynchus claytoni* d. *Longidorus elongatus*
7. This nematode is known as 'Cotton root-knot nematode
   a. *Meloidogyne incognita acrita* b. *M.incognita indicus*
   c. *M. arenaria acrita* d. *M.hapla indicus*
8. The symptom caused by *Xiphinema* in onion root is
   a. Fire fly blast b. Fish hook c. Dry wilting d. Stubby root
9. *Ditylenchus dipsaci* is also known by this name
   a. Bulbous nematode b. Teasel nematode c. Decay nematode d. Persian nematode
10. The first plant parasitic nematode recorded
    a. Root knot nematode b. Citrus nematode
    c. Wheat ear cockle nematode d. Cyst nematode
11. 'Ear cockle disease' is locally known as
    a. Gegla b. Sehun c. Mamni d. a + b + c
12. This is also known as 'Sugarcane stylet nematode'
    a. *Pratylenchus sacchari* b. *Tylenchorhynchus martini*
    c. *Hoplolaimus sacchari* d. *Hetrodera marioni*
13. This nematode always attacks above plant parts
    a. *Pratylenchus coffeae* b. *Rotylenchulus reniformis*
    c. *Meloidogyne graminicola* d. *Anguina tritici*
14. High moisture/ humidity favours the attack of which nematode
    a. *Anguina tritici* b. *Heterodera* c. *Meloidogyne* d. b and c
15. Ear cockle nematode may remain viable in galls in quiescent state for
    a. 10 years b. 20 years c. 26 years d. 40 years

16. The larvae of *Anguina tritici* in wheat galls can survive up to
    a. 28 years under dry conditions only
    b. 9-14 years under moist conditions only
    c. a and b
    d. Only for one year
17. Earcockle disease of barley can be eradicated by
    a. Use of certified seed
    b. Removal of galls by water floatation
    c. Crop rotation
    d. Winnowing
18. Yellow ear rot/ Tundu disease of Wheat is caused by
    a. *Anguina tritici*
    b. *Corynebacterium tritici*
    c. *a and b*
    d. *Heterodera avenae*
19. In general, Tundu disease is more severe at
    a. Low temperature and low relative humidity
    b. High temperature and low relative humidity
    c. Low temperature and high relative humidity
    d. High temperature and high relative humidity
20. Annual rye grass toxicity is caused by
    a. *Anguina millefolii*
    b. *Subanguina radicicola*
    c. *Anguina balsamophila*
    d. *Anguina agrostis*
21. The galls on the stem of Aeroptilon picris (orchids) is caused by
    a. *Anguina tritici*
    b. *A. picridis*
    c. *Heterodera picridis*
    d. *A. orchidi*
22. The leaf gall on Cynodon transvalensis is caused by
    a. *Anguina tumefaciens*
    b. *A. picridis*
    c. *A. agrostis*
    d. *A. tritici*
23. 'Maple root knot nematode' is
    a. *Meloidogyne graminicola*
    b. *Meloidogyne ovalis*
    c. *Meloidogyne indicus*
    d. *Meloidogyne mapli*
24. The symptom caused by *Aphelenchoides besseyi* in rice is called
    a. Incipient wilting
    b. Yellowing
    c. Stunting
    d. Firefly blast
25. Hot water treatment is the effective method of control for
    a. *Aphelenchoides besseyi*
    b. *Globodera rostochiensis*
    c. *Hirschmanniella oryzae*
    d. *Pratylenchus coffeae*
26. *Aphelenchoides composticola* infects
    a. Potato
    b. Strawberry
    c. Mushroom
    d. Fungi
27. This pathogen can reproduce on mushroom also
    a. *Aphelenchoides besseyi*
    b. *Aphelenchoides coffeae*
    c. *Aphelenchoides ritzemabosi*
    d. a + c
28. Number of *Aphelenchoides composticola* required to kill mushroom mycelium
    a. 1/100 g compost
    b. 10/100 g compost
    c. 100/100 g compost
    d. 1000/100 g compost
29. Nematode associated with Cauliflower disease of Strawberry
    a. *Ditylenchus destructor*
    b. *Aphelenchoides fragariae*
    c. *Aphelenchoides ritzemabosi*
    d. *Aphelenchoides composticola*

30. *Ditylecnhus destructor* is also called as
    a. Potato scab nematode
    b. Potato wilt nematode
    c. Potato tuber nematode
    d. Potato spot nematode
31. A short valve which connects the stoma with the oesophagus is known as
    a. Melastom b. Locostom c. Merastom d. Telostom
32. The term 'Stone cell" is related to this nematode
    a. Root-knot b. Spiral c. Sheath d. Cyst
33. The nematode disease not prevalent in India
    a. Red rot b. Red ring c. Molya d. Ear cockle
34. Red ring disease of coconut is mostly localized in
    a. Coastal peninsular India
    b. Sub- Saharan Africa
    c. Central and South America
    d. South – East Asia
35. The red ring nematode along with the diseased tissues is transmitted by
    a. *Nephotettix virescens*
    b. *Aphis gossypii*
    c. *Bemisia tabaci*
    d. *Rhynchophorus palmarum*
36. "Pine wilt" disease caused by *Bursaphelenchus xylophilus* appeared in epidemic form in
    a. Germany b. Japan c. China d. France
37. In Japan pine wilt disease is more severe on
    a. Black pine b. Red pine c. a and b d. White pine
38. 'A difference in form or structure as males and females of the same species' is known as
    a. Sexual dimorphism b. Sex determination c. Sex ratio d. Sexual polymorphism
39. 'Ufra' refers to
    a. Shriveled grain formation
    b. Ears fail to emerge from boot leaf
    c. Inflorescence fails to form grains
    d. Ears become twisted
40. In 'swollen ufra' the
    a. Panicles fails to emerge
    b. Panicles emerge but are distorted
    c. Panicles are sterile
    d. b + c
41. The symptoms noticed in 'ripe ufra'
    a. Panicles fails to emerge
    b. Panicles emerge but are distorted
    c. Panicles are sterile
    d. a + c
42. This nematode is considered economically important on bulbous ornamentals
    a. *Aphelenchoides fragariae*
    b. *Ditylenchus destructor*
    c. *Ditylenchus dipsaci*
    d. *Aphelenchoides ritzemabosi*
43. This enlarged portion of the female reproductive system functions as a reservoir in receving and holding spermatozoa from the male
    a. Oviduct b. Uterus c. Ovary d. Spermatheca
44. A stylet which is derived through evolutionary development from a stoma is termed as
    a. Stomatostyle b. Stomastyle c. Stomodeum d. Stonchiostyle

45. Brown ring of Daffodils and Hyacinth is due to
 a. *Radopholus similis* b. *Ditylenchus dipsaci*
 c. *Rhadinaphelenchus cocophilus* d. *Pratylenchus daffodili*

46. The most effective method for the control of Stem and Bulb nematode is
 a. Hot water treatment of bulbous ornamentals b. Flooding
 c. Resistant varieties d. b + c

47. One of the most important seed borne nematode diseases of crops is
 a. *Anguina tritici* b. *Meloidogyne incognita*
 c. *Trichodorus christei* d. *Pratylenchus coffeae*

48 This is not a fungivorous nematode
 a. *Aphelenchoides avenae* b. *Hoplolaimus indicus*
 c. *Aphelenchoides myceliophagus* d. *Aphelenchoides composticola*

49. The systematic grouping of organisms according to their natural relationships is known as
 a. Systematics b. Classification c. Taxonomy d. Phylogeny

50. *Nacobbus batatiformis* is also called as
 a. Hairy gall nematode b. Incomplete root-knot nematode
 c. True hairy nematode d. Hairy cyst nematode

ANSWERS KEY FOR XXXII

Sl. No.	Ans	Sl. No.	Ans	Sl. No.	Ans	Sl. No.	Ans	Sl. No.	Ans
1	d	11	d	21	b	31	d	41	d
2	d	12	b	22	a	32	a	42	c
3	a	13	d	23	b	33	b	43	d
4	a	14	a	24	d	34	c	44	a
5	c	15	c	25	a	35	d	45	b
6	b	16	c	26	c	36	b	46	a
7	a	17	b	27	b	37	c	47	a
8	b	18	c	28	d	38	a	48	b
9	b	19	a	29	b	39	d	49	c
10	c	20	d	30	c	40	a	50	a

XXXIII.

1. In the nematode cuticle, a circular groove or arc whose plane is perpendicular to the longitudinal axis is known as
   a. Longitudinal striation b. Transverse alae
   c. Longitudinal alae d. Transverse striation
2. Sensory organs of touch in nematodes are called as
   a. Tactoreceptors b. Tactocarriers c. Tactotaxis d. Tactovectors
3. The phenomenon where a violent reaction to parasitic attack resulting in sudden death of invaded tissues providing a barrier against further invasion, is known as
   a. Hydrotaxis b. Hypertonic c. Hypersensitive d. Hyposensitive
4. The situation in which head region of a nematode is delimited by a depression or a constriction is termed as
   a. Cephalic head b. Raised head c. Head onset d. Head offset
5. Nematodes, in general are also known as
   a. Round worms b. Eel worms c. Tape worms d. a and b
6. Smallest nematode reported is
   a. *Paralongidorus maximus* b.*Paratylenchus projectus*
   c. *Pratylenchus coffeae* d. *Longidorus elongatus*
7. The nematode cuticle is basically
   a. 1-layerd structure b. 2-layerd structure c. 3- layerd structure d. Multilayered
8. Platymyarian cells are usually of this shape
   a. Flat b. U shaped c. Round d. Rod shaped
9. This is the chief constituent of chitin
   a. Sucrose b. Fatty acids c. Gucosamine d. Glucosediamine
10. Foregut also is referred to as
    a. The intestinal region of the alimentary tract
    b. The oesophageal region of the alimentary tract
    c. Reproductive system
    d. The glandular portion of the alimentary canal
11. 'Filiform tail' in nematodes means
    a. Flat tial tip b. Pointed tail tip c. Rounded tail tip d. Tapering tail tip
12. Nematodes usually undergo first moult
    a. Before hatching b. After hatching c. During hatching d. Not specific
13. Eggs are laid in gelatinous matrix in
    a. *Meloidogyne incognita* b. *Globodera*
    c. *Rotylenchulus* d. a and c

14. This system is not well developed in nematodes
    a. Nervous b. Reproductive c. Excretory d. Muscular
15. Infective stage of *Ditylenchus* is
    a. Second stage juvenile b. Pre adult female
    c. Fourth stage juvenile d. First stage juvenile
16. This term indicates 'capable of growth outside the body'.
    a. Ectogenous b. Extrogenous c. Extragenous d. Epigenous
17. Stomatostylets are present in
    a. Tylenchids b. Rhabditids c. Trichodorids d. a + b
18. Most common type of reproduction in nematodes
    a. Bisexual b. Parthenogenesis c. Hermophroditism d. Intersex
19. A vulval flap is also known as
    a. Epiptygma b. Ectotygma c. Vulvtigma d. Vulvotoigma
20. Presence of coarse annules is the distict feature of this family
    a. Criconemtidae b. Hoploaimidae c. Pratylenchidae d. Tylenchidae
21. A nerve that extends from the rectal commissure to the rectum is termed as
    a. Ventrorectal nerve b. Dorsorectal nerve c. Laterorectal nerve d. Transorectal nerve
22. Sugar beet cyst nematode was reported in
    a. 1859 by Kuhn b. 1889 by H. Schacht c. 1859 by H. Schacht d. 1959 by H. Schacht
23. The conical perineal region of cyst forming nematodes is termed as
    a. Cyst fnestrae b. Cyst cone c. Cyst pore d. Cyst offset
24. Potato cyst nematode was first reported in India by Jones in
    a. 1967 b. 1951 c. 1961 d. 1971
25. A bundle of nerve fibres connecting ganglia are known as
    a. Commissure b. Neuron c. Nerve fibrilla d. Commissuron
26. A type of muscle cell in which the muscle fibres completely surround the sarcoplasm is referred to as
    a. Platymyarian b. Circomyarian c. Coelomyarian d. Polymyarian
27. In nematodes, exchange of gases takes place through
    a. Hypodermis b. Cuticle c. Microvilli d. Oesophagous
28. The shape of coelomyarian cells is
    a. Round b. U shaped c. Rods d. Pleomorphic
29. All plant parasitic nematodes possess this organ
    a. Ovary b. Three oesophageal glands
    c. Stylet d. b and c
30. In monodelphic prodelphic nematode, the ovary is
    a. Anteriorly directed b. Posteriorly directed
    c. One anterior and one posterior d. Curved

31. Nematodes with two ovaries are known as
    a. Didelphic b. Ophistodelphic c. Diovarian d. Diorchic
32. A mass of resistant desiccated nematode larva frequently found on bulb crops is
    a. Dauer larva b. Gall c. Cyst d. Desiccated nematode
33. Growing root knot nematodes get nourishment from
    a. Giant cells b. Parenchymatous cells c. Feeding cells d. Host cells
34. Non-infective stage of Meloidogyne is
    a. II stage juvenile and adult female b. Adult male
    c. IV stage juvenile d. b and c
35. The dead heteroderid nematodes containing viable eggs are called
    a. Dauer larva b. Cysts c. Egg sac d. Egg-mass
36. The phylum Nematoda is broadly divided into
    a. Two classes b. Four classes c. Three classes d. d.Five classes
37. 'Nematode wool' is characteristic in
    a. *Radopholus* b. *Hoplolaimus* c. *Meloidogyne* d. *Ditylenchus*
38. A change of position toward or away from a source of chemical stimulation is known as
    a. Chemotaxis b. Chemostatis c. Chemorecptyle d. Chemokinesis
39. Technically, 'Caudal pore' is known as the
    a. Spingernet b. Calvarium c. Cephalic setae d. Spinneret
40. 'Perennial pattern' is not used for species identification in this nematode
    a. Cyst b. Root knot c. Root lesion d. a and c
41. This nematode is the causal agent of 'toppling-over' disease in banana
    a. Burrowing nematode b. Cyst nematode c. Lesion nematode d. Spiral nematode
42. This nematode is a true sedentary endoparasite in its habitat
    a. *Anguina tritici* b. *Globodera rostochiensis*
    c. *Pratylenchus penetrans* d. *Xiphinema index*
43. Appearance of light necrotic lesions on root tips of rice is due to
    a. *Hirschmaniella oryzae* b. *Melodogyne graminicola*
    c. *Aphelenchoides besseyi* d. *Aphelenchoides oryzicola*
44. This is the most commonly used bacterial bioagent against nematodes
    a. *Pseudomonas fluorescens* b. *P. solanacearum*
    c. *Pasteuria penetrans* d. *Erwinia indica*
45. Most effective enemy plant of plant parasitic nematodes is
    a. Crossandra b. Mustard c. Asparagus d. Crotalaria
46. This is the infective stage in Sheath nematode
    a. II stage juvenile b. III stage juvenile c. Adult male d. IV stage juvenile
47. The most preferred host of reniform nematode is
    a. Onion b. Cotton c. Groundnut d. Tomato

48. This nematode induces specialized feeding cells in infected host tissues
    a. *Tylenchulus semipenetrans* b. *Meloidogyne hapla*
    c. *Rotylenchulus reniformis* d. a + b + c
49. The most commonly used and commercially available chemical against phytonematodes on several crops is
    a. Carbofuran 5G b. Carbofuran 3G c. Carbofuran 10G d. Carbofuran 2G
50. Of late, this nematode is gaining economic importance in most rice growing regions of the country
    a. *Meloidogyne javanica* b. *M.graminicola*
    c. *M.incognita* d. *Ditylenchus dipsaci*

ANSWERS KEY FOR XXXIII

Sl. No.	Ans	Sl. No.	Ans	Sl. No.	Ans	Sl. No.	Ans	Sl. No.	Ans
1	d	11	d	21	b	31	a	41	a
2	a	12	a	22	c	32	a	42	b
3	c	13	d	23	b	33	a	43	a
4	d	14	c	24	c	34	d	44	a
5	d	15	c	25	a	35	b	45	b
6	b	16	b	26	b	36	a	46	a
7	c	17	a	27	b	37	d	47	b
8	a	18	a	28	a	38	a	48	b
9	c	19	a	29	d	39	d	49	b
10	b	20	a	30	a	40	d	50	b

XXXIV

1. The average length of the survival of *Radopholus similis* in soil without host
   a. 12 months b. 6 moths c. 2 months d. 24 months
2. In general, avermectins affect nematodes by inhibiting
   a. Neuron b. Cuticle c. Stylet d. Oesophagous
3. Heavy soils can also support the survival of this nematode
   a. *Trichodorus christei* b. *Meloidogyne incognita*
   c. *M.hapla* d. *Xiphinema index*
4. This zoosporic fungus is quite effective against nematodes but not easy to culture
   a. *Pochonia* b. *Bacillus* c. *Catenaria* d. *Purpureomycellium*
5. This bacterium can manage phytonematodes efficiently
   a. *Pasteuria abbeti* b. *P. thornei* c. *P. electon* d. *P. dubaei*
6. This rhizobacterium alters root exudates that affect hatching of nematode juveniles
   a. *Bacillus mucori* b. *Erwinia radiobactor*
   c. *Pseudomonas solanacearum* d. *Agrobacterium radiobactor*
7. This soil bacterium has been reported to be antagonistic to phytonematodes
   a. *Clostridium butyricum* b. *C. bacterium*
   c. *Bacillus necotinae* d. *B. begoniae*
8. The strains SAN 415, CR 371 and Dipel refer to this antagonistic microbe
   a. *Purpureocillium lilacinum* b. *Atkinson*
   c. *Bacillus thuringiensis* d. *T. Goody*
9. The ducts of the subventral oesophageal glands in preparasitic juveniles contain granules of this diameter (in micro meter)
   a. 0.6-0.8 b. 1.0-2.0 c. 0.2-0.5 d. 2.2-3.5
10. The term 'oxyuroid' esophagus refers to
    a. Bullet shaped b. Bulboid c. Spherical d. Tubular
11. Quantitiative estimation of a biologically active substance by testing its effect under standardized conditions on living organisms or the effect of one organism upon other is known as
    a. Biosecirity b. Biosafety c. Biogenesis d. Bioassay
12. The number of rectal gland cells in many nematodes varies from
    a. 1-3 b. 4-6 c. 3-6 d. 5-8
13. The exudation of the gelatinous matrix in root-knot nematode is associated with a rhythmic contraction of the
    a. Lateral depressor ani muscle b. Subventral depressor ani muscle
    c. Ventral depressor ani muscle d. Dorsal depressor ani muscle
14. Cereal cyst nematode can exhibit maximum activity at the soil pH of
    a. 7.5-11.0 b. Less than 3.0 c. 5.0-7.5 d. 3.0-4.0

15. Basal knobs are located at
    a. Anterior to the stylet
    b. Posterior to the stylet
    c. Lateral to the stylet
    d. Ventral to the stylet
16. The rate growth Meloidogyne javanica is said to be also related to the
    a. degree of stress it is subjected
    b. temperature exposed
    c. atmospheric CO_2
    d. high moisture
17. *Ditylenchus dipsaci* infecting the heads of teasel was first discovered by
    a. W. carter (1943)
    b. Atkinson (1892)
    c. Neal (1894)
    d. Julius Kuhn (1857)
18. A wasting away or arrested development is referred to as
    a. Antrophy
    b. Neotrophy
    c. Atrophy
    d. Distrophy
19. In order to know whether a nematode is involved in a certain poor growth of the crop, this sampling type needs to be adapted
    a. Diagnostic sampling
    b. Quantitative sampling
    c. Stratified sampling
    d. Systematic sampling
20. In this type of extraction technique, nematodes may die due to lack of oxygen
    a. Cobb's sieving
    b. Fenwick can
    c. *Modified Baermann funnel*
    d. Incubation
21. The method followed commonly for assessing colonization of mycorrhizal in galled tissues of *Meloidogyne hapla* is
    a. Stem clearing & staining technique
    b. Root clearing & staining technique
    c. Foliage clearing & staining technique
    d. Root dissection technique
22. Most optimum temperature for *Aphelenchiodes* besseyi is
    a. 28-370C
    b. 12-220 C
    c. 18-210 C
    d. 23-300 C
23. Seed treatment with a slurry of biocontrol agents mixed with an organic carrier is referred to
    a. Seed organic priming
    b. Bio seed priming
    c. Seed biopriming
    d. Biopriming of seed
24. 'Cervival alae' are located in this region of nematodes
    a. Neck
    b. Intestine
    c. Rectum
    d. Spicule
25. 'Entonem', an International entomopathogenic nematode commercial product belongs to
    a. *Heterorhabditis bacteriophora*
    b. *H. megidis*
    c. *Steinernema carpocapsae*
    d. *S. feltiae*
26. *Aphelenchoides fragariae* on straw berry was first recorded 1891 by
    a. J.f dastur (1881)
    b. J. Ritzemaboss (1891)
    c. Mankau (1801)
    d. Duddington (1871)
27. *Heterodera goettingiana* on pea roots was reported in 1892 by
    a. C. Liebscher
    b. T. Goodey
    c. Steiner
    d. Poinar
28. Brown leaf disease on Asplenium nidus is incited by
    a. *Aphelenhoides ritzemabosi*
    b. *Aphelenhoides besseyi*
    c. *Aphelenhoides fragariae*
    d. *Aphelenhoides nidum*

29. Spermatozoa of phytonematodes are by
    a. Flagellate, spherical
    b. Non-flagellate, spherical
    c. Flagellate, conical
    d. Non-flagellate, square
30. The chromosomes in oogonial divisions of these nematodes usually are not discrete
    a. *Radopholus, Criconema*
    b. *Globodera, Xiphinema*
    c. *Pratylenchus, Paratylenchus*
    d. *Heterodera, Pratylenchus*
31. Avermectins affect the uptake of oxygen by nematode juveniles at 0.05 ppm by
    a. 18 %
    b. 85 %
    c. 61 %
    d. 40 %
32. The term 'bifarious' means
    a. Oriented in two direcitons
    b. Having two branches
    c. Hvaving two lobes
    d. Two branches meeting at one point
33. A chromosome which is not a sex chromosome is named as
    a. Setosome
    b. Autosome
    c. Sctosome
    d. Cytosome
34. Use of set of sieves for extraction o fnematodes was first developed by
    a. Mcbath
    b. Chitwood
    c. Hue
    d. N. A. Cobb
35. Pratylenchus *brachyurus* is easily parasitized by
    a. Pasteuria thornei
    b. *P. labia*
    c. *P. claytori*
    d. *P. phylogi*
36. The striations on the cuticle of juvenile of Meloidogyne javanica averages about
    a. 10 nm
    b. 5 nm
    c. 20 nm
    d. 45 nm
37. Under stress, a major ecological adaptation in most nematodes is absence of production of
    a. males
    b. females
    c. Stewart
    d. Paramanov
38. Multiplication of oogonia in nematodes occurs in this zone of the ovary
    a. Central, epithelial
    b. Apical, epithelial
    c. Apical, germinal
    d. Central germinal
39. "Wout's medium" is commonly used to culture this form of nematodes
    a. Free-living
    b. Entomopathogenic
    c. Predacious
    d. Mycophagous
40. This genus closely resembles reniform nematode
    a. *Criconemoides*
    b. *Pratylenchus*
    c. *Anguina*
    d. *Tylenchulus*
41. In centrifugation technique to study the Pasteuria penetrans spore attachment onto the nematode, the centrifuge should be run at
    a. 5000 rpm for 15 min
    b. 1000 rpm for 15 min
    c. 5000 rpm for 5 min.
    d. 1000 rpm for 30 min.
42. King's B medium is most commonly used to culture
    a. *Pseudomonas fluorescens*
    b. *Bacillus macerans*
    c. *Trichoderma harzianum*
    d. *Pochonia chlamydosporia*
43. For anaesthetization of nematodes, this can be used as a good relaxant
    a. Potassium iodide
    b. FAA
    c. Dichloroethyl ether
    d. Lactophenol
44. For calibration of ocular micrometer in Camera Lucida studies, this formula is used
    a. 1 mm=(Axi) /B
    b. 1 mm=(Azi) /N
    c. 1 mm=(AB) /Z
    d. 1 mm=(Axii) /B

45. Earlier name of *Pasteuria penetrans* is
    a. *Clostridium penetrans*
    b. *Pasteuria claytori*
    c. *P. phylogei*
    d. *Dubosquia penetrans*
46. *Glomus fasciculatum* is an
    a. Epimycorrhiza b. Mesomycorrhiza c. Ectomycorrhiza d. Endomycorrhiza
47. The normal dosage of neem cake recommended per sq. m. to manage phytonematodes in crops
    a. 100 g b. 50 g c. 250 g d. 500 g
48. *Gliocladium virens* can also parasitize this stage of nematode
    a. J2 b. Egg c. Adult female d. Adult male
49. This heterotrophic bacterium has shown antagonistic effect against root-knot nematodes
    a. *Bacillus magari*
    b. *B. nlsln*
    c. *Pseudomonas truncatum*
    d. *Azatobacter chroococcum*
50. The chemical name 22,23-dihydroavermectin B1 refers to this compound
    a. Avymectin b. Slepmectin c. Ivermectin d. Bevermectin

ANSWERS KEY FOR XXXIV

Sl. No.	Ans	Sl. No.	Ans	Sl. No.	Ans	Sl. No.	Ans	Sl. No.	Ans
1	b	11	d	21	b	31	c	41	c
2	a	12	c	22	d	32	a	42	a
3	d	13	d	23	c	33	b	43	c
4	c	14	c	24	a	34	d	44	a
5	b	15	b	25	d	35	a	45	d
6	d	16	a	26	b	36	c	46	d
7	a	17	d	27	a	37	a	47	a
8	c	18	c	28	c	38	c	48	b
9	a	19	a	29	b	39	b	49	d
10	b	20	c	30	d	40	d	50	c

XXXV

1. Nematodes basically, are
   a. Bilaterally symmetrical
   b. Laterally symmetrical
   c. Radially symmetrical
   d. Triilaterally symmetrical
2. The term 'Pseudocoelom' denotes
   a. Body cavity lined by mesodermal epithelium on both sides
   b. Body cavity formed as a result of gastrulation
   c. Body cavity filled with fluid
   d. Body cavity not lined by mesodermal epithelium on both sides
3. This structure helps in maintaining the shape of nematodes
   a. Pseudocoelom
   b. Muscle
   c. Entire cuticle
   d. Annulations of the cuticle
4. Nematodes are strictly
   a. Diploblastic b. Triploblastic c. Tetrablastic d. Polyblastic
5. Nematode body shape, in general is typically
   a. Vermiform b. Pleomorphic c. Round d. Rod shaped
6. A cell of the epithelium that forms the gonoduct wall and is located at the distal end of the ovary and to be differentiated from the terminal cell, is known as
   a. Epicell b. Cephalic cell c. Apical cell d. Distal cell
7. Adult females of citrus nematode are always
   a. Spherical shaped b. Melon shaped c. Bottle shaped d. Kidney shaped
8. Cyst nematode females are strictly
   a. Kidney shaped b. Melon shaped c. Tubular shaped d. Spherical shaped
9. The long axis from head to tail is termed as
   a. Tansoanterior axis b. Lateroanterior axis c. Transoposterior axis d. Anteroposterior axis
10. 'Asteriform' indicates this shape of a structure
    a. Pea shaped b. Star shaped c. Wheel shaped d. Pin shaped
11. The general length of a nematode may range
    a. 0.01 mm to 10.0 mm b. 0.01mm to 11.0 mm c. 0.2 mm to 10.0 mm d. 0.3 mm to 11.0 mm
12. The body width of nematode may vary from
    a. 0.01 mm to 0.5 mm
    b. 0.1 mm to 0.7 mm
    c. 0.2 mm to 0.9 mm
    d. 0.3 mm to 1.0 mm
13. This term indicates 'similar in function but differing in origin and structure' of an organ
    a. Homologous b. Analogous c. Paralogous d. Heterologous
14. Nematode with a long stylet
    a. *Belanolaimus longicaudatus*
    b. *Longidorus elingatus*
    c. *Radopholus similis*
    d. *Xiphinema americanum*

15. 'Aestivation' alos refers to
    a. Locomotion of nematodes in clayey soils
    b. Locomotion of nematodes in dry soils
    c. Dormancy during unfavourable conditions
    d. Excretion of waste products
16. 'Flask' shaped structures are also known as
    a. Flagellate b. Ampulliform c. Pyriform d. Periform
17. All the natural openings of nematodes are present on
    a. Ventral side b. Dorsal side
    c. Laterel side d. Not confined to any particular side
18. Exchange of gases in nematodes takes place through
    a. Hypodermis b. Cuticle c. Microvilli d. Stylet
19. The feeding stylet and copulatory spicules are formed from
    a. Hypodermis b. Outer body tube c. Cuticle d. Tail
20. Under high pressure the nematode body maintains its shape with the aid of the
    a. Cuticle b. Fluid filled interior c. Muscles d. Hypodermis
21. The structures which helps in strengthening of the cuticle
    a. Punctuations b. Transverse striations
    c. Longitudinal markings d. Caudal alae
22. Punctations are visible in
    a. *Rugose wall pattern of Heterodera* b. *Perineal patern of Meloidogyne hapla*
    c. *Scales of Criconematids* d. a and b
23. Transverse striations are pronounced in
    a. Tylenchidae b. Heteroderidae
    c. Criconematicidae d. a + b + c
24. The perineal in Meloidogyne and rogues wall pattern of Heterodera are
    a. Modification of punctuations b. Modification of longitudinal markings
    c. Modification of transverse markings d. Modification of cross markings
25. The following structures, in general aids taxonomists in species identification
    a. Caudal alae b. Ridges c. Cervical alae d. Longitudinal alae
26. The presence of alae in the anterior part of the body around the neck is known as
    a. Cervical alae b. Caudal alae
    c. Longitudinal alae d. Transverse alae
27. Caudal alae which are restricted to males as coupulatory bursa are present in the
    a. Anterior region b. Ventral region c. Lateral region d. Posterior region
28. The nematode cuticle is basically
    a. Opaque b. Thin and simple c. Transparent d. Translusent

29. Cortical layer of nematode comprises of
    a. Keratin b. Dermin c. Tannin d. Paraffin
30. Chemically median layer consists of
    a. Lipids b. Collagen c. Esterase d. Glycogen
31. The nematode cuticle is secreted by
    a. Matrix b. Fibrilar layer c. Cortex d. Hypodermis
32. Lorge amount of glycogen is stored in
    a. Cuticle b. Cloaca c. Hypodermis d. Buccal cavity
33. Nematode muscle cells are situated in the
    a. Intercaudal zone of hypodermis b. Body cavity
    c. Adjacent to the cuticle d. Matrix layer
34. The main function of hypodermis is
    a. Serves as exoskeleton for attachment of muscles
    b. Secretion of cuticle
    c. To accommodate longitudinal nerves and excretory canals
    d. Storage of reserve food material
35. Hypodermal glands act as
    a. Osmotic regulators b. Ionic regulators c. Both a and b d. Either a or b
36. The caudal glands of the hypodermis which secrete adhesive substances for anchoring of nematodes are found in
    a. Lateral region b. Ventral region c. Tail region d. Dorsal region
37. Sensory organs like amphids, phasmids and deirids are associated with the
    a. Hypodermal glands b. Caudal glands c. Oesophgeal glands d. Rectal glands
38. In nematodes the longitudinal running hypodermal chords are
    a. One b. Three c. Four d. Five
39. This is not a unique feature of nematode muscles
    a. Spindle shape
    b. Oblique striation
    c. Innervations processes joining nerve chords
    d. Presence of z-bundles
40. Protein predominantly present in thick mycofilaments of nematodes is
    a. Tropomyosin b. Myosin c. Actin d. Paramyosin
41. Stimulation of random movement induced by a chemical agency is termed as
    a. Chemotrophy b. Chemokinet c. Chemokinensis d. Chemosinensis
42. In the outer body tube of the nematode the muscle layer is responsible for
    a. Locomotion b. Secretion c. Respiration d. Digestion
43. The outer shell or covering of a nematode egg is termed as
    a. Dorion b. Dendron c. Goggion d. Chorion

44. 'Collaret' refers to the
   a. A band like cuticular extension of the cephalic region opening anteriorly
   b. A cup like cuticular extension of the cephalic region opening anteriorly
   c. A band like cuticular extension of the caudalic region opening posteriorly
   d. A ball like cuticular extension of the cephalic region opening posteriorly
45. Trangular shape is also expressed as
   a. Askoid b. Dentoid c. Deltoid d. Pleomoid
46. The position on the nematode body situated 30^0 laterally from the dorsomedian line and perpendicular to the anteroposterior axis is referred to as
   a. Dorsoventral b. Venreolateral c. Laterodorsal d. Dorsolateral
47. Technically, vas deferens is known as
   a. Spermus dactic b. Ductus spiculae c. Spermus defens d. Ductus Spermatic
48. Ectopore is a synonym of the
   a. Excretory pore b. Anal pore c. Cuticular pore d. Intestinal pore
49. Platymyarian cells are usually
   a. Circomyarian b. Coelomyarian c. Meromyarian d. Polymyarian
50. The portion of the muscle cell differentiated into muscle fibres is known as
   a. Plasmofibre b. Fibroplasm c. Fibroventron d. Fibroentron

ANSWERS KEY FOR XXXV

Sl. No.	Ans	Sl. No.	Ans	Sl. No.	Ans	Sl. No.	Ans	Sl. No.	Ans
1	a	11	c	21	a	31	d	41	c
2	d	12	a	22	d	32	c	42	a
3	c	13	b	23	d	33	a	43	d
4	b	14	d	24	c	34	b	44	a
5	a	15	c	25	b	35	c	45	c
6	c	16	b	26	a	36	c	46	d
7	c	17	a	27	d	37	a	47	d
8	d	18	b	28	c	38	c	48	a
9	d	19	c	29	a	39	d	49	c
10	b	20	a	30	b	40	b	50	b

XXXVI

1. Number of stages in the life cycle of a phytonematode
   a. 2 stages b. 4 stages c. 6 stages d. 8 stages
2. The number of juvenile stages found in a nematode is
   a. One b. Two c. Three d. Four
3. Nematodes moult
   a. Once b. Twice c. Thrice d. Four times
4. Nematodes usually undergo first moult
   a. Before hatching b. After hatching c. During hatching d. Not specific
5. The pre-hatch moult may also be absent in this group of nematodes
   a. Parasitic b. Saprophytic c. Marine d. Free living
6. Juveniles of *Xiphinema index* are reported to emerge from the egg before
   a. Second moult b. First moult c. Thitd moult d. Fourth moult
7. In blastula stage the cells are arranged to form a fluid sphere which is bounded by
   a. Single layers of cells b. Two layers of cells
   c. Three layers of cells d. Four layers of cells
8. In gastrula stage the early embryo consists of an open mouthed sac like body with
   a. No layer of cells b. Single layer of cells c. Two layer of cells d. Four layer of cells
9. The layer present in typical nematode egg shell
   a. Lipid layer b. Chitinuous layer c. Vitelline layer d. a + b + c
10. The layer which offers resistance to nematode eggs under adverse environmental conditions
    a. Chitinuous layer b. Lipid layer c. Vitelline layer d. Inner layer
11. Chitin is present only in
    a. Egg shell b. Cuticle c. Spicule d. Stylet
12. In phytonematodes embryonic development has been well studied in
    a. *Radopholus similis* b. *Ditylenchus dipsaci*
    c. *Meloidogyne incognita* d. a and b
13. The process of pushing ventral surface of nematode embryo into the interior is known as
    a. Epiboly b. Endoly c. Ectoboly d. Endopoly
14. Artificial hatching agent to hatch potato cyst nematode is
    a. Anhydrotetronic acid b. Picrolinic acid
    c. Sodium metavandate d. a + b + c
15. Following acts as the hatching factor and is ofund in root exudates of soybean
    a. $H_{18}C_{20}O_6$ b. $H_{14}C_{24}O_8$ c. $H_{18}C_{24}O_6$ d. $H_{10}C_{20}O_{12}$
16. There is a definite relationship between the onset of moulting and this nematode activity
    a. Feeding habit b. Genus c. Size d. Parasitism

17. The most characteristic feature of post embrytonic development in phytonematodes is
    a. Ecdysis b. Moulting c. a and b d. Hatching
18. Pre-hatch moulting has not been noticed in
    a. *Tylenchulus semipenetrans* b. *Meloidogyne incognita*
    c. *Rotylenchulus reniformis* d. *Aphelencoides besseyi*
19. The juvenile ensheathed by the cuticle of previous stage is called
    a. Cyst b. Neonate larva c. Old larva d. Dauer larva
20. A mass of resistant dessicated juvenile frequently found on bulb crops is
    a. Dauer larva b. Firefly blast c. Cyst d. Desiccated nematode
21. The immature stage of nematodes are actually termed as
    a. Larva b. Moult c. Juvenile d. Neont
22. The infective stage in sedentaryendoparasitic nematodes is
    a. J1 stage b. J2 stage c. J3 stage d. Adult male
23. This stage of embryo formation is not found in all species of nematodes
    a. Blastula b. Psedoblatula c. Gastrula d. Coelograstula
24. Eggs are laid within the female body in
    a. *Meloidogyne incognita* b. *Globodera*
    c. *Heterodera* d. b + c
25. The rate of egg laying in *Radopholus similis* and *Aphelenchoides ritzemabosi* is
    a. 4 & 2 b. 2 & 4 c. 4 & 8 d. 6 & 4
26. The fecundity rate of ectoparasites and migratory endoparasites ranges between
    a. 5-10 b. 10-25 c. 25-35 d. 30-45
27. The fecundity rate of sedentary endoparasites and semi endoparasites ranges between
    a. 50-100 b. 100-250 c. 50-500 d. 300-450
28. In general, the feeding by most phytonematodes stimulates the development of a
    a. Host cells b. Syncytium c. Nurse cells d. Feeding tubes
29. Development of a specific giant cell by *Meloidogyne arenaria* exclusively involves this mechanism
    a. Repeated mitosis without cytokinesis b. Repeated mitosis with cytokinesis
    c. Repeated meiosis without cytokinesis d. Repeated meiosis with cytokinesis
30. Developing cyst nematodes get nourishment from
    a. Giant cells b. Nurse cells c. Syncytia d. Host cells
31. This genus has the special types of nurse cell system which serves as source of continuous supply of nutrients to the nematodes
    a. *Rotylenchus* b. *Pratylenchus* c. *Anguina* d. *Rotylenchulus*
32. The second stage juveniles infective in
    a. Cyst nematode b. Root knot nematode c. a & b d. Stem & bulb nematode
33. Infective stage in all species of *Meloidogyne* is
    a. Second stage juvenile b. First stage juvenile
    c. Fourth stage juvenile d. Varies with the speceis

34. The term 'acaudal' refers to
    a. Absence of testes
    b. Absence of a tail
    c. Absence of an egg sac
    d. Absence of ovary
35. The condition wherein 'lips are fused to give a smooth or nearly smooth contour' is known as
    a. Malgamated lip
    b. Overlapping lip
    c. Superficial lip
    d. Amalgamated lip
36. After removal from the host, cysts normally turn into this colour
    a. Dark brown
    b. Black
    c. Light brown
    d. Orange brown
37. Acid phosphatase has been frequently reported from the following nematode
    a. Root knot
    b. Cyst
    c. Dagger
    d. Lesion
38. The hydrolytic enzyme produced by potato cyst nematode
    a. Cellulose
    b. Invertase
    c. Alkaline phosphatase
    d. Alpa-galactosidase
39. In this nematode there may be a decrease in size as they pass develop from the juvenile stage to adults
    a. *Tylenchulus semipenetrans*
    b. *Globodera pallida*
    c. *Meloidogyne javanica*
    d. *Pratylenchus penetrans*
40. The cells formed due to the feeding by Tylenchulus semipenetrans are typically known as
    a. Transfer cells
    b. Gaint cells
    c. Nurse cells
    d. Syncytia
41. "Nurse cells" are the modification of these tissues
    a. Cortex mesophyll
    b. Cortex-epithelial
    c. Cortex-parenchyma
    d. Vascular
42. These cells are absent in nematode
    a. Hypodermal
    b. Flame cells
    c. Mesodermal cells
    d. Endodermal cells
43. A large egg sac at the posterior end of female is present in the cyst nematode
    a. *Heterodera avenae*
    b. *H. sorghi*
    c. *H. cajani*
    d. *H. zeae*
44. Phytonematodes generally contain more
    a. Proteins
    b. Enzymes
    c. Polyphenols
    d. Lipids
45. The major energy reserve in phytonematodes is
    a. Lipids
    b. Carbohydrates
    c. Proteins
    d. Triacylglycerols
46. The mechanism of host finding in nematodes was proposed by
    a. Zukerman
    b. Jenson
    c. Linford
    d. a and b
47. The most accepted theory of chemotaxis based upon interactions between molecules emanating from host and nematode was proposed by
    a. Kuhn, 1959
    b. Klink et al., 1970
    c. Zukerman & Jenson, 1984
    d. Van gundy & Rackman, 1961
48. In general, soil-inhabiting nematodes require ample supply of
    a. CO_2
    b. O_2
    c. H_2O
    d. N
49. The cavity or chamber of the amphid which contains the sensilla is termed as
    a. Amphidial sac
    b. Amphidial pouch
    c. Amphidial lobe
    d. Amphidial gland

50. 'Cells situated along the esophagus, two opposite each esophageal lumen radius and one opposite each sector for a total of nine' are known as

a. Arcuate cells b. Arcade cells c. Radial cells d. Lumen cells

ANSWERS KEY FOR XXXVI

Sl. No.	Ans	Sl. No.	Ans	Sl. No.	Ans	Sl. No.	Ans	Sl. No.	Ans
1	c	11	a	21	c	31	d	41	a
2	d	12	c	22	b	32	c	42	b
3	d	13	a	23	d	33	a	43	c
4	a	14	d	24	d	34	b	44	d
5	d	15	b	25	a	35	d	45	a
6	b	16	c	26	b	36	c	46	d
7	a	17	b	27	c	37	b	47	c
8	c	18	c	28	d	38	d	48	b
9	d	19	b	29	b	39	a	49	b
10	b	20	c	30	c	40	c	50	b

XXXVII

1. Sometimes delay in hatching of cysts is mainly due to accumulation of

a. CO_2 b. O_2 c. H_2O d. H_2S

2. This nematode can sense and get more attracted to H_2S

a. *Radopholus similis* b. *Ditylenchus dipsaci*
c. *Meloidogyne incognita* d. None of these

3. During aerobic starvation usually nematodes utilize these stored nutrients

a. Carbohydrates b. Proteins c. Lipids d. a + b + c

4. The optimum temperature required for growth and development of phytonematodes is in the range of

a. 5-30° C b. 10-25° C c. 15-35° C d. 20-35°C

5. The upper temperature limit for development of *Meloidogyne* and *Globoderar*

a. 5-15°C b. 15-30° C c. 20-30° C d. 25-35° C

6. The optimum temperature required for egg production in root knot nematodes varies between

a. 15-20° C b. 25-32° C c. 35-37° C d. Above 37° C

7. *Globodera rostochiensis* develops and produces eggs most rtapidly at

a. 10° C b. 14° C c. 18° C d. 21° C

8. 'A coelomocyte capable of absorbing foreign material from body fluids and storing it in crystalline form' is known as

a. Coelocyte b. Arthrocyte c. Acoelocyte d. Crystocyte

9. Nematodes inhabiting the above ground parts of plants as well as soil inhabiting nematodes are resistant to

a. Desiccation b. Moisture c. Temperature d. CO_2

10. Suspended animation of nematode due to desiccation is called

a. Anhydrobiosis b. Anoxybiosis c. Cryptobiosis d. Anabiosios

11. Nematodes normally shrink in this solution

a. Hypotonic b. Hypertonic c. Alkaline d. Salt

12. Nematodes quickly become turgid and active in this solution

a. Hypotonic b. Isotonic c. Alkaline d. a and b

13. The juvenile of *Anguina tritici* in wheat galls can survive up to

a. 28 years under dry conditions only b. 9-14 years under moist conditions only
c. a and b d. Only few months

14. *Ditylenchus dipsaci* have been revived from infected plant material after

a. 23 years of storage b. 25 years of storage c. 28 years of storage d. 32 years of storage

15. Hatching and development of nematodes is inhibited under

a. High osmotic pressure b. Low osmotic pressure
c. High moisture d. Low moisture

16. Following opined that at pH 6 there is maximum hatching of Globodera rostochiensis
   a. Duggan b. Seshadri c. Williams d. Ellenby
17. 'Ammonium ions are responsible for hatching of *Meloidogyne incognita* eggs' was reported by
   a. Oteifa b. Ross c. Castaner d. Kimpinski
18. The stage at which *Anguina tritici* survives dormancy is
   a. First stage larva b. Second stage larva c. Pre adult d. Adults
19. In *Ditylenchus* and *Aphelenchoides* the resistant stages are
   a. Second stage larva b. Pre adult c. Adults d. b and c
20. Dormant stages of nematodes are particularly resistant to the action of
   a. Fungicides b. Nematicides c. Herbicides d. Soil fauna
21. The process of destruction of a cell by internal agents, is known as
   a. Endolysis b. Photolysis c. Circolysis d. Autolysis
22. The coelomyarian cells are virtually a storage depot for
   a. Glycogen b. Lipids c. Proteins d. Starch
23. Muscle cells are connected to each other by means of
   a. Alpa- 1, 4 glycosidic linkage b. Peptide linkage
   c. Cytoplasmic bridges d. a + b
24. Muscles serving as ovijectors are found in this nematode organ
   a. Vulva b. Spicules c. Testis d. Uterus
25. This is the chemical transmitter, which contracts muscles in nematodes
   a. Amine b. Chlorogenic acid
   c. Pentose monophosphate d. Lactose dinine
26. The stomodeum includes
   a. Buccal cavity b. Pharynx c. Mouth and lips d. a + b + c
27. In Nematodlogy the term 'bank' means
   a. Group of peripheral transverse muscle cells
   b. Group of oesophageal transverse muscle cells
   c. Group of peripheral longitudinal muscle cells.
   d. Group of peripheral lateral and dorsal muscle cells
28. In Secernentea the stylet is derived from the fusion of stomatal lining, which is known as
   a. Stomatostylet b. Odontostylet c. Onchiostylet d. Stromatostylet
29. In Adenophorea the stylet is thought to be derived from a tooth which is termed as
   a. Onchiostylet b. Odontostylet c. Stomatostylet d. Strongystylet
30. Nematodes select a spot on the host tissue before penetration, which is known as
   a. Penetration site b. Accession site c. Feeding spot d. Feeding site
31. In dorylaimids stylets develop from
   a. Oesophagus b. Buccal cavity c. Mouth d. Median bulb

32. In all phytonematodes, adult males are
    a. Parasitic and saprophytic
    b. Predaceous
    c. Saprophytic
    d. Parasitic
33. Stylet is very long and strong in
    a. *Pratylenchus* b. *Criconema* c. *Xiphinema* d. *Meloidogyne*
34. Anchor sahped stylet knob is present in
    a. *Longidorus* b. *Xiphinema* c. *Hoplolaimus* d. *Criconema*
35. In general, all phytonematodes possess
    a. Two-part oesophagus
    b. Three-part oesophagus
    c. Four-part oesophagus
    d. Non-specific
36. Oesophagus is the largest part of stomodeum and found between
    a. Stoma and intestine
    b. Stoma and rectum
    c. Intestine and rectum
    d. Cardia and median bulb
37. Lumen in nematodes is divided into
    a. Four parts b. Two parts c. Three parts d. Multiple parts
38. The oesopageal lumen in nematodes is
    a. Asymmetrically b. Triradiate symmetry c. Bilaterally symmetry d. Radial symmetry
39. In Dorylaimodea the oesophagus is divided into
    a. Anterior procorpus b. Posterior corpus c. Isthmus d. a and b
40. In Tylenchida the oesophagus is divided into
    a. Procorpus b. Posterior bulb c. Isthmus d. a + b + c
41. Royal 300 strain- Antipolis is a commercial product of this bioagent
    a. *Arthrobotrytis superb*
    b. *A. irregularis*
    c. *A. robusta*
    d. *A. oligospora*
42. The commercial nematode formulation used for insect control
    a. Royal 300 b. DD 135 c. Royal 350 d. Biocon
43. Nematode affecting insects is
    a. *Stenernema carpocapsae*
    b. *Hoplolaimus seinhorsti*
    c. *Xiphinema carpocapsae*
    d. *Stenernema indicae*
44. Nematode affecting mosquito
    a. *Trichodorus teres*
    b. *Stenernema glaseri*
    c. *Romamomermus cultivorax*
    d. *Heterorhabditis indicus*
45. Nematode used for controlling mosquito larva
    a. Dorylaimids b. Mermithids c. Tylenchids d. Aphelenchids
46. The principal sterol identified from nematodes is
    a. Campesterol b. Lathosterol c. Phytosterol d. Cholestanol
47. Usually, nematode resistant cultivars contain more quantity of
    a. Free polyphenol b. Phenolic oxidase c. Polyphenol oxidase d. Conjugated polyphenol

48. Presence of these in plants analogous to antibodies in animals, responds to nematode infection
a. Agglutin b. Collagenin c. Collagen d. Agglutinins

49. Resistance to cyst nematodes is based on
a. Inhibition of hatching b. Inactivation
c. Prevention of entry d. Inhibition of female development

50. Best testing sites for resistance screening purpose
a. Natural endemics b. Natural epidemics
c. Moderate to high incidence d. Uniform disease incidence

ANSWERS KEY FOR XXXVII

Sl. No.	Ans	Sl. No.	Ans	Sl. No.	Ans	Sl. No.	Ans	Sl. No.	Ans
1	c	11	b	21	d	31	a	41	a
2	d	12	d	22	a	32	c	42	b
3	a	13	c	23	c	33	c	43	a
4	b	14	a	24	b	34	d	44	c
5	c	15	c	25	b	35	b	45	b
6	a	16	d	26	d	36	a	46	d
7	c	17	a	27	c	37	c	47	b
8	b	18	b	28	a	38	b	48	d
9	a	19	d	29	b	39	d	49	b
10	b	20	b	30	d	40	d	50	c

XXXVIII

1. The most predominant nematode genus worldwide that acts as a virus vector is
   a. *Longidorus elonagtus*
   b. *Xiphinema index*
   c. *Trichodorus christei*
   d. *Paralongidorus xyli*
2. Phosphate glucose has been most commonly employed in identifying this nematode
   a. Banana race of *Radopholus*
   b. Citrus race of *Radopholus*
   c. *Xiphinema index*
   d. *Tylenchorhynchus claytoni*
3. Phylogenic tree obtained by alignment of the sequences of internal transcribed spacers of ribosomal DNA gene complex (ITS of rDNA) for the four races of *Meloidogyne incognita* indicate that
   a. Race 3 and race 4 are most related
   b. Race 3 and race 2 are most related
   c. Race 2 and race 4 are most related
   c. Race 1 and race 4 are most related
4. *Hero* gene confers resisitnace to this nematode
   a. *Globodera* isolated from tomato
   b. *Heterodera* from Pigeon pea
   c. *Anguina* from barley
   d. *Radopholus* from banana
5. This is the first nematode resistance gene to be cloned from sugar beet, which confers resistance against the sugar beet cyst nematode
   a. Hs1 pro-2
   b. Hs2 pro-2
   c. Hs1pro-1
   d. Hs1 pro-4
6. This is a protein or other compound from the nematode that is recognized by the host R-gene
   a. Elicitor
   b. Effector
   c. Encoded protein
   d. Uncoded protein
7. Mi-1, Gpa2 and Hero A are members of the NBS-LRR class of plant R-genes that do not contain
   a. N-terminal toll-interleukin receptor-like (TIR) domain
   b. N-diagonal TIR domain
   c. M- terminal TIR domain
   d. N-lateral TIR domain
8. In nematode-resistant tomatoes, this is not allowed to develope
   a. Feeding site
   b. Nematode growth
   c. Nematode reproduction
   d. Neither the feeding site nor the nematode
9. Plants are defined as resistant to nematodes when they show reduced levels of
   a. Penetration
   b. Reproduction
   c. Movement
   d. Symptoms
10. This is one of the best characterized nematode resistance genes, which confers resistance to several root-knot nematode species in tomato.
    a. Mi
    b. Mii
    c. Mi1
    d. MI
11. *Meloidogyne hapla* reproduces by
    a. Facultative mitootic parthenogenesis
    b. Meiotic parthenogenesis
    c. Facultative meiotic parthenogenesis
    d. Mitootic parthenogenesis
12. The two models, plant and animal that have potential for providing insights into phytonematode interactions are

a. *Arabidopsis thaliana and earthworm* b. *A. thaliana and Caenorhabditis elegans*
c. *Wild tomato and Caenorhabditis elegans* d. *A. thaliana and wucheria bancroftii*

13. During parasitism by cyst nematode, due to the activation of the following, syncytium is formed.
a. Cytokinin signaling b. Auxin signaling c. Gibberlin signaling d. Ethylene signaling

14. Hypersensitive response due to nematode infestation normally leads to
a. Increase in peroxidase activity and decreases lignin deposition
b. Decrease in peroxidase activity and increase in lignin deposition
c. Increase in peroxidase activity and lignin deposition
d. Derease in peroxidase activity and lignin deposition

15. *Mae* and *Mag* genes are related to
a. *M. javanica* b. *M.arenaria* c. *M.hapla* d. *M.graminicola*

16. Much of the nematode body composition is
a. Proteins & carbohydrates b. Proteins & fats
c. Carbohydrates & fats d. Collagen & chitin

17. *Meloidogyne arenaria, M. incognita, M. javanica* and *M. hapla* represent up to this much of root-knot nematode in cultivated soils
a. 95% b. 70% c. 60% d. 50%

18. In trees, the sampling depth for *Xiphinema* or *Longidorus* may be up to
a. 10 cm b. 40 cm c. 20 cm d. 30 cm

19. This method is based on differences in weight and rate of sedimentation of soil particles.
a. Baermann's Funnel b. Cobb's sieving c. Elutriation d. Waring blender

20. For capturing small juveniles of *Meloidogyne*, this sieve mesh is more suitable
a. 38 -45 µm b. 60-85 µm c. 00-150 µm d. More than 150 µm

21. This nematode prefers moving up instead of crawling down during extraction
a. *Anguina tritici* b. *Cricoenma spp.* c. *Dolichodorus spp.* d. *Turbatrix aceti*

22. Schuiling centrifuge extraction apparatus is adopted for the extraction of
a. Eggs b. Egg-mass c. Cysts d. Galls

23. The members of this genus are heavy and normally settle quickly during extraction
a. *Longidorus* b. *Meloidogyne* c. *Pratylenchus* d. *Paratylenchus*

24. This technique is used for the extraction of nematodes from soil, river silt, manure and litter
a. Waring bender b. Oostenbrink elutriator
c. Sugar floatation d. Cobb's sieving

25. This nematode is extremely sensitive to mechanical damage or friction
a. *Meloidogyne hapla* b. *Heterodera avenae* c. *Trichodorus teres* d. *Pratylenchus coffeae*

26. The specific gravity of living nematodes is
a. 1.084 b. 1.000 c. 2.084 d. 2.000

27. The cysts of the following nematode are relatively small in size
a. *Globodera rostochiensis* b. *Globodera pallida*
c. *Heterodera carotae* d. *H. glycines*

28. Killing of nematodes is done by heating at this temperrture

a. 25-30°C b. 65-90°C c. 35-45°C d. 50-55°C

29. Detailed studies on the development of reproductive system in juvenile stages can be made by using 1 % of this stain

a. Aceto carmine b. Butyric carmine c. Aceto butanol d. Aceto blue

30. This nematdode was subjected to Federal Quarantine in United States during 1954

a. *Meloidogyne javanica* b. *Anguina tritici*

c. *Heteodera glycines* d. *H.schachtii*

31. Pangola grass has high amounts of this nematoxic compound

a. Catechol b. Pyrocatechol c. Acetic amine d. hydrogen amine

32. This acts as a general chemotactic attractant for most phytonematodes

a. CO b. CO_2 c. NO d. O_2

33. Most common type of reproduction in root-knot nematodes is

a. Bisexual b. Intersex c. Hermophroditism d. Parthenogenesis

34. *Panagrellus redivivus* is strongly attracted to cell-free filtrates of culture media of fungi, which secrete the following chemo-attractants

a. Amino acids b. Linolic acid c. Esters or fatty acids d. Acetic acid

35. The chromosome number of *M. incognita*

a. 2n = 32–38 b. 2n = 44 c. n = 18 d. 2n = 54–58

36. This organ plays a vital role in regular removal of material by defaecation in actively feeding nematodes

a. Pseudocoelom b. Intestine c. Oesophagous d. Rectal glands

37. L-glutamic acid is a specific attractant for

a. *Globodera pallida* b. *Tylenchulus semipenetrans*

c. *Ditylenchus dipsaci* d. *Anguina tritici*

38. This protein is involved in ionic and osmotic regulation in root-knot nematodes

a. P–Cl cotransporter protein b. Cl –K cotransporter protein

c. K–Cl cotransporter protein d. P-K cotransporter protein

39. The majority of nitrogenous waste product in nematodes is

a. Ammonia b. Nitrogen c. H_2S d. Alkaloids

40. This is a major component of the gelatinous matrix in the root-knot nematode egg mass

a. Chitin b. Glycoprotein c. Collagen d. Amine

41. Once the nematode stylet has entered the host cell, granules in the dorsal esophageal gland move forward in the cytoplasmic extension and accumulate in this organ

a. Ampulla b. Lumen c. Gland cell d. Amphid

42. The modified host cells by *Rotylenchulus reniformis* are referred to

a. Giant cells b. Syncytia c. Nurse cells d. Mother cells

43. Several multinucleate giant cells are induced by

a. *Meloidogyne spp.,* b. *Rotylenchulus reniformis*

c. *Heterodera spp.* d. *Globodera spp.*

44. Cyst nematodes select this cell to modify into syncytial cell
    a. Outer cortical cell b. Inner cortical cell
    c. Periferal cortical cell d. Parenchyma
45. This enzyme is involved in the breakdown of the excretory cell and the release of the fluids.
    a. Amylase b. Cellulase
    c. Pectinase d. Leucine aminopeptidase
46. By heating proteases and chitinase enzymes to 80° C for 10 min, nematicidal effectiveness
    a. Reduces b. Increases c. Stabilises d. Neutralises
47. For soil solarization, these types of plastic sheets are most preferable
    a. Opaque b. Translucent c. Black coloured d. Thin, transparent
48. Urea and ammonia are effective against nematodes at
    a. 0.01-0.02 % b. 0.03 – 0.04% c. 0.3-0.4 % d. 0.1-0.2 %
49. This hypothesis states that addition of organic matter to soil stimulates natural enemies of nematodes and reduce the nematode numbers
    a. Wang hypothisis b. Stirling hypothisis c. Linford hypothesis d. Riegal hypothesis
50. During soil organic amendments, chitin increases levels of these fungi in soil
    a. Chitinolytic fungi b. Linolytic fungi c. Acetolytic fungi d. Saprophytic fungi

ANSWERS KEY FOR XXXVIII

Sl. No.	Ans	Sl. No.	Ans	Sl. No.	Ans	Sl. No.	Ans	Sl. No.	Ans
1	b	11	c	21	d	31	b	41	a
2	a	12	b	22	c	32	b	42	d
3	c	13	a	23	a	33	d	43	a
4	a	14	c	24	b	34	c	44	b
5	c	15	b	25	c	35	a	45	d
6	a	16	b	26	a	36	b	46	a
7	a	17	a	27	c	37	a	47	d
8	d	18	b	28	b	38	c	48	b
9	b	19	c	29	a	39	a	49	c
10	a	20	a	30	c	40	b	50	a

XXXIX

1. Increased crop yields observed with organic soil amendments are due to the enhanced activities of these microfauna

a. Fungivores and bacterivores
b. Protozoans
c. Actinomycetes
d. Predaceous

2. Problems from *Meloidogyne* spp. are common in the soils with this organic matter content

a. <10%
b. < 20%
c. < 2%
d. <30%

3. Major nematode problem in avocado is

a. *Pratylenchus*
b. *Radopholus*
c. *Anguina*
d. *Aphelenchoides*

4. *Caenorhabditis elegans* also has a small number of genes for a multicellular organism, about

a. 10,000
b. 19,000
c. 25,000
d. 5000

5. There are four bands of muscles in nematodes, two are dorso-lateral and two

a. Ventro-lateral
b. Dorso-ventral
c. Dorso-lateral
d. Latero-ventral

6. Digestive system of nematodes have two openings and in form of

a. Muscules
b. Digestive tract
c. Oesophagous
d. Alimentary canal

7. 'Nematoda' actually means

a. Pointed bodies
b. Pointed ends
c. Pointed cells
d. Peaked heads

8. Undulating waves of contraction and relaxation of muscles helps nematodes in

a. Excretion
b. Digestion
c. Locomotion
d. Respiration

9. Sexual reproduction where male gametes and female gametes fuse to form the zygote is termed

a. Amphimixis
b. Anastomoses
c. Areolation
d. Amixis

10. Small sequence fragment of DNA sequence used as a unique identifier to separate organisms into species is known as

a. Bacillary code
b. Bar code
c. Annuli code
d. Zonal code

11. The opposite of the term 'suppressive soil' is

a. Conducive soil
b. Stimulative soil
c. Favourable soil
d. Synergistic soil

12. This is the anterior part of pharynx

a. Metacorpus
b. Polycorpus
c. Corpus
d. Anticorpus

13. Denticles refer to the minute

a. Spine like structures
b. Amphid like structures
c. Por like structure
d. Teeth-like structures

14. Trilaminar outermost part of the cuticle and the first layer to be laid down during moulting is

a. Epicuticle
b. Ectocuticle
c. Procuticle
d. Exoxuticle

15. Nematode movement resulting from undirected responses to changes in stimulus intensity is known as

a. Kinetics
b. Kinesis
c. Kinensis
d. Kinens

16. This is the zone of the cuticle, located immediately beneath the cortical zone

a. Middle zone
b. Medium zone
c. Median zone
d. Semi zone

17. Short sequences of 1–6 base nucleotide repeats with highly variable length distributed widely throughout the genome are termed
    a. Microsatellite b. Tinysatellites c. Macrosatellites d. Minisatellites
18. A molecule that is labelled and used to detect another molecule (target) through its specific affinity is
    a. Marker probe b. Tracer probe c. Molecular probe d. Local probe
19. This is a gene derived from the same copy of a gene in the most recent common ancestor.
    a. Orthoprobe gene b. Orthologous gene c. Orthrocant gene d. Orthrolie gene
20. This enzyme breaks galacturon sugar polymers
    a. Pectate lyase b. Pectate lipase c. Cellulytic lyase d. Cellulolipase
21. This is a non-functional member of a gene family
    a. Agene b. Pseudogene c. Neogene d. Non gene
22. Locomotion of the nematode resulting from orienting the body to the stimulus direction refers to
    a. Tactic b. Migration c. Taxis d. Penetration
23. Transverse striae are the shallow transverse grooves restricted to the
    a. Cortical zone b. Meristematic zone c. Epideral zone d. Xylem zone
24. Hemizonid is normally located in this side of the nematode body
    a. Dorsal b. Ventral c. Lateral d. Dorsolateral
25. Enzymes that break non-covalent linkages between cellulose chains are
    a. Effectors b. Proctors c. Promons d. Expansins
26. *Aphelenchoides fragariae* is of Quarantine importance on this crop in Europe
    a. Narcissus b. Chrysanthemum c. Aster d. Strawberry
27. In fast green staining during histopathological studies, the dose of fast green used is
    a. 0.5 % b. 0.1% c. 5 % d. 1%
28. For staining root-knot nematode egg-mass, the stain used is
    a. Lactophenol b. Acid fuchsin c. Cotton blue d. Phloxine B
29. Baker's Rapid Method is adopted for preparation of
    a. Permanent nematode mount b. Temporary nematode mount
    c. Staining of nematodes in plant tissue d. Staining of cysts
30. 'Green commandos' is the brand name of the following bioagent
    a. *Pasteuria penetrans* b. *Pseudomonas fluorescens*
    c. *Steinernema carpocapsae* d. *Trichoderma harzianum*
31. Collagen is the major structural component of this part of nematodes
    a. Intesine b. Stylet c. Cuticle d. Oesophagous
32. The total lipid content in phytonematodes is
    a. 15-41 % b. 11-64% c. 1.4-4% d. 5-10%
33. Numerical methods of analysis used to create diagrams of overall similarity among species is known as
    a. Phenogram b. Phylogram c. Phylogeny d. Phenogeny

34. This was the first of the molecular techniques to be applied in Nematology
    a. PCR b. Protein electrophoresis
    c. RFLP d. RAPD
35. In a phylogenetic tree, a group of taxa that belong to the same branch is called a
    a. Clad b. Node c. Unit d. Cluster
36. *Meloidogyne* species are active in soils with moisture levels at this much field capacity
    a. 40-60% b. 10-20% c. 20-30% d. 60-80%
37. The chromosome number in Radopholus similis is
    a. n=10 b. n=4 c. n=12 d. n=16
38. Water is a very good hatching agent for this species of Heterodera
    a. *H.schachtii* b. *H.glycines* c. *H.cajani* d. *H.avenae*
39. 'Granek's ratio' is used in the morphometry of this nematode
    a. Root-knot b. Burrowing c. Cyst d. Seed gall
40. 'Sudden Death Syndrome' (SDS) caused by *Fusarium solani* is sometimes associated with
    a. *Heterodera glycines* in soybean b. *Meloidogyne incognita* in soybean
    c. *Radopholus similis* in banana d. *Anguina tritci* in wheat
41. The host range of *G. rostochiensis* and *G. pallida* include
    a. Potato, brinjal and tomato b. Potato, aubergine and tomato
    c. Potato, chilli and tomato d. potato, aubergine and brinjal
42. Members in the following families are unique among nematodes in parasitizing above-ground plant parts
    a. Anguinidae and Aphelenchoididae b. Tylenchidae and Longidoridae
    c. Aphelenchoididae and Hoplolaimidae d. Pratylenchidae and Aphelenchoididae
43. *Pratylenchus brachyurus* is widely associated with
    a. Grapes b. Papaya c. Citrus d. Banana
44. Rod-shaped sperm, known as 'spermatids' are characteristic feature of
    a. *Longidorus attenuates* b. *Xiphinema index*
    c. *Hoplolaimus indicus* d. *Radopholus similis*
45. 'Yellows disease' in black pepper is mainly due to
    a. *Meloidogyne incognita* b. *Pratylenchus coffeae*
    c. *Radopholus similis* d. *Tylencchorhynchus indicus.*
46. Eggs of *Hirschmaniella oryzae* are deposited
    a. Within the roots b. Outside the roots c. Within leaf sheath d. On the stem
47. *Dolichodorus* is a
    a. Migratory endoparasite b. Migratory ectoparasite
    c. Sedentaty ectoparasite d. Migratory endoparasite
48. The genus *Helicotylenchus* is one of the largest genera in the tylenchs, which contains more than
    a. 100 species b. 50 species c. 200 species d. 160 species

49. The most common species of *Rotylenchus* is

a. *robustus* b. *indicus* c. *reniformis* d. *laurentinus*

50. *Meloidogyne mali* infects this crop

a. Barley b. Almond c. Apple d. Cabbage

ANSWERS KEY FOR XXXIX

Sl. No.	Ans	Sl. No.	Ans	Sl. No.	Ans	Sl. No.	Ans	Sl. No.	Ans
1	a	11	a	21	b	31	c	41	b
2	c	12	c	22	c	32	b	42	a
3	a	13	d	23	a	33	a	43	c
4	b	14	a	24	b	34	b	44	d
5	a	15	b	25	d	35	d	45	c
6	d	16	c	26	a	36	a	46	a
7	b	17	a	27	b	37	b	47	b
8	c	18	c	28	d	38	c	48	d
9	a	19	b	29	a	39	c	49	a
10	b	20	a	30	c	40	a	50	c

XL

1. Cysts are formed mostly by the tanning of the female cuticle due to the
   a. Polyphenol oxidase b. Polyphenol hydrogenase
   c. Polyphenol esterase d. Pectinase
2. The glands overlap the intestine lateroventrally in
   a. *Radopholus similis* b. *Anguina tritici* c. *Heterodera schachtii* d. *H.avenae*
3. 'Potato early dying' syndrome is due to *Verticillium dahliae* and this nematode
   a. *Pratylenchus penetrans* b. *Globodera rostochiensis*
   c. *G. pallida* d. *Meloidogyne hapla*
4. Common name of this nematode is 'shethoid nematode'
   a. *Hemicycliophora* b. *Hemicriconemoides* c. *Criconema* d. *Scutellonema*
5. A 400 mesh sieve has this pore aperture
   a. 25 microns b. 200 micros c. 150 microns d. 38 microns
6. TAF fixative mainly contains
   a. Triethanoamine b. Triethylamine c. Trimethylamine d. Trimethanoamine
7. The best staining method to show gonads, oesophageal glands and nerve ring is
   a. Silver nitrate b. Gold chloride c. Acetic orcein d. Phloxine B
8. The total number of nerve cells present in nematodes is
   a. 500 b. 100 c. 250 d. 350
9. High molecular weight chemical compounds like polypeptides and proteins pass out from the nematode body through
   a. Anus b. Vulva c. Rectum d. Excretory pore
10. The average thickness of an egg of *Meloidogyne javanica* is
    a. 10 mili micron b. 30 mili micron c. 20 mili micron d. 50 mili micron
11. The nematode cuticle is shed in one piece in this nematode
    a. *Paratylenchus nanus* b. *Meloidogyne hapla* c. *Radopholus similis* d. *Criconema parentus*
12. Microrespirometry is used to measure this activity in nematodes
    a. CO_2 consumption b. O_2 consumption c. O_2:CO_2 d. CO_2 release
13. When the moisture content of soil approaches the wilting point for plants, the nematode activity in soil
    a. Inceases b. Decreses
    c. Not affected d. Varies with the nematode species
14. During shortage of food availability, the nematode sex ratio is shifted toward this stage
    a. Females b. J2 c. Males d. J4
15. This nematode can survive well in both sandy and clay soils
    a. *Meloidogyne spp.* b. *Cricoenma spp.*
    c. *Anguina tritici* d. *Tylenchulus semipenetrans*

16. This nematode is a semi-endoparasitic in nature
   a. *Nacobbus* b. *Radopholus similis* c. *Pratylenchus coffeae* d. *Trichodorus christei*
17. This stage in *Bursaphelenchus xylophilus* is known as 'dauer larvae'
   a. J_2 b. J_1 c. J_3 d. J_4
18. Chemicals that cause interactions between organisms are called
   a. Semiochemicals b. Interochemicals c. Intrachemicals d. Sanochemicals
19. This compound present in the amphidial secretions in J_2 of *Meloidogyne incognita* is important in chemoreception
   a. O-glycans b. N-acetylglucosamine
   c. N-acetylgalactosamine d. Glucose
20. A stack of contractile units in muscle fiber is termed
   a. Meromere b. Circomere c. Platomere d. Sarcomere
21. This group of nematodes, in general, has a greater amount of lipid content than animal parasitic nematodes
   a. Phytopathogens b. Free-living c. Bacterial feeders d. Predacious
22. Hatching in cyst nematodes is associated with the release of this ompound, which is found in high concentrations in the perivitelline fluid
   a. Acetyl-CoA b. N-acetylglucosamine c. Trehalose d. Tricarboxylic acid
23. All phytonematodes have a small body radius and a
   a. Large surface area to volume ratio b. Small surface area to volume ratio
   c. Medium surface area to volume ratio d. Non-variable large surface area to volume ratio
24. When this nematode attacks crops plants, no clear visible disease symptoms are seen
   a. *Pratylenchus coffeae* b. *Tylenchorhynchus claytoni*
   c. *Anguina tritici* d. *Helicotylenchus indicus*
25. These glands are responsible for secreting the gelatinous matrix to protect eggs
   a. Oesophageal b. Rectal c. Hypodermal d. Cervical
26. This nematode was earlier linked with the 'mentek' disease in rice
   a. *Meloidogyne graminicola* b. *Hirschmaniella oryzae*
   c. *Ditylenchus angustus* d. *Heterodera oryzicola*
27. This nematode is called 'filarial worm'
   a. *Wuchereria bancroftii* b. *Ascaris lumbricoides*
   c. *Taenia solium* d. *Dracunculus medinesis*
28. In this nematode, excretory pore secrets gelatinous matrix
   a. *Meloidogyne hapla* b. *M. arenaria*
   c. *Tylenchulus semipenetrans* d. *Rotylenchulus reniformis*
29. Post uterine sac is absent in this suerfamily
   a. Criconematoidea b. Tylenchoides c. Aphelenchoidea c. Diplogasteroidea

30. Rugose wall pattern is a characteristic feature of this genus
   a. *Radopholus* b. *Anguina* c. *Rhadinaphelenchus* d. *Heterodera*
31. The first Postgraduate Nematology course in Southeast Asia was held at
   a. Indian Agricultural Research Institute, New Delhi
   b. Aligarh Muslim University, Aligarh
   c. a and b
   d. Tamil Nadu Agricultural University, Coimbatore
32. Post graduate courses in Nematology were started at Indian Agricultural Research Institute, New Delhi in the year
   a. 1968 b. 1960 c. 1969 d. 1965
33. This rice variety is resistant to white tip disease
   a. IR 60 b. Sona Masuri c. IR 8 d. IR 5
34. Race-2 of *Meloidogyne incognita* is present in this State
   a. Karnataka b. Assam c. Punjab d. Arunachal Pradesh
35. In general, the pH of the intestine of a nematode ranges from
   a. 3-7.5 b. 4 -8.2 c. 4-6.8 d. 5-8.2
36. Some species of these two genera often live in the tunnels made by bark beetles and transported on the bodies of beetles
   a. *Pratylenchus and Aphelenchus* b. *Anguina and Dolichodorus*
   c. *Tylenchus and Aphelenchoides* d. *Turbatrix and Nacobbus*
37. Mucopolysaccharides are mostly produced by nematodes attacking
   a. Animals b. Plants c. Insects d. Birds
38. This nematode organ stores glycogen
   a. Oesophagus b. Intestine c. Hypodermis d. Cuticle
39. *Heterodera cajani* was reported in India for the first time by
   a. Koshy b. Das c. Thitumalachar d. Nayar
40. The cuticle of nematode, in general, is
   a. Completely permeable b. Completely impermeable
   c. Both permeable and impermeable d. Neither permeable and impermeable
41. The brown colour of the matured cyst of *Heterodera* is mainly due to the activity of
   a. Phenol oxidase b. Polyphenol oxidase
   c. Triphenol oxidase d. Diphenol oxidase
42. *Rhabditis terrestris* often occurs as a resting larva in
   a. Earthworm b. Plant roots c. Insect larva d. Weed hosts
43. This is excreted by some nematodes, which sometimes is metabolically produced and used in their metabolism
   a. CO_2 b. O_2 c. N d. H_2O

44. The potato steroid glycoalkaloids α-solanine and α-chaconine stimulate the hatch of

a. *G. rostochiensis* b. *G.pallida* c. a + b d. *Meloiodgyne incognita*

45. A ventro-lateral commissure of the nervous system in the hypodermis found in some nematodes is termed

a. Ganglion b. Nerve ring c. Hemizonid d. Phasmid

46. The nematode cuticle is primarily composed of

a. Proteins b. Lipids c. Polyphenols d. Chitin

47. Pigeonpea variety Pusa -23 is resisitant to this nematode

a. *Heterodera cajani* b. *Meloidogyne incognita*

c. *M. javanica* d. a and b

48. Nematodes, in general, differ from other animals in having only these muscles in the body wall

a. Transverse muscle b. Cross muscle c. Longitudinal muscle d. Cephalic muscle

49. The terms 'annelid-like, caterpillar- like and looping progression' in nematodes refer to

a. Excretion b. Respiration c. Digestion d. Locomotion

50. The total lipid content of the plant-parasitic nematodes generally ranges from

a. 11% to 67% of dry weight b. 9-28 %

c. 15-37 % d. 22-88%

ANSWERS KEY FOR XL

Sl. No.	Ans	Sl. No.	Ans	Sl. No.	Ans	Sl. No.	Ans	Sl. No.	Ans
1	a	11	a	21	a	31	a	41	b
2	c	12	b	22	c	32	c	42	a
3	a	13	b	23	a	33	d	43	a
4	b	14	c	24	b	34	a	44	c
5	d	15	d	25	b	35	c	45	c
6	a	16	a	26	b	36	c	46	a
7	b	17	c	27	a	37	a	47	b
8	c	18	a	28	c	38	c	48	c
9	d	19	c	29	a	39	a	49	d
10	b	20	d	30	d	40	c	50	a

XLI

1. *Criconemoides* exhibits this type of movement
   a. Caterpillar-like b. Annelid like c. Looping type d. Circular type
2. 'Soldier' is the brand name of this bio-agent
   a. *Heterorhabditis indica* b. *Steinernema carpocapsae*
   c. *Trichoderma harzianum* d. *Bacillus putida*
3. The habitat of *Tylenchorhynchus* is
   a. Endoparasitic b. Ectoparasitc c. Semi-endoparasitic d. a + b
4. Entomopathogenic nematodes belong to this order
   a. Tylenchida b. Dorylaimida c. Rhabditida d. Strongylida
5. FORTRAN computer program was developed mainly to simulate
   a. Nematode soil sampling strategies b. Nematode population strategies
   c. Nematode distribution strategies d. Nematode movement strategies
6. *Mae* and *Mag* resistance genes are related to
   a. Tomato-Meloidogyne arenaria b. Peanut-M. arenaria
   c. Rice-M.graminicola d. Maize-M. arenaria
7. *Mi* gene does not confers resistance to
   a. *M.incognita and M. javanica*
   b. *M.incognita, M. javanica and M. arenaria*
   c. *M.exigua and M.graminicola.*
   d. *M.incognita, M. javanica, M. hapla and M. arenaria*
8. A cysteine proteinase inhibitor from rice, oryzacystatin I (OC-I), completely inhibits the proteolytic activity of all stages of
   a. *M. hapla* b. *M. incognita* c. *M.arenaria* d. *M.javanica*
9. Total assembly length of Meloidogyne arenaria genotype A2-O is
   a. 248.05 Mb b. 274.01 Mb c. 284.05 Mb. d. 250.02 Mb
10. These are small molecules acting as pheromones in the social behavior of nematodes
    a. Azacarosides b. Ascarosides c. Abcarosides d. Acetocarosides
11. 'IMP' stands for
    a. Indian *Meloidogyne* Project b. International *Meloidogyne* Project
    c. International nematode Management Project d. Indian root-knot Management Project
12. This term is concerned with the methods and principles by which organisms are arranged in the form of classification
    a. Macrotaxonomy b. Microtaxonomy c. Eutaxonomy d. Pseudotaxonomy
13. Major genera under the family Meloidogynidae are
    a. *Meloidogyne, Meloinema and Meloidodera* b. *Meloidogyne, Nacobbus and Meloidodera*
    c. *Meloidogyne, Meloinema and Meloidoderella* d. *Meloidogyne, Meloinema and Mononchus*

14. The anal opening of the female root-knot nematode is located between the tail terminus and the
   a. Rectum b. Vulva c. Vagina d. Spermatheca
15. The gelatinous egg matrix in Meloidogyne cosists of proteins, enzymes and a
   a. Mucomonosaccharide b. Mucodisaccharide
   c. Mucoside d. Mucopolysaccharide
16. Generally second stage juveniles accumulate just behind root cap at the region of
   a. Cell elongation b. Cell formation c. Cell multiplication d. Cell widening
17. Formation of a knot on the root by root-knot nematode involves hypertrophy of this tissue
   a. Pericycle b. Cortical parenchyma c. Cambium d. Stele parenchyma
18. The nematoxic compound 'thionemone' is present in the root exudates of
   a. Marigold b. Mustard c. Neem d. Asparagus
19. Trichostrongylids have proven useful in the ivestigaiton of this activity of nematodes
   a. Ecdysis b. Locomotion c. Excretion d. Hatching
20. *Panagrellus redivivus* is a
   a. Predatory nematode b. Free living nematode
   c. Mycophagous d. Bacterial feeder
21. The *Mi*-1.2 gene is closely related to another tomato gene for resistance against
   a. *Pseudomonas putida* b. *P. syringae*
   c. *Ralstonia solanacearum* d. *Rhizoctonia solani*
22. The eggs of *Meloidogyne* species are approximately of this micrometer dimension
   a. 95 × 40 b. 100 × 25 c. 50 × 50 d. 80 × 100
23. An interactive computer simulation programme 'FORTRAN' is developed for
   a. *Heterodera avenae-wheat* b. *Radopholus similis -banana*
   c. *Heterodera schachtii-sugar beet* d. *Anguina tritici-wheat*
24. This is the original serine proteinase inhibitor identified from cowpea
   a. CpT b. CP ti c. CPT d. CpTi
25. The oviduct in a female root-knot nematode consists of this number of staggered rows of cells
   a. Two b. Three c. Six d. Four
26. The initial generic name of root-knot nematode in 1879 was
   a. *Anguillula indica* b. *Anguillula marioni*
   c. *Heterodera radicicola* d. *Anguillulina radicicola*
27. In sedentary endoparsitic nematodes, this sterol type is more abunadant than cholesterol
   a. Sitosterol b. Phytosterol c. Campsterol d. Lathosterol
28. This is the most important nematode stimulus in soil
   a. Moisture b. pH c. Temperature d. Soil solution
29. Avermectins are derived from the mycelia of
   a. *Streptococcus avermitilis* b. *Pseudomonas avermitilis*
   c. *Bacillus avermitilis* d. *Streptomyces avermitilis*

30. Both sex of this nematode can produce pheromones

a. *Rhabditis pellio* b. *Cylindrocarpus longistoma*

c. *Cephalobes persignis* d. *Mononchus spp.*

31. The thickness of the surface coat of phytonematodes, in general, is about

a. 2-10 nm b. 2-50 nm c. 5-30 nm d. 5-90 nm

32. The surface carbohydrate detected on the cuticle of Meloidogyne incognita is

a. Salicylic acid b. Mannose c. Galactose d. N-acetylgalactosamine

33. Root diffusates are essential for the hatching of eggs of this nematode

a. *Meloidogyne hapla* b. *M.javanica*

c. *M. incognita* d. *Globodera rostochiensis*

34. The permeability characteristics of the nematode eggshell is derived mainly from

a. Lipid layer b. Crystalline layer c. Middle layer d. External layer

35. Presence of this compound in the fuids within the eggs of golden cyst nematode of potato plays a major role in egg hatching

a. Glucose b. Fructose c. Mannose d. None of these

36. Glycinoeclepins act as egg hatching factors in this nematode

a. Soybean cyst b. Soybean root-knot c. Soybean lesion d. Pea cyst

37. This nematode is one of the premiere model systems for ageing research

a. *Paracanthonchus olgae* b. *Caenorhabditis elegans*

c. *Oncholaimus scanicus* d. a + c

38. One of the characteristics of nematode senescence is age pigment, which is also known as

a. Lipofucosine b. Lipofuscin c. Lipoalamin d. Lipomalanin

39. This phytonematode can be cultured axenically

a. *Criconema mutabile* b. *Hirschmaniella oryzae*

c. *Hoplolaimus indicus* d. *Bursaphelenchus cocophilus*

40. The movement of a nematode sperm is

a. Amoeboid b. Gliding c. Jumping d. Not specific

41. Crop damage prediction with a minimum of measurement and monitoring is characteristic of this model system

a. Multiple point b. Simulation model c. Critical point d. Synoptic approach

42. This is the first fresh water nematode to be reported

a. *Trilobus longus* b. *Turbatrix aceti*

c. *Aphelenchoides indicus* d. *Seinura calnus*

43. *Alternanthera philoxeroides* is commonly found associated with

a. Herbs b. Shrubs

c. Aquatic vascular plants d. Forest trees

44. This is known to be highly cellulolytic and is a major industrial microbial source for commercial cellulases and other cell walldegrading enzymes.

a. *Trichoderma reesei* b. *T.harzianum* c. *T.viride* d. *T.reasin*

45. This compound present on the surface coat of the nematode cuticle, reduces active oxygen plant defenses.
    a. Saractivators
    b. Glutathione peroxidase
    c. Glutathione oxidase
    d. Glutathione dioxide
46. This enzyme is particularly involved in biosynthesis of phytoalexin
    a. Glutamase
    b. Peroxidase
    c. Polyphenoloxidase
    d. Phenylalanine ammonia-lyase
47. During 2007-2009, this nematode was added as a new quatantine pest
    a. *Rhadinaphelenchus cocophilus*
    b. *Ditylenchus dipsasi*
    c. *Anguina tritici*
    d. *Radopholus similis*
48. *Heterodera glycines* is currently on the list of pests regulated by this country under Schedule II of the Plant Proteciton Regulations
    a. Japan
    b. Argentina
    c. Canada
    d. India
49. 'NAPPO' stands for
    a. National Plant Protection Organization
    b. National Pest Protection Organization
    c. National Protection of Pests under Acts
    d. National Advanced Plant Protection Organization
50. In India, this nematode on this crop is of quarantine importance
    a. *Meloidogyne incognita –Beet root*
    b. *Radopholus similis-Banana*
    c. *Ditylenchus destructor-Potato*
    d. *Rhadinaphelenchus cocophilus-coconut*

ANSWERS KEY FOR XLI

Sl. No.	Ans	Sl. No.	Ans	Sl. No.	Ans	Sl. No.	Ans	Sl. No.	Ans
1	b	11	b	21	b	31	c	41	c
2	a	12	a	22	a	32	b	42	a
3	d	13	c	23	c	33	d	43	c
4	c	14	d	24	d	34	a	44	a
5	a	15	b	25	a	35	d	45	b
6	b	16	a	26	b	36	a	46	d
7	c	17	b	27	b	37	b	47	b
8	a	18	c	28	c	38	b	48	c
9	c	19	a	29	d	39	d	49	a
10	b	20	b	30	a	40	a	50	c

XLII

1. The longevity of *Radopholus similis* without food is for about
   a. Three months b. Six months c. One month d. One year
2. The abbreviation "CBD" stands for
   a. Convention on Biological Diversity b. Conference on Bio Diversity
   c. Consortium on Biosafety d. Convention on Biosecurity
3. The location of the World Trade Organization is
   a. London b. Greece c. Italy d. Geneva
4. Nanotechnology is the understanding and control of matter at dimensions between
   a. 1-100 nm b. 1-1000 nm c. 1-10 nm d. 1-100 mm
5. The term 'Bionanotechnology" consists of nanaotechnology and
   a. Biochemistry b. Biology c. Biotechnology d. Bioscience, in general
6. The eggs of the following nematode show great diversity in their size, shape and markings
   a. Phytoparasitic b. Animal feeding c. Free- living d. Insect feeding
7. This nematode can be revived after storage in liquid nitrogen
   a. *Anguina tritici* b. *Meloidogyne graminicola*
   c. *Caenorhabditis elegans* d. *Creconema*
8. The possibility of using nematodes for management of phytonematodes was first suggested by
   a. Cobb b. Barber c. Seinhorst d. Thorne
9. Medium containing crude materials which supply most of the dietery requirements is referred to
   a. Holidic medium b. Oligidic medium c. Meridic medium d. Ketonic medium
10. This is a nucleic acid based molecular technique used in Nematology
    a. PCR b. DIBA c. ELISA d. PAGE
11. The "Synoptic method" is related to this aspect of nematodes
    a. Reproduction b. Excretion c. Crop loss assessment d. Surveying
12. High extraction efficiency of Criconemella can be achieved by the following techninque
    a. Baermann funnel b. Cobb's sieving
    c. Centrifugal floatation + Elutriation d. Elutriaiton
13. Floatation-sieving technique is mainly used to extract large sized nematodes like
    a. *Xiphinema* b. *Meloidogyne* c. *Anguina* d. *Paratylenchus*
14. This nematode genus is difficult to eradicate with fumigants
    a. *Radopholus* b. *Pratylenchus* c. *Paratrichodorus* d. *Meloidogyne*
15. *Pratylenchus brachyurus* is frequently found associated with this crop
    a. Brinjal b. Pigeon pea c. Groundnut d. Potato
16. The expanded form of 'SWNT' is
    a. Single-walled Carbon Nanotube b. Single-walled Copper Nanotube

c. Simple walled Carbon Nanotune d. Straight walled Copper Nanotube

17. Trizole reagent is used to isolate this molecule from nematodes

a. DNA b. Total RNA c. Total DNA d. Proteins

18. The most strongest and important members of the hollw 1-D nanosystem are

a. Carbon nanotubes b. Carbon nanofibres c. Onion-like carbons d. Nanodiamond

19. In histopathological technique, the thickness of the sections for very fine cellular details should be

a. 1-2 micrometer b. 3-5 micro meter c. 5-7 micro meter d. 2-4 milimeter

20. The life cycle of myceliophahous nematodes may be completed in

a. Few weeks b. Few months c. Non-specific d. Few days

21. Ro1, Ro2 and Ro3 are the pathotypes of this nematode

a. *Radopholus similis* b. *Rotylenchulus reniformis*

c. *Globodera rostochiensis* d. *Rotylenchus spp.*

22. Distribution of *Meloidogyne graminicola* is mostly in this zone

a. Tropical b. Temperate c. Cosmopolitan d. b + c

23. This species lacks egg-masses, if present, it is devoid of eggs

a. *Heterodera zeae* b. *H.mothi* c. *Meloidogyne exigua* d. *M. indica*

24. The modified 'Gall Index Test' is used in this study related to nematodes

a. Crop loss assessment b. Nematicide testing

c. Survival d. Reproduction

25. This nematode is also known as 'purple nematode'

a. *Anguina agrostis* b. *Aphelenchoides fragaeriae*

c. *Hoplolaimus indicus* d. *Dolichodorus spp.*

26. Viehmeyer tube and oakfield tube are employed in this technique of nematodes

a. Micrometry b. Histopathology c. Sampling d. Drawing

27. The most appropriate sampling time for nematodes from tomato rhizosphere is

a. August-October b. September-October c. May-August d. June-September

28. Plant parasitic nematodes are specialized by the stylet and the following structure, which are considered the most significant evolutionary adaptations for plant parasitism

a. Esophageal glands b. Excretory pore c. Hemizonids d. Phasmids

29. Compost incorporation into soil has this effect on soil density

a. Increases b. Decreases

c. Non-specific d. Intitally increases, later decreases

30. Sudangrass can be used in this form to manage phytonematodes

a. Mulching b. Biofumigant c. Antagonisitc crop d. Cover crop

31. This layer of nematode cuticle is osmiophilic in nature.

a. Inner b. Outer c. Middle d. b + c

32. The relative impermeability of nematode eggs to water soluble substances is due to the

a. Outer protein layer b. Thin inner layer

c. Thick chitinous layer d. Middle layer

33. This appears to function as the major circulating blood sugar in nematodes
a. Trehalose b. Glucose c. Sucrose d. Maltose

34. Absorption of food materials in nematodes takes place usually by
a. Cuticle b. Intestine c. Oesophagous d. Hypodermis

35. The pH of the cells in the intestine of most nematodes, usually
a. Neutral b. Acidic c. Alkaline d. Varies

36. The glycogen is stored mainly in the hypodermis, the intestine and in the epithelial cells of
a. Circulatory system b. Reproductory organs c. Nervous organs d. Pseudocoelm

37. Glucose-6-phosphate dehydrogenase is specifically present in this phytonematode
a. *Ditylenchus* b. *Meloidogyne* c. *Radopholus* d. *Heterodera*

38. Trace amounts of volatile acids may also be produced during the metabaolic activities in
a. *Seinura* b. *Globodera* c. *Caenorhabditis* d. *Criconema*

39. Tricarboxylic acid cycle is completely absent in
a. *Paratylenchus* b. *Turbatrix aceti* c. *Panagrellus* d. *Caenorhabditis*

40. In many free-living stages of animal parasitic nematodes, fat is stored in the
a. Lumen of the intestine b. Oesophagous
c. Cuticle d. Ovaries

41. Haemoglobin is the only respiratory pigment present in this form of nematodes
a. Free-living b. Animal parasitic c. Plant parasitic d. Predatory

42. The cuticle of nematode is
a. Completely impermeable b. Completely permeable
c. Inert d. Varies with the surrounding

43. The colour of the cyst in Heterodera cajani changes from white to dark due to the activity of
a. Peroxidase b. Phenol oxidase c. Polyphenol oxidase d. Diphenol oxidase

44. *Ditylenchus dipsaci* can survive this temperature, if water is removed from the body by previous immersion in concentrated salt solutions.
a. –80° C b. –40° C c. –20° C d. –5° C

45. Although the shape of the nematode egg varies, they are essentially
a. Bullet shped b. Ovoid c. Circular d. Spiral

46. Stylet is actuallay, a hollow structure with a
a. Sublongitudinal orifice b. Subterminal orifice
c. Subdorsal orifice d. Subventral orifice

47. The chitinous layer of nematode egg shell also contains
a. Proteins b. Fats c. Amino acids d. Polysaccharides

48. The eggs of this nematode will hatch only in soil containing host roots
a. *Meloidogyne hapla* b. *Pratylenchus coffeae*

c. *Anguina tritici* d. *Heterodera goettingiana*

49. Major hosts of reniform nematode are
    a. Cowpea & Caster b. Carrot & Raddish c. Tomato & Brinjal d. Banana & Citurs
50. The stylet of *Meloidogyne* juvenile makes this number of thrusts to puncture the egg shell to emerge out during hatching
    a. 20-40 per second b. 70-90 per minute c. 45-65 per minute d. 50-60 per second

ANSWERS KEY FOR XLII

Sl. No.	Ans	Sl. No.	Ans	Sl. No.	Ans	Sl. No.	Ans	Sl. No.	Ans
1	b	11	c	21	a	31	a	41	b
2	a	12	c	22	a	32	b	42	d
3	d	13	a	23	d	33	a	43	c
4	a	14	c	24	b	34	b	44	a
5	c	15	c	25	a	35	d	45	b
6	c	16	a	26	c	36	b	46	b
7	c	17	b	27	d	37	a	47	a
8	a	18	a	28	a	38	c	48	d
9	b	19	b	29	b	39	b	49	a
10	a	20	b	30	d	40	a	50	b

XLIII

1. The egg shell of *Globodera rostochiensis* consisits this amount of ash

   a. 10% b. 3% c. 20% d. 15%

2. The egg gelatinous matrix of *Meloidogyne javanica* also contains

   a. Glycoprotein b. Polyprotein c. Aminoprotein d. Lipoprotein

3. Most preferred part of root system by phytonematodes is

   a. Secondary roots b. Root tips c. Primary roots d. Tertiary roots

4. The nematode cuticle secreted by the hypodermis is not

   a. Inactive b. Active c. Sensitive d. Protective

5. Optimum temperature required for the maximum hatching of eggs of *Heterodera avenae* is

   a. 30° C b. 20-25°C c. 36°C d. 10-18°C

6. Root-knot nematode on black pepper was reported for the first time from Kerala, India by

   a. Butler-1901 b. Koshy- 1906 c. Ayyar-1926 d. Nair-1966

7. This genus an example for migratory semi-endoparsitic nematode

   a. *Rotylenchulus* b. *Tylenchulus* c. *Helicotylenchus* d. *Pratylenchus*

8. In several parts of the world including India, Paratylenchus projectus is a mjaor problem on

   a. Tube rosel b. Cabbage c. Black pepper d. Grapes

9. The leghaemoglobin content of cowpea is drastically reduced when root-knot nematode interacts with

   a. *Rhizctonia* b. *Rhizobium* c. *Sclerotium* d. *Fusarium*

10. Biointensive Integrated Nematode Management is based on an understanding of

    a. Nematode biology b. Nematode physiology

    c. Nematode ecology d. Nematode morphology

11. The cheapest, practically feasible and safest method to manage nematodes is

    a. Fallowing b. Flooding c. Soil solarization d. Crop rotation

12. This cropping sequence is most appropriate to suppress *Meloidogyne javanica* in brinjal

    a. Cotton-wheat-cotton b. Cotton-cabbage-cotton

    c. Cowpea-cotton-cotton d. Gourd-cotton-cowpea

13. This sweet potato variety is a resisitant trap crop for root-knot nematode

    a. Sree Nandini b. Sree Ratna c. Sree Bhadra d. Sree Vardhin

14. This is an example for antagonistic crop in nematode management

    a. Sesame b. Cowpea c. Crorolaria d. Cabbage

15. In diploblastic animals, this major layer is absent

    a. Epoderm b. Ectoderm c. Endoderm d. Mesoderm

16. The revised name of Paecilomyces lilacinus is

    a. *Purpureomycillis lilacinus* b. *Purpureocillium lilacinum*

    c. *Purpureomycillium lilacinum* d. *Purpureocillium lilacinus*

17. Pseudomonas stutzeri is effective against
    a. *Meloidogyne incognita- Black pepper*
    b. *Radopholus similis-Black pepper*
    c. *Xiphinema index-Grapevine*
    d. *Anguina tritici-Oat*
18. *Glomus macrocorpus* is effective against
    a. *Pratylecnhus coffeae-Coconut*
    b. *Meloidogyne incognita-Okra*
    c. *Radopholus similis-Coconut*
    d. *Rotylenchulus reniformis-Cotton*
19. *Acaulospora bireticulata* is effective against the burrowing nematode in coconut, which is a
    a. Bacterial bioagent
    b. Opportunistic fungus
    c. Mycorrhiza
    d. Actinomycetes
20. 'Mycorimix' is an anti-nematode commercial formulation of this origin
    a. Entomopathogenic nematode
    b. Bacterium
    c. Mycorrhiza
    d. Fungus
21. Cultivar 'Kufri Dewa' is resisitant to
    a. *Globodera rostochiensis in potato*
    b. *G.pallida in potato*
    c. *Meloidogyne incognita in potato*
    d. *M.incognita in sweet potato*
22. The monograph on nematodes by this nematologist is regarded as the "Bible of Nematologists"
    a. Micoletzky
    b. de Man
    c. Paramonov
    d. Grasse
23. Somatic musculature in nematodes is a layer of the following shaped muscle cells attached to the hypodermis.
    a. Spindle
    b. Spherical
    c. Elongated
    d. Irregular
24. Hypodermis is characterized by the presence of the following number of chords
    a. Five
    b. Two
    c. Four
    d. Six
25. This technique can amplify and quantify nucleic acids simultaneously
    a. Quantitative PCR
    b. Progressive PCR
    c. Consisitent PCR
    d. Determinent PCR
26. In general, the external body surface of most nematodes is
    a. Dry
    b. Oily
    c. Watery
    d. Varies
27. In soils with high organic content, the damage caused by cyst nematodes is normally
    a. High
    b. Low
    c. Moderate
    d. Varies
28. *Tylenchorhynchus* is known as 'stunt nematode' because
    a. Short body length
    b. Causes stunted growth in plants
    c. Short stylet
    d. Short tail
29. These three genera belong to one family
    a. *Radopholus, Hirschmaniella and Pratylenchus*
    b. *Radopholus, Hirschmaniella and Paratylenchus*
    c. *Meloidogyne, Heterodera and Radopholus*
    d. *Meloidogyne, Heterodera and Anguina*
30. Unlike other species of Meloidogyne, the head region of Meloidogyne hapla is
    a. Rounded and smooth
    b. With 1-3 complete annulaitons
    c. With 1-3 incomplete annulaitons
    d. Cylindrical and smooth
31. After heating, nematodes of this genus become straight or slightly curved but not spiral
    a. *Helicotylenchus*
    b. *Hoplolaimus*
    c. *Rotylenchus*
    d. a + b

32. Very large phasmids opposite each other in the tial region are the characteristic feature of this genus
    a. *Longidorus* b. *Paralongidorus* c. *Scutellonema* d. *Dolichodorus*
33. This nematode is of economic importance on banana in most tropical countries
    a. *Pratylenchus* b. *Hoplolaimus* c. *Meloidogyne* d. *Helicotylenchus*
34. In this superfamily, the procorpus and the metacarpus are fused
    a. Criconematoidea b. Heteroderoidea c. Aphelenchoidea d. Tylenchoidea
35. This genus cannot be collected from soil in Baermann funnel funnel
    a. *Xiphinema* b. *Paratylenchus* c. *Criconemoides* d. *Anguina*
36. The revised name for the genus Macroposthonia
    a. *Trichodorus* b. *Paratrichodorus* c. *Criconemoides* d. *Hoplolaimus*
37. The common name "ring nematode" is in reference to the
    a. Ring shape of the nematode b. Ring-like coarse annulations
    c. Ring shaped tail d. Ring shaped mouth
38. This genus has two cuticles
    a. *Dolichodorus* b. *Hoplolaimus* c. *Belonolaimus* d. *Bursaphelenchus*
39. Female nematodes have a long slender stylet with rounded knobs, but males have less developed or without stylet in the following genus
    a. *Paratylenchus* b. *Nacobbus* c. *Aphelenchoides* d. *Tylenchorhynchus*
40. *Xiphinema diversicaudatum* occurs more frequently on
    a. Rose b. Fig c. Woody plants d. Fruits
41. Clay soils favour this nematode
    a. *Meloidogyne javanica* b. *Anguina tritici*
    c. *Xiphinema index* d. *Heterodera avenae*
42. *Ditylenchus myceliophagus* is morphologically very closely related to
    a. *Ditylenchus dipsaci* b. D.*destructor*
    c. D. *angustus* d. *Aphelenchoides besseyi*
43. Species of this genus are also favourbale hosts for *Aphelenchoides composticola*
    a. *Fusarium* b. *Alternaria* c. *Botrytis* d. *Pythium*
44. *Meloidogyne artiella* is a major pest on
    a. Forest tress b. Cabbage c. Okra d. Betel vine
45. Watermelon is relatively a poor host for
    a. *Meloidogyne incognita* b. *M.hapla*
    c. *M.arenaria* d. *M.javanica*
46. Stylet knobs are not set-off in this species
    a. *Meloidogyne hapla* b. *M.incognita* c. *M.arenaria* d. *M. javania*
47. *Pratylenchus vulnus* usually, is a serious pest on
    a. Vegetables b. Fruits c. Plantation crops d. Pulses

48. 'Maize-mustard-menthol mint' cropping sequence can effectively manage root-knot nematodes on

a. Maize b. Mustard c. Mint d. a + b

49. Major damage caused by *Meloidogyne graminicola* on rice is

a. Non-transplantable seedlings b. Wilting

c. Stunting d. Yellowing

50. Cultivars 'Shringar' and 'Suvasini' of this crop are resisitant to *Meloidogyne incognita*

a. Champaka b. Jasmine c. Tube rose d. Rose

ANSWERS KEY FOR XLIII

Sl. No.	Ans	Sl. No.	Ans	Sl. No.	Ans	Sl. No.	Ans	Sl. No.	Ans
1	a	11	d	21	c	31	b	41	c
2	b	12	a	22	d	32	c	42	d
3	b	13	c	23	a	33	d	43	a
4	a	14	a	24	b	34	a	44	b
5	d	15	d	25	b	35	c	45	b
6	b	16	b	26	b	36	c	46	c
7	c	17	a	27	b	37	b	47	b
8	b	18	c	28	b	38	b	48	a
9	b	19	c	29	a	39	a	49	a
10	c	20	c	30	a	40	c	50	c

MATCH THE FOLLOWING

1.

A	B
1. Bulbiformin	a. *Pseudomonas fluorescens*
2. *Bt* toxin	b. *Clostridium butyricum*
3. Phenazin	c. *Streptomyces avermitilis*
4. H_2S	d. *Bacillus subtilis*
5. Butyric acid	e. *Streptomyces anulatus*
6. Valinomycin	f. *Bacillus thuringiensis*
7. Avermectin	g. *Desulfovibrio desulfurican*
8. Alpha diethienyl	h. *Arthrobotrys robusta*
9. Alpha terthienyl	i. *Meloidogyne graminicola*
10. Azadirachtin	j. *Meloidogyne exigua*
11. Predaceous fungus	k. *Tagetes erecta*
12. Egg-mass within roots	l. *Neem*
13. Egg-mass outside roots	m. *Meloidogyne coffeicola*
14. Southern root-knot nematode	n. *Tagetes patula*
15. Javanese root-knot nematode	o. *Meloidogyne hapla*
16. False root-knot nematode	p. *Meloidogyne enterolobii*
17. Northern root-knot nematode	q. *Meloidogyne arenaria*
18. Guava	r. *Meloidogyne javanica*
19. Coffee	s. *Meloidogyne incognita*
20. Peanut root-knot nematode	t. *Nacobbus aberrans*

ANSWERS 1

1. d	11. h
2. f	12. i
3. a	13. j
4. g	14. s
5. b	15. r
6. e	16. t
7. c	17. o
8. n	18. p
9. k	19. m
10. l	20. q

2.

A	B
1. *Belonolaimus*	a. Golden cyst nematode
2. *Criconema*	b. Pea cyst nematode
3. *Globodera palida*	c. Citrus nematode
4. *Heterodera avenae*	d. Stubby root nematode
5. *Heterodera goettingiana*	e.Sheath nematode
6. *Dolichodrus*	f. Sting nematode
7. *Tylenchulus semipenctrans*	g.Lesion nematode
8. *Globodera rostochiensis*	h. Ring nematode
9. *Trichodorus*	i. Pale cyst nematode
10. *Hemicycliophora*	j. Awl nematode
11. *Hoplolaimus*	k. Cereal cyst nematode
12. *Paratylenchus*	l. Reniform nematode
13. *Rotylenchus*	m.Yam nematode
14. *Xiphinema*	n. Pine wood nematode
15. *Paralongodorus*	o.Stunt nematode
16. *Scutellonema*	p. Needle nematode
17. *Bursaphelenchus*	q.Dagger nematode
18. *Rotylenchulus*	r. Spiral nematode
19. *Pratylenchus*	s.Pin nematode
20. *Tylenchorhynchus*	t.Lance nematode

ANSWERS 2

1. f	11. t
2. h	12. s
3. i	13. r
4. k	14. q
5. b	15. p
6. j	16. m
7. c	17. n
8. a	18. l
9. d	19. g
10. e	20. o

3.

A	B
1. Migratory endoparasite	a. Sheath nematode
2. Migratory ectoparasite	b. Citrus nematode
3. Sedentary endoparasite	c. Cyst nematode
4. Sedentary ectoparasite	d. Stubby root nematode
5. Semi endoparasite	e. Lesion nematode
6. Seed borne	f. Stem and bulb nematode
7. Nematode wool	g. Chrysanthemum foliar nematode
8. Migratory ecto-and endoparasites	h. Wheat gall nematode
9. Insect as a vector	i. Golden cyst nematode
10. Cavities and tunnels in roots	j. Pin nematode
11. Quarantine nematode	k. Coconut red ring nematode
12. Smallest phytonematode	l. Burrowing nematode
13. Opisthodelphic	m. Lesion nematode
14.Optional sense organ	n. Amphid
15. Longest phytonematode	o. Root-knot nematode
16. Didelphic, prodelphic	p. Dagger nematode
17. Compulsory sense organ	q. Gubernaculum
18. Stylet	r. Ring nematode
19. Long tapering tail	s. Phytonematode
20. Poor plant parasite	t. Saprophytic nematode

ANSWERS 3

1. e	11. i
2. d	12. j
3. c	13. p
4. a	14. q
5. a	15. m
6. h	16. o
7. f	17. n
8. g	18. s
9. k	19. t
10. l	20. r

4.

A	B
1. *Ditylenchus myceliophagus*	a. Sticky lateral branches
2. *Monacrosporium drechsleri*	b. Sticky nets
3. *Monacrosporium eudermatum*	c. Sticky knobs
4. *Arthrobotrys robusta*	d. Non-constricting rings
5. *Dactylaria candida*	e. *Agaricus*
6. *Dactylaria brochophaga*	f. Constricting rings
7. *Xiphinema index*	g. Tomato black ring virus
8. *Longidorus elongates*	h. Strawberry latent ring spot virus
9. *Paralongidorus maximus*	i. Tobacoo rattle virus
10. *Trichodorus viruliferus*	j. Arabis mosaic virus
11. *Glomus fasciculatum*	k. Black palm weevil
12. *Rhadinaphelenchus cocophilus*	l. Citrus root-knot nematode
13. *Steinernema carpocapsae*	m. Endomycorrhizal fungus
14. *Meloidogyne exigua*	n. Cereal root-knot nematode
15. *Meloidogyne naasi*	o. Entomopathogenic nematode
16. *Meloidogyne indca*	p. False root-knot nematode
17. *Meloidodera spp.*	q. Cotton root-knot nematode
18. *Nacobbus spp.*	r. Cystoid nematode
19. *Meloidogyne incognita acrita*	s. Coffee root-knot nematode
20. *Meloidogyne brevicauda*	t. Tea root-knot nematode

ANSWERS 4

1. e	11. m
2. c	12. k
3. a	13. o
4. b	14. s
5. d	15. n
6. f	16. l
7. j	17. r
8. g	18. p
9. h	19. q
10.i	20. t

5.

A	B
1. Ripe ufra	a. *Anguina tritici*
2. Toppling-over	b. *Anguina tritici +Clavibacter tritici*
3. Ear-cockle	c. *Aphelenchoides besseyi*
4. Ear Rot	d. *Radopholus similis*
5. White tip of leaf	e. *Ditylenchus angustus*
6. Onion bloat	f. *Meloidogyne hapla*
7. Excessive root branching	g. *Ditylenchus dipsaci*
8. Coarse root	h. *Xiphinema species*
9. Stubby root	i. *Trichodorus christei*
10. Curly root tip	j. *Belonolaimus longicaudatus*
11. Rotting of fleshy parts	k. *Rhadinaphelenchus cocophilus*
12. Necrosis of host tissue	l. *Scutellonema bradys*
13. Leaf galls	m. *Pratylenchus brachyurus*
14. Leaf lesions	n. *Meloidogyne graminicola*
15. Root lesions	o. *Anguina millefoli*
16. Cysts on roots	p. *Tylenchulus semipenetrans*
17. Egg-mass within root tissue	q. *Aphelenchoides ritzemabosi*
18. Egg-mass outside root tissue	r. *Heterodera glycines*
19. Die-back	s. *Meloidogyne javanica*
20. Dead/devitalized buds	t. *Aphelenchoides fragariae*

ANSWERS 5

1. e	11. l
2. d	12. k
3. a	13. o
4. b	14. q
5. c	15. m
6. g	16. r
7. f	17. n
8. j	18. s
9. i	19. p
10. h	20. t

6.

A	B
1. Monodelphic	a. *Heterodera*
2. Monodelphic, ophistodelphic	b. *Xiphinema*
3. Sexual dimorphism	c. *Radopholus*
4. Didelphic	d. *Pratylenchus*
5. Outstretched ovaries	e. *Tylenchulus*
6. Intersex	f. *Meloidogyne*
7. Sex reversal	g. *Ditylenchus*
8. Sphincter muscle	h. *Nacobbus*
9. Cystoid nematode	i. *Paralongidorus*
10. False root-knot nematode	j. *Xiphinema coxi*
11. Funnel shaped amphids	k. *Meloidodera*
12. Auto fluorescent guiding sheath	l. Cloaca
13. Tooth-like stylet	m. *Dactylella candida*
14. Predaceous nematode	n. *Globodera*
15. Constricting rings	o. *Trichodorus*
16. Syncytia	p. *Mononchus*
17. Anchor shaped stylet	q. *Paratrichodorus*
18. Asymmetrical head in males	r. *Criconematids*
19. Found in 100 cm soil depth	s. *Rotylenchus*
20. Most primitive form of plant parasite	t. *Criconema*

ANSWERS 6

1. d	11. i
2. b	12. j
3. e	13. o
4. a	14. p
5. c	15. m
6. g	16. n
7. f	17. r
8. l	18. t
9. k	19. s
10. h	20. q

7.

A	B
1. Immersion oil in 100x objective	a. permanent mount
2. Dichloro-triethly-ether	b. Histopathological slide mount
3. Lactophenol	c. anaeathetizes nematode
4. Glycerine	d. prevents the bending of light rays
5. Glyceel	e. aluminum slide hole diameter
6. Meyer's alanine	f. semi-permanent mount
7. Vulval cone top	g. cover slip sealing
8. 10 mm	h. cutting en-face view of nematodes
9. 50-110 micron	i. cyst nematode mount
10. Gelatin+phenol jelly	j. Inactivating temperature
11. Lactophenol-Acid fuchsin	k. temperature for root invasion by sugar beet cyst nematode
12. 15-30^0 C	l. thickness of cover slip
13. 25-30^0 C	m. found in galled tissues
14. 10-20^0 C	n. staining plant tissue
15. 30-40^0 C	o. favorable temperature for nematodes
16. Keto acids	p. temperature for activity of potato cyst nematode
17. Taylor	q. reported races in M.arenaria
18. Seinhorst	r. Egg hatching factor
19. Sodium hypochlorite	s. reported races in M.incognita
20. Taylor and Sasser	t. reported races in Ditylenchus dipsaci

ANSWERS 7

1. d	11. n
2. c	12. o
3. f	13. k
4. a	14. p
5. g	15. j
6. b	16. m
7. i	17. s
8. e	18. t
9. l	19. r
10. h	20.q

8.

A	B
1. Leaf galls	a. *Tylenchulus semipenetrans*
2. Grass galls	b. *Nacobbus srendipictus*
3. Non-gall forming	c. *Ditylenchus radidicola*
4. Stubby and malformed roots	d. *Rhadinaphelenchus cocophilus*
5. Galls on rose	e. *Aphelenchoides fragariae*
6. Galls on wheat	f. *Trichodorus christei*
7. Galls on lemon	g. *Anguina millefoli*
8. Longer time host attachment	h. *Helicotylenchus multicinctus*
9. False galls on tomato	i. *Hemicycliophora arenaria*
10. Elliptical necrotic lesions on roots	j. *Xiphienma diversicaudatum*
11. Root rots	k. *Anguina agrostis*
12. Insect transmitted nematode	l. *Tagetes*
13. Celery	m. *Tylenchorhynchus claytoni*
14. Allyl isithiocyanate	n. *Tylenchorhynchus vulgaris*
15. Tobacco	o. *Sctellonema bradys*
16. 'Kalahasti' melody in ground nut	p. *Fusarium wilt + Belonolaimus*
17. Chickpea	q. Mustard
18. Alpha terthienyl	r. *Tylenchorhynchus brevilineatus*
19. 'Fish hook' roots	s. *Tylenchorhynchus brassicae*
20. Cauliflower	t. *Xiphinema on onion*

ANSWERS 8

1. g	11. o
2. k	12. d
3. e	13. p
4. f	14. q
5. j	15. m
6. c	16. r
7. i	17. n
8. a	18. l
9. b	19. t
10. h	20. s

9.

A	B
1. Root-knot nematode and Fusarium wilt complex in cotton	a. Melendez and Powell
2. Sting nematode and Fusarium wilt complex in cotton	b. Pitcher and Crosse
3. Root-knot nematode and Fusarium wilt complex in tobacco	c. Hewitt, Raski and Goheen
4. Root-knot nematode and bacterial wilt complex in tomato	d. Atkinson
5. *Aphelenchoides ritzemabosi* and *Clavibacter tritici complex*	e. Linford *et.al.,*
6. Nematodes as virus vectors	f. Holdeman and Graham
7. Organic matter in soil reduces nematode population	g. Hunger
8. Carbon di sulphide as soil fumigant	h. Carter
9. Chloropicrin as nematicide	i. Halogenated hydrocarbon
10. Dichloropropane dichloropropene as nematicide	j. Mathews
11. Ethylene dibromide as nematicide	k. EDB
12. Methyle bromide	l. Julius Kuhn
13. Chemical with outstanding solubility in water	m. DBCP
14. Chemical which does not hydrolyze readily, persist longer	n. Christie
15. Dow fume	o. Fensulphothion
16. Compound with high vapour pressure	p. Methyl bromide
17. Nemaphos	q. DD
18. Dasanit	r. Aldicarb
19. Nemacur	s. Phenamiphos
20. Temik	t. Thionazin

ANSWERS 9

1. d	11. n
2. f	12. i
3. a	13. p
4. g	14. q
5. b	15. k
6. c	16. m
7. e	17. t
8. l	18. o
9. j	19. s
10. h	20. r

10.

A	B
1. Thimet	a. Metham sodium
2. Furadon	b. Entomopathogenic nematode
3. Vapam	c. To control mosquito larva
4. Royal -300	d. Resistant to *Globodera rostochiensis*
5. DD 135	e. Phorate
6. *Rhomanomermis cultivorax*	f. Resistant to *Heterodera avenae*
7. Kufri Swarna potato variety	g. *Arthtobotrytis robusta*
8. Mermithid	h. Carbofuran
9. Rajkiran barley variety	i. Resistant to dagger nematode
10. Hisar Lalit tomato variety	j. Nematode affecting mosquito
11. Dogridge grape root stock	k. Arabis mosaic
12. Trifoliate orange root stock	l. Resistant to root-knot nematode
13. NETU virus	m. *Trichodrus christei*
14. Strawberry latent ring spot virus	n. Resistant to *Tylenchulus semipenetrans*
15. Mulberry ring spot virus	o. *Xiphinema*
16. Tomato black ring virus	p. *Tobacco ring spot*
17. NEPO virus	q. Longidorus martini
18. Tobacco rattle virus	r. Pea early browning
19. Nematode that remains for longer time at one feeding site	s. *Paralongidorus maximus*
20. Symptoms of shock and recovery phases	t. *Longidorus elongatus*

ANSWERS 10

1. e	11. i
2. h	12. n
3. a	13. r
4. g	14. s
5. b	15. q
6. j	16. t1
7. d	17. k
8. c	18. m
9. f	19. o
10. l	20. p

11.

A	B
1. Root-knots	a. Citrus
2. Root-knot infested tissues	b. Grape
3. Number of races in *M. incognita*	c. Tamil Nadu
4. Number of races in *M. arenaria*	d. Hypertrophy & hyperplasia
5. *Meloidogyne indica*	e. West Bengal
6. *Meloidogyne incognita*	f. Tea
7. *Meloidogyne arenaria on ground nut*	g. Tryptophan
8. *Rotylenchulus renoformis on cotton*	h. 04
9. *Aphelenchoides besseyi on tube rose*	i. Andhra Pradesh
10. *Meloidogyne brevicauda*	j. Hairy and witches' broom symptoms
11. *Anguina tritici*	k. 02
12. *Nacobbus spp.*	l. Sugar beet
13. *Heterodera delvi*	m. Cowpea
14. *Non-leguminous host of Heterodera cajani*	n. Tomato and brinjal
15. *Heterodera schachtii*	o. Seed borne nematode
16. *Heterodera vignae*	p. Finger millet (Ragi)
17. *Pratylenchus indicus*	q. Sesamum
18. *Pratylenchus penetrans*	r. Mentek disease in rice
19. *Hirschmaniella oryzae*	s. Sugar cane
20. *Hoplolaimus indicus*	t. Fruits in nursery stage

ANSWERS 11

1. d	11. o
2. g	12. j
3. h	13. p
4. k	14. q
5. a	15. l
6. b	16. m
7. i	17. n
8. c	18. t
9. e	19. r
10. f	20. s

12.

A	B
1. Rye grass toxicity	a. Aphelenchoides besseyi
2. Stem galls on orchids	b. Aphelenchoides fragariae
3. Leaf galls on Cynodon transvalensis	c. Rhadinaphelenchus cocophilus
4. Firefly blast symptoms on rice	d. Anguina picridis
5. Mushroom	e. Ditylenchus dipsaci
6. Cauliflower disease	f. Anguina tumefaciens
7. Central and South America	g. Anguina agrostis
8. Bulbous ornamentals	h. Aphelenchoides composticola
9. Lesion nematode on Maize	i. Pratylenchus indicus
10. Lesion nematode on Coffee	j. Hirschmaniella oryzae
11. Lesion nematode on Rice	k. Hoplolaimus galeatus
12. Lesion nematode on nursery fruit trees	l. Pratylenchus zeae
13. Interveinal necrosis in rice	m. Helicotylenchus multicinctus
14. Miti-miti disease of Colocasia esculenta	n. Tylenchorhynchus vulgaris
15.Cotton	o. Pratylenchus coffeae
16. Banana	p. Pratylenchus penetrans
17. Tobacco	q. Hirschmaniella miticausa
18. Pearl millet	r. Mesocriconema xenoplax
19. Peach Tree Short Life disease in USA	s. Criconema
20. Gradual die back of walnut	t. Tylenchorhynchus claytoni

ANSWERS 12

1. g	11. i
2. d	12. p
3. f	13. j
4. a	14. q
5. h	15. k
6. b	16. m
7. c	17. t
8. e	18. n
9. l	19. r
10. o	20. s

13.

A	B
1. Esterase and Malate dehydrogenase	a. *Heterodera*
2. Alkaline phosphatase	b. *Radopholus*
3. Superoxide dismutase	c. *Rotylenchulus*
4. Phosphate glucose	d. *Heterodera schachtii*
5. Hs1pro1	e. *Heterorhabditis*
6. Hi	f. *Globodera pallida*
7. Gpa2	g. Reduces the fecundity of *Meloidogyne*
8. Gro1-4	h. Slow growth of *G. tabaccum*
9. CpTi	i. *Meloidogyne incognita*
10. Cystine proteinase inhibitor	j. Slow growth of female of *Heterodera schachtii*
11. Serine protease	k. *Globodera rostochiensis*
12. *Rhg1*	l. *Helicotylenchus multicinctus*
13. *Cre loci*	m. Abnormal sexual development nematodes
14. *CCII*	n. *Heterodera glycines*
15. ACHE-I-7.1	o. *Heterodera avenae*
16.LEV-I-7.1	p. Disrupts chemosensory function of invading J2s
17.nAChRbp	q. Reduced reproduction of nematodes coupled with disruption of chemosensory function of invading J2s
18.*CCII + nAChRbp*	r. Disorientation of invading J2s
19. *CeCPI + PjCHI-1*	s. Reduced chitin content and retardation in embryogenesis in the nematode egg
20. *CpTI + Oc-IΔD86*	t. Chemodisruption of J2s

ANSWERS 13

1. c	11. j
2. e	12. n
3. a	13. o
4. b	14. l
5. d	15. r
6. i	16. t
7. f	17. p
8. k	18. q
9. g	19. s
10. h	20. m

14.

A	B
1. Rmc1	a. *Heterodera avenae*
2. H1	b. Pepper
3. Cre1	c. *Meloidogyne chitwoodi*
4. Me3	d. Potato
5. Rmc1	e. *G. rostochiensis*
6. Ma	f. Soybean
7. Rhg1 and Rhg4	g. *Heterodera sacchari*
8. Hero A	h. Plum
9. Hsa-1^{Og}	i. *Globodera pallida*
10. Mi-1	j. Wheat
11. Cre3	k. 40-90 eggs/female
12. Fecundity of root-knot nematode	l. Tomato
13. Fecundity of citrus nematode	m. 14-20 eggs/female
14. Fecundity of lesion nematode	n. 300-900 eggs/female
15. Fecundity of lance nematode	o. 10-32 eggs/female
16. Fecundity of reniform nematode	p. 30-40 eggs // female
17. Fecundity of burrowing nematode	q. Amphimictic
18. Fecundity of foliar nematode	r. 50-75 eggs/female
19. Rice root nematode	s. Parthenogenetic
20. Stem and bulb nematode	t. 20-40 eggs/female

ANSWERS 14

1. d	11. j
2. e	12. n
3. a	13. k
4. b	14. p
5. c	15. m
6. h	16. r
7. f	17. o
8. i	18. t
9. g	19. q
10. l	20. s

15.

A	B
1. Oxyuroidea	a. Diplogasterida
2. Cylindrocorporidae	b. Pratylenchidae
3. Hoplolaimina	c. Hemicycliophoroidea
4. Nacobbinae	d. Rhabditida
5. Criconematina	e. Dolichodoroidea
6. Seinuridae	f. Dorylaimoidea
7. Xiphinematinae	g. Aphelenchoidea
8. Longidoridae	h. Aphelenchoidoidea
9. Aphelenchinae	i. Longidoridae
10. Enoplia	j. Diphtherophorina
11. Belondiridae	k. Cephalobina
12. Trichodoridae	l. Dorylaimina
13. Chromadoridae	m. Adenophorea
14. Panagrolaomidae	n. Spiruria
15. Strongylidae	o. Tylenchoidea
16. Ascarididae	p. Holplolaimina
17. Tylodoridae	q. Chromadorida
18. Anguinidae	r. Strongylida
19. Rotylenchulidae	s. Tylenchuloidea
20. Paratylenchidae	t. Anguinoidea

ANSWERS 15

1. d	11. l
2. a	12. j
3. e	13. q
4. b	14. k
5. c	15. r
6. h	16. n
7. i	17. o
8. f	18. t
9. g	19. p
10. m	20. s

16.

A	B
1. Oostenbrink elutriator	a. Plant tissues
2. Baunacke method	b. Active nematodes
3. Blender centrifugal flotation method	c. Soil
4. Kort extraction apparatus	d. Bamboo splinter
5. Baermann's funnel	e. Cysts dry extraction
6. Extraction fluid	f. Formalin (4-5%)
7. Nematode anaesthetizer	g. Glass wool
8. Nematode picking	h. $MgSO_4$
9. Fixative	i. Cysts wet extraction
10. Nematode permanent mount	j. Sodium-azide
11. To avoid crushing of nematodes	k. Lacto phenol-cotton blue
12. Sealing the mount	l. Glycerine-gelatine
13. Perineal pattern	m. Glycerine
14. Mounting nematode head	n. Lugol's solution
15. Mounting cysts	o. Glyceel
16. Root staining	**p. Cason's fluid**
17. Utra thin sections	q. Perspex slides
18. Staining microtome sections	r. Glycerine jelly
19. Gelatin powder	s. Glycerine alcohol
20. Clearing solution	t. Microtome

ANSWERS 16

1. c	11. g
2. e	12. o
3. a	13. k
4. i	14. l
5. b	15. q
6. h	16. n
7. j	17. t
8. d	18. p
9. f	19. r
10. m	20. s

17.

A	B
1. Rearing of *Turbatrix aceti*	a. Malt (Difco)
2. Rearing of *Panagrellus redivivus*	b. Body length ÷ greatest width
3. Rearing of *Rhabditis* spp	c. 50% (v/v) acetic acid
4. L	d. Body length ÷ body width at the anus
5. a	e. Oatmeal
6. b	f. Overall length of the posterior ovary from vulva x 100 ÷ body length
7. b'	g. Total body length in microns or millimetres
8. c	h. NA
9. c'	i. Body length ÷ oesophagus length from the lips to the end of the oesophageal glands
10. G1	j. Body length ÷ oesophagus length
11. G2	k. Body length ÷ tail length
12. V	l. Overall length of the anterior ovary from vulva x 100 ÷ body length
13. 60 mesh	m. 38 micro meter
14. 100 mesh	n. PDA
15. 200 mesh	o. 250 micro meter
16. 400 mesh	p. Distance of the vulva from the lips x 100 ÷ body leng
17. White trap	q. 150 micro meter
18. Rearing of mycophagus nematodes	r. Carrot disc
19. Rearing of bacteriovorus nematodes	s. Entomopathogenic nematodes
20. Culturing of burrowing nematode	t. 75 micro meter

ANSWERS 17

1. c	11. f
2. e	12. p
3. a	13. o
4. g	14. q
5. b	15. t
6. j	16. m
7. l	17. s
8. k	18. n
9. d	19. h
10. l	20. r

18.

A	B
1. R	a. Annuli between labial disc and first annule after excretory pore
2. R st	b. Annuli between vulva and anus
3. R oes	c. Annuli between tail terminus and vulva
4. R ex	d. *Meloidogyne*
5. R V	e. Total number of body annuli
6. R an	f. Annuli between labial disc and base of stylet
7. R Van	g. Annuli between labial disc and oesophago-intestinal valve
8. Amphidelphic	h. *Hemicriconemoides*
9. Tail with mucro	i. Annuli from anus to tail terminus
10. Dorsal overlapping	j. *Hirschmaniella*
11. Monodelphic	k. *Helicotylenchus*
12. Female with additional cuticular sheath	l. Aphelenchids
13. Spicules arcuate	m. *Heterodera*
14. Spicules arcuate or rose-thorn shaped	n. *Radopholus similis*
15. Rounded cysts without vulval cone	o. *Tylenchulus*
16. Lemon shaped cysts with vulval cone	p. *Pratylenchus*
17. Kidney shaped female	q. *Hemicycliophora*
18. Long necked /bottle shaped female	r. Tylenchids
19. Tail cylindrical or conoid	s. *Rotylenchulus*
20. Stylet with anchor shaped knobs	t. *Globodera*

ANSWERS 18

1. e	11. d
2. f	12. h
3. g	13. r
4. a	14. l
5. c	15. t
6. i	16. m
7. b	17. s
8. k	18. o
9. j	19. p
10. n	20. q

19.

A	B
1. *Heterodera avenae*	a. Perineal tubercles large and clustered
2. *Heterodera cajani*	b. Egg mass absent /devoid of eggs
3. *Globodera rostochiensis*	c. Bifenestrate
4. *Globodera pallida*	d. Lateral lines
5. *Heterodera zeae*	e. Ambifenestrate
6. *Meloidogyne javanica*	f. Dorsal arch low with smooth striae
7. *Meloidogyne incognita*	g. Perineal tubercles discrete
8. *Meloidogyne graminicola*	h. Head cap low, slopping posteriorly
9. *Meloidogyne hapla*	i. Cotton
10. *Meloidogyne arenaria*	j. Dorsal arch high
11. Deltapine	k. Tobacco
12. Pusa Komal	l. Temperate
13. Pusa 23	m. Pepper
14. HS 101	n. Groundnut
15. Pusa Bold	o. Cowpea
16. Florunner	p. Pearl millet
17. CH-1	q. Paddy
18. Early California Wonder	r. Tomato
19. NC 95	s. Mustard
20. IR-64	t. Castor

ANSWERS 19

1. c	11. i
2. e	12. o
3. a	13. p
4. g	14. r
5. b	15. s
6. d	16. n
7. j	17. t
8. f	18. m
9. l	19. k
10. h	20. q

20.

A	B
1. *Aphelenchoides besseyi*	a. 55^0 C for 20 minutes
2. *A. ritzemabosi*	b. 43^0 C for 240 minutes
3. *Radopholus similis*	c. 46.7^0 C for 10 minutes
4. *Pratylenchus penetrans*	d. 52^0 C for 10 minutes
5. *Ditylenchus dipsaci*	e. 54^0 C for 10 minutes
6. *Tylenchulus semipenetrans*	f. 47.8^0 C for 15 minutes
7. *Anguina tritici*	g. 51^0 C for 7.5 minutes
8. *Meloidogyne spp.in ginger*	h. Cowpea
9. *Ditylenchus myceliophagus*	i. Asparagus
10. Trap crop	j. *Carthanus tinctorius*
11. Cucurbitacin	k. 55^{0}C for 10 minutes
12. Aspargic acid	l. Irradiation (48,000 to 96, 000 r)
13. Isomeric polyacetalene	m. Mustard
14. Thionemone	n. Bitter cucumber
15. Antagonistic crop	o. Galactose is the surface receptor
16. Meloidogyne javanica	p. Mannose is the surface receptor
17. Meloidogyne incognita	q. Salicylic acid is the surface receptor
18. Globodera pallida	r. Glucose is the surface receptor
19. Globodera rostochiensis	s.. Margosa
20. Xiphinema index	t.. N-acetylglucosamine is the surface receptor

ANSWERS 20

1. d	11. n
2. f	12. i
3. a	13. j
4. g	14. s
5. b	15. m
6. c	16. q
7. e	17. r
8. k	18. p
9. l	19. t
10. h	20. o

21.

A	B
1. Tear gas	a. Phenamiphos
2. Nemagon	b. Phorate
3. Nemaphos	c. Methomyl
4. Nemacur	d. DD
5. Mocap	e. Thionazin
6. Thimet	f. Chloropicrin
7. Temik	g. Cadusaphos
8. Furadon	h. DBCP
9. Vydate	i. Triazaphos
10. Lannate	j. Ethoprop
11. Telone	k. Dichlorofenthion
12. Dow fume	l. Carbosulfan
13. Basamid	m. Carbofuran
14. Rugby	n. Oxamyl
15. Hostothion	o. Demeton
16. VC-13	p. Aldicarb
17. Systox	q. Dazomet
18. Marshall	r. Terbuphos
19. Counter	s. Phosalone
20. Zolone	t. EDB

ANSWERS 21

1. f	11. d
2. h	12. t
3. e	13. q
4. a	14. g
5. j	15. i
6. b	16. k
7. p	17. o
8. m	18. l
9. n	19. r
10. c	20. s

22.

A	B
1. Rajendra Annapurna	a. Musk Melon
2. Pusa Parvathi	b. Chilli
3. C-152	c. Sesamum
4. Phule G-5	d. Potato
5. Karimunda	e. Soybean
6. TMV-10	f. Rice
7. Arka anamika	g. Cotton
8. TNAU-119	h. Cowpea
9. Pusa jwala	i. Brinjal
10. Kufri Swarna	j. Black pepper
11. KBSH-31	k. Pigeon pea
12. Shilajeet	l. Rajmah
13. Gold star	m. Tomato
14. Surat Dwarf	n. Chickpea
15. NC 79	o. Sunflower
16. Karnataka Hybrid	p. Sweet potato
17. Nemagold	q. Okra
18. Pioneer 511 A	r. Tobacco
19. TKM-9	s. Groundnut
20. Pant A-10	t. Maize

ANSWERS 22

1. i	11. o
2. l	12. e
3. h	13. a
4. n	14. g
5. j	15. r
6. s	16. m
7. q	17. p
8. c	18. t
9. b	19. f
10. d	20. k

23.

A	B
1. *Purpureocillium lilacinum*	a. Non-constricting rings
2. *Pasteuria penetrans*	b. Trichodermin
3. *Arthrobotrys oligospora*	c. Rhizobacterium
4. *Catenaria*	d. Mycorrhizal fungus
5. *Trichoderma viride*	e. Adhesive knobs
6. *Stylopage*	f. Swallowing type of predacious nematodes
7. ***Pseudomonas fluorescens***	g. Zoosporic
8. *Glomus fasciculatum*	h. Egg-parasitic
9. *Streptomyces avermitilus*	i. Suction capped predacious nematode
10. *Dactylella doedycoides*	j. Three dimensional adhesive nets
11. *Bacillus thuringiensis*	k. Constricting rings
12. *Tripyla*	l. Sporozoa
13. *Dorylaimid*	m. Parasitic protozoa
14. *Dactylaria candida*	n. Obligate parasite
15. *Diplogaster*	**o. Non-modified adhesive hyphae**
16. *Cauliculosporium penetrans*	p. Stylet armed predacious nematode
17. *Mononchus macrostoma*	q. Avermectin
18. *Dactylella brachophaga*	r. Delta endotoxin
19. *Hypsibius myrops*	s. Turbellarian
20. *Adenoplea*	t. Tardigrade

ANSWERS 23

1. h	11. r
2. n	12. f
3. j	13. p
4. g	14. a
5. b	15. i
6. o	16. m
7. c	17. l
8. d	18. k
9. q	19. t
10. e	20. s

24.

A	B
1. *Anguina tritici*	a. Life cycle is 45-48 days
2. *Ditylenchus dipsaci*	b. *Arthrobotrys robusta*
3. *Pratylenchus spp.*	c. *Pochonia chlamydosporia*
4. *Hirschmaniella oryzae*	d. Life cycle is 21-27 days
5. *Aphelenchoides besseyi*	e. Tomato resistant variety
6. *Rotylenchulus reniformis*	f. Life cycle is 5 months
7. *Tylenchulus semipenetrans*	g. Brinjal resistant variety
8. *Pratylenchus coffeae*	h. Banana resistant variety
9. *Radopholus similis*	i. Life cycle is 30 days
10. Royal 300 or Antipolis	j. Pigeon pea resistant variety
11. Nematofagin-BL	k. Life cycle is one year
12. Melo Con WG	l. *Arthrobotrys oligospora*
13. IIHR-Biovert	**m. Cauliflower resistant variety**
14. Trichorich	n. Life cycle is 54-65 days
15. NTDR-1	o. Tobacco resistant variety
16. Pusa Purple Long	p. Life cycle is 35-50 days
17. Pusa Snow Ball	q. *Trichoderma harzianum*
18. Yelakki Bale	r. Life cycle is 17-28 days
19. ICPL-95	s. *Purpureocillium lilacinum*
20. Anand 145	t. Life cycle is 6-8 weeks

ANSWERS 24

1. f	11. l
2. k	12. s
3. n	13. c
4. i	14. q
5. p	15. e
6. r	16. g
7. t	17. m
8. a	18. h
9. d	19. j
10. b	20. o

25.

A	B
1. *hgctl*, AF498244	a. *Globodera rostochiensis*
2. *gp ^c p-i*	b. *Meloidogyne javanica*
3. *pMiDuoxl*, DQ082753	c. Isopropyl amid phosphate
4. *Gr-eng-l*, AF004523	d. Oxime carbamate
5. *MjTIS-11*	e. Fast, far-reaching
6. *Pv010*	f. Monothiophosphate
7. Phenamiphos	g. *Heterodera glycines*
8. Fensulphothion	h. *Meloidogyne incognita*
9. Phorate	i. Leaf is nematicidal
10. Oxamyl	j. Globodera pallida
11. Methyl bromide	k. Organic phosphate
12. Methylisothiocyanate	l. *Pratylenchus vulnus*
13. Amaranth	m. Entire plant is nematicidal
14. Marigold	n. Seed meal is nematicidal
15. Mustard	o. Fruit peal is nematicidal
16. Hemp	p. Root is nematicidal
17. Cotton	q. Seed is nematicidal
18. Sour orange	r. Shoot is nematicidal
19. Safflower	s. Optimal movement
20. Soybean	t. Flower is nematicidal

ANSWERS 25

1. g	11. e
2. j	12. s
3. h	13. i
4. a	14. p
5. b	15. q
6. l	16. m
7. c	17. n
8. f	18. o
9. k	19. t
10. d	20. r

26.

A	B
1. Duer larvae	a. Root-knot nematode
2. Giant cells	b. Good host for burrowing nematode
3. Dormant adult female	c. **Cauliflower disease complex**
4. Finger millet	d. Stem & bulb nematode
5. Arecanut	e. 'Kalahasty 'melody
6. Strawberry	f. Seed gall nematode
7 Ground nut	g. Reniform nematode
8. 'Tundu'	h. Non-host for root-knot nematode
9. Castor	i. Cereal cyst nematode
10. Rice	j. Rice root nematode
11. Grapes	k. Seed gall nematode
12. Ectoparasite	l. Rice root-knot nematode
13. Endoparasite	m. Citrus nematode
14. Semi-endoparasite	n. Lesion nematode
15. Barley	o. Pine wood nematode
16. Egg-mass inside the gall	p. Dagger nematode
17. Foliar nematode	q. Needle nematode
18. Coffee	r. Burrowing nematode
19. Pine	s. Coconut red ring nematode
20. Virus vector	t. Stubby root nematode

ANSWERS 26

1. d	11. p
2. a	12. t
3. i	13. r
4. h	14. m
5. b	15. k
6. c	16. l
7. e	17. s
8. f	18. n
9. g	19. o
10. n	20. q

27.

A	B
1. Sodium hypochlorite acid fuchsin technique	a. Fixation step
2. New Blue R	b. Ribbon mounting step
3. TBA (tertiary butyl alcohol)	c. *Meloidogyne incognita*
4. Paraffin	d. Nematode staining in plant tissues
5. FAA	e. *Meloidogyne hapla*
6. Haupt's adhesive	f. Dead nematode staining
7. High arch, zig-zag patterns	g. *Meloidogyne arenaria*
8. Lateral lines present	h. Dehydration step
9. Rounded to flattened arch, forking striae	i. Best time for sampling in corn
10. Striae smooth or wavy	j. *Meloidogyne javanica*
11. Sodium hypochlorite	k. Best time for sampling in tomato
12. Fenwick can technique	l. Embedding step
13. The Oosternbrink elutriation method	m. Best time for sampling in potato
14. Mist chamber technique	n. 200 cc
15. Waring blender method	o. 5 g
16. August-October	p. Extraction of cysts
17. June-September	q. Extraction of cysts and juveniles of cyst nematodes
18. May-August	r. Extraction of nematodes from soil
19. Soil sample quantity.	s. Extraction of nematode both from soil and roots
20. Root sample quantity	t. Extraction of nematodes from plant parts

ANSWERS 27

1. d	11. p
2. f	12. q
3. h	13. r
4. l	14. s
5. a	15. t
f. b	16. i
7. c	17. k
8. j	18. m
9. g	19. n
10. e	20. o

28.

A	B
1. *Bradyrhizobium japonicum*	a. Excretory pore at the level of esophagus-intestinal junction valve
2. *Rhizobium trifoli*	b. Excretory pore at level or posterior to spear knobs
3. *Radopholus*	c. Excretory pore near base of isthamus
4. *Pratylenchus*	d. Excretory pore above the level of intestinal junction valve
5. *Meloidogyne*	e. Transmits Pea early browning virus
6. *Tylenchorhynchus*	f. Excretory poreposteriorly in a protruberance anterior to the vulva
7 *Hoplolaimus*	g. Excretory pore posterior to median bulb
8. *Tylenchulus*	h. Controls burrowing nematode on banana
9. *Rotylenchulus*	i. Controls root-knot nematode on bean
10. *Rhizobium leguminosarum*	j. Excretory pore opposite to esophageal intestinal junction
11. *Rhizobium phaseoli*	k. Controls reniform nematode on cotton
12. *Glomus fasciculatum*	l. Controls root-knot nematode on clover
13. *Glomus etunicatum*	m. Controls root-knot nematode on blackgram
14. *Glomus intraradices*	n. *Meloidogyne incognita* on mungbean
15. *Gigaspora marginata*	o. *Heterodera glycines* in soybean
16. *Glomus mossae*	p. *Meloidogyne javanica* on white clover
17. *Xiphinema basiri*	q. *Meloidogyne hapla* on hairy vetch
18. *Longidorus attenuatus*	r. Transmits Cowpea mosaic virus
19. *Trichodorus cyclindricus*	s. Transmits Lettuce ring spot virus
20. *Paratrichodorus anemones*	t. Transmits Tobacco rattle virus

ANSWERS 28

1. o	11. n
2. p	12. h
3. a	13. i
4. j	14. k
5. b	15. l
6. c	16. m
7. d	17. r
8. f	18. s
9. g	19. t
10. q	20. e

29.

A	B
1. *Meloidogyne incognita*	a. Interacts with *Phytophthora parasitica* var. *nicotianae* on tobacco
2. *Aphelenchoides fragariae*	b. Interacts with *Fusarium oxysporum* f.spp.*vasinfectum* on cotton
3. *Helicotylenchus dihystera*	c. Interacts with *Corynebacterium insidiosum* on lucerne
4. *Meloidogyne javanica*	d. Interacts with *Agrobacterium tumefacians* on raspberry
5. *Criconemella xenoplex*	e. Interacts with *Agrobacterium tumefacians* on peach
6. *Meloidogyne spp.*	f. Interacts with *Pseudomonas syringe* on plum
7. *Meloidogyne hapla*	Interacts with *Ralstonia solanaceraum* on brinjal
8. *Ditylenchus dipsaci*	h. Interacts with *Corynebacterium fasciens* on strawberry
9. *Rotylenchulus reniformis*	i. Interacts with *Pseudomonas caryophylli* on carnation
10. *Meloidogyne incognita acrita*	j. Interacts with *Pseudomonas marginata* on gladiolus
11. *Radopholus similis*	k. Interacts with *Rhizoctonia solani* on wheat
12. *Ditylenchus destructor*	l. Interacts with *Fusarium oxysporum* on soybean
13. *Ditylenchus dipsaci*	m. Interacts with *Cylindrocarpon radicicola* on tulip
14. *Pratylenchus penetrans*	n. Interacts with *Dilophospora alopercuri* on wheat
15. *Heterodera avenae*	o. Interacts with *Phytophthora infestans* on potato
16. *Anguina tritici*	p. Interacts with *Fusarium oxysporum* f.spp. *tracheiphylum* on *cowpea*
17. *Meloidogyne* spp.	q. Interacts with *Botrytis alli* on onion
18. *Heterodera glycines*	r. Interacts with *Fusarium pisi* on pea
19. *Meloidogyne* spp.	s. Interacts with *Fusarium oxysporum* f.spp. *cubensei* on banana
20. *Meloidogyne* spp.	t. Interacts with *Fusarium oxysporum* f.spp. *dianthi* on carnation

ANSWERS 29

1. g	11. s
2. h	12. o
3. i	13. q
4. j	14. r
5. f	15. k
6. e	16. n
7. d	17. t
8. c	18. l
9. b	19. m
10. a	20. p

30.

A	B
1. *Meloidogyne incognita* ETL	a. 1-5/500 g soil
2. *Heterodera schachtii* ETL	b. 50-1300/100 cc soil
3. *Globodera rostochiensis* ETL	c. 20-240/100 cc soil
4. *Ditylenchus dipsaci* ETL	d. More than 200 nematodes/200 cc soil
5. *Pratylenchus penetrans* ETL	e. 5-10 cysts/100 cc soil
6. *Hemicycliophora* spp. ETL	f. More than 500/500 g soil
7. *Belonolaimus longicaudatus* ETL	g. Methyl isothiocyanate
8. *Trichodorus* spp. ETL	h. 4-10 eggs/g soil
9. *Heterodera carotae* ETL	i. 25/500 cc soil
10. Fumigants	j. Irreversible binding of acetylcholinesterase
11. Methyl bromide	k. 31 cysts/100 g soil
12. Organophosphates	l. Refractive, with nerve fibres
13. Organocarbamates	m. Located in posterior part of body
14. Avermectins	n. Encircles isthmus
15. Hemizonid	o. Encircles spicules
16. Deirids	p. Alkyl halides
17. Phasmids	q. Located near tail
18. Caudalids	r. Inhibits neurons
19. Nerve ring	s. Reversible binding of acetylcholinesterase
20. Bursa	t. Small protuberances in lateral field

ANSWERS 30

1. d	11. p
2. h	12. j
3. k	13. s
4. a	14. r
5. f	15. l
6. b	16. t
7. i	17. m
8. c	18. q
9. e	19. n
10. g	20. o

31.

A	B
1. *Dolichodorus*	a. Pratylenchidae
2. *Ditylenchus*	b.Heteroderidae
3. *Tylenchorhynchus*	c. Hoplolaimidae
4. *Hirschmaniella*	d. Criconematidae
5. *Scutellonema*	e.Anguinidae
6. *Globodera*	f. Longidoridae
7. *Rhadinaphelenchus*	g. Paratylenchidae
8. *Hemicriconemoides*	h. Dolichodoridae
9. *Tylenchulus*	i. Nacobbinae
10. *Xiphinema*	j. Dorylaimida
11. *Hemicycliophora*	k. Belonolaimidae
12. *Cacopaurus*	l. Tylenchida
13. *Bursaphelenchus*	m. Secernentea
14. Subfamily with single ovary	n. Tylenchulidae
15. Subfamily with hard tough female cyst	o. Aphelenchoididae
16.Orrder with odontostylet	p. Adenophorea
17.Order with three part oesophagous	q. Heteroderinae
18.Class with amphids in labial position	r. Rotylenchinae
19.Class with presence of caudal glands	s. Aphelenchidae
20.Sub family of Rotylenchus	t. Hemicycliophorinae

ANSWERS 31

1. h	11. t
2. e	12. g
3. k	13. s
4. a	14. i
5. c	15. q
6. b	16. j
7. o	17. l
8. d	18. m
9. n	19. p
10. f	20. r

32.

A	B
1. Resembling genus of *Tylenchorhynchus*	a. *Hirschmanella*
2. Resembling genus of *Radopholus*	b. *Criconemoides*
3. Resembling genus of *Hoplolaimus*	c. *Heterodera*
4. Resembling genus of *Helicotylenchus*	d. *Dolichodorus*
5. Resembling genus of *Pratylenchus*	e. *Tylenchulus*
6. Resembling genus of *Criconemoides*	f. Dolichodoridae
7. Resembling genus of *Hemicycliophora*	g. *Tylenchus*
8. Resembling genus of *Belonolaimus*	h. *Longidorus*
9. Resembling genus of *Meloidogyne*	i. Anguinidae
10. Resembling genus of *Heterodera*	j. Heteroderidae
11. Resembling genus of *Rotylenchulus*	k. Pratylenchidae
12. Resembling genus of *Xiphinema*	l. Belonolaimidae
13. Family with high, set-off, round and trapezoid labia	m. *Rotylenchus*
14. Family with 4, 6 or more lateral fields	n. Tylenchidae
15. Family with long, well developed stylet	o. *Criconema*
16. Family with three part oesophagous, ventral overlapping	p. *Helicotylenchus*
17. Family with conoid to pointed tail	q. Hoplolaimidae
18. Family with males monarchic, leptoderan bursa	r. *Scutellonema*
19. Family with slender and robust body	s. *Globodera*
20. Genus with very big phasmid	t. *Radopholus*

ANSWERS 32

1. g	11. e
2. a	12. h
3. p	13. q
4. m	14. i
5. t	15. f
6. o	16. k
7. b	17. j
8. d	18. n
9. c	19. l
10. s	20. r

33.

A	B
1. *CpTISpTI-1*	a. Potato
2. *PIN2*	b. Rice
3. *Oc-IΔD86*	c. Maize
4. *CeCP1*	d. Promoter
5. *CCII*	e. Arabidopsis
6. GNA	f. Tomato
7. *Cry6A*	g. Transcription factor
8. nAChRbp	h. Cowpea
9. *Rhg4*	i. *M. javanica effector*
10. *Mi*-1.2	j. *Colocasia esculenta*
11. TUB-1	k. *M. incognita* calreticulin
12. *Pdf2.1*	l. Sugar beet
13. *At1g64110)*	m. Avr protein
14. *WRKY33*	n. Venom allergen-like protein
15. *Mj-NULG1a*	o. *Galanthus nivalis*
16. *Mi-CRT*	p. Heterodera avenae
17. *Hs4F01*	q. Soybean
18. Ha-annexin	r. Brinjal
19. *map-1.2*	s. Plantain
20. *Gr-Vap1*	t. *Bacillus thuringiensis*

ANSWERS 33

1. h	11. r
2. a	12. d
3. b	13. e
4. j	14. g
5. c	15. i
6. o	16. k
7. t	17. l
8. s	18. p
9. q	19. m
10. f	20. n

34.

A	B
1. Polyacetylene	a. Marigold
2. Alpha chaconine	b. Soybean
3. Alpha terthienyl	c. *Gossypium hirsutum*
4. Pyrocatechol	d. Juvenile hormone
5. Nimbidin	e. Lima bean
6. Glyceollin	f. Vertebrate steroid hormone
7. Terpenoid aldehydes	g. *Eragrostis curvula*
8. Phaseolin	h. Sterol inhibitor
9. Coumestrol	**i. Safflower**
10. Phenyl Phenalenones	j. Nematode stimulus
11. Medicarpin	**k. Sex specific mating pheromone**
12. Gossypol	l. Neem
13. Ecdysteroids	m. Protein synthesis inhibitor
14. Farnesyl methyl ether	n. Macrocyclic lactone
15. Progresterone	o. *Phaseolus vulgaris*
16. Azasteroid	p. Cotton
17. Carbon dioxide	q. Polyhydroxylated derivatives of cholesterol
18. Abamectin	r. Alfalfa
19. Actidione	s. Potato sprouts
20. Ascaroside	t. *Musa* spp.,

ANSWERS 34

1. i	11. r
2. s	12. c
3. a	13. q
4. g	14. d
5. l	15. f
6. b	16. h
7. p	17. j
8. o	18. n
9. e	19. m
10. t	20. k

35.

A	B
1. Synoptic Model	a. Model involves repeated measurement of the disease progress
2. Simulation Model	b. E(Y)=Ymax{1-(1-m)Pi/(c+Pi)}
3. Multiple point Model	c. Breaks cutin
4. Critical point Model	**d. Degrade highly esterified pectins**
5. Brown Model	**e. Catalyzes demethylesterification of cell wall**
6. Oostenbrink Model	f. E (Y)=a+b.Pi
7. Seinhorst Model	g. Breaks lignin
8. Elston Model	h. Catalyze reactions that break alpha 1,4-glycosideic bonds
9. MELSIM	i. Breaks starch
10. Cutinase	j. Reduces chorismate
11. Pectin methyl esterase	k. Lytic enzyme
12.Polygalactouronase	l. Model with classical mutivanate statistical version of the systems analysis
13. Pectinase	m. E (Y)=a+b(log Pi)
14.Pectin lyase	n. Defense enzyme
15. Ligninase	o. Breaks lipids
16.Lipase	p. Model that predicts losses based on knowledge of damage function
17.Amylase	q. Meloidogyne-grapevine system
18. Chorismate mutase	r. $y=m+(1-m)Z^{Pi-T}$
19. Chitinase	s. Breaks pectic substances
20.Peroxidase	t. Model deals with relative rates of growth of pest and plant populations

ANSWERS 35

1. l	11. e
2. t	12. h
3. a	13. s
4. p	14. e
5. f	15. g
6. m	16. o
7. r	17. i
8. b	18. j
9. q	19. k
10. c	20. n

36.

A	B
1. *Ditylenchus destructor* on potato releases	a. Amylase
2. *Ditylenchus dipsaci* on onion releases	b. 3-bonded carbon nanopartcile
3. *Globodera rostochiensis* on potato releases	c. Alkaline Phosphatase
4. *Meloidogyne javanica* on cuttings & tubers releases	**d. Also called as carbon nanoflake**
5. *Radopholus similis* on banana releases	e. Cellulase
6. *Tylenchulus semipenetrans* on citrus releases	f. New class of biodegradable macromolecule
7. Nanodiamond	g. Hollow 1-D nanosystem
8. Fullerene	h. Produces highly reactive hydroxyl radicles
9. Onion-like carbon	i. Polygalacturonase pectinase
10. Carbon nanofibre	j. 10-50 nm
11. Carbon nanotube	k. Also called 'Halloysite'
12. Carbon nanowall	l. Most studied and utilized nanoparticle for biosystem
13. Dendrimer	**m. Helps in the target-specific delivery of proteins**
14. Chitosan	n. Zero dimensional carbon nanomaterial
15. Copper nanoparticle	o. Biodegradable and non-toxic nanoparticle
16. Quantun dots	p. Endo-polygalacturonase Alkaline Phosphatase
17. Nanofabrication	q. Filamentous carbon nanostructure
18. Clay nanotube	r. Polyhedral nanoparticle
19. Nanosilver	**s. Creates artificial plant parts**
20. Silica nanoparticle	t. Acid Phosphatase

ANSWERS 36

1. i	11. g
2. p	12. d
3. a	13. f
4. t	14. o
5. e	15. h
6. c	16. j
7. b	17. s
8. n	18. k
9. r	19. l
10. q	20. m

37.

A	B
1. Apargine- Mannitol agar	a. Histopathological studies
2. Oostenbrink technique	b. Cyst extraction
3. 60 mesh sieve	c. Temporary nematode mounts
4. Maize- meal culture	d. Only for endoparasitic nematodes
5. Carrot disc culture	e. Active nematodes extraction
6. Formalin acetic acid	f. Nematode killing agent
7. Mistifier technique	g. Cover slip sealer
8. Hanging blood drop method	h. Large sized nematodes extraction
9. Baerman's funnel technique	i. Nematode stain
10. Fenwick can method	j. Aphelenchs culturing
10. Bromophenol	k. Only for soil nematodes
12. Haem gum stain	l. *Ditylenchus myceliophagus*
13. Nigon's medium	m. Entomopathogenic nematodes
14. Paraffin embedding	n. Fixative
15. Malt agar	o. Nematode extraction from large number of samples
16. Fleming's solution	p. Root stain
17. Zut	q. Bacteriovorus nematodes
18. Waring Blender technique	r. *Radopholus similis*
19. Iodine	s. Foliage stain
20. Aluminium slides	t. Rhabditids culturing

ANSWERS 37

1. q	11. p
2. k	12. s
3. h	13. t
4. j	14. a
5. r	15. l
6. n	16. i
7. o	17. g
8. m	18. d
9. e	19. f
10. b	20. c

38.

A	B
1. Callus tissue technique	**a. Nematodes appear living after fixation**
2. Excised root tissue technique	b. Differentiates living and dead nematodes
3. Alexander Y. Ryss's technique	**c. Shrinkage of the specimen after fixation**
4. TAF (Triethanolamine fixative)	d. To stain gonads, nerve ring and oesphageal glands
5. Formalin-glacial fixative	e. To stain nematode lipids
6. Ditlevsen's fixative	f. To stain nematode chromosomes
7. Acetic Orcein	g. To stain nematode nucleic acids
8. Gold chloride	h. To stain faint surface nematode strucutres
9. Silver nitrate	i. *Ditylenchus dipsaci*
10. New blue R	**j.No drying of nematodes after fixation**
11. Phloxin-B	k. To stain nematode proteins
12. Picrocarmine	l. To stain nematode neurosecretions
13. PAS (Periodic acid-Schiff) technique	**m. Nematode killing and fixing using two Eppendorf** tubes
14. Sudan black B	n. To stain esterase in nematodes
15. Gallocyanin-chromatin technique	o. To stain egg-masses
16. Mercuric-bromophenol blue technique	p. To stain reproductory system in nematodes
17. Aldehyde-fuchsin technique	q. To stain nematode carbohydrates
18. Potassium ferricyanide +Potassium ferrocyanide	r. *Heterodera schactii* on sugar beet
19. Ammoniacal silver nitrate	s. To stain acid phosphatase in nematodes
20. Dimeethyl formamide with naphthol AS-BI phosphate	t. To stain peripheral nervous system of nematodes

ANSWERS 38

1. i	11. o
2. r	12. f
3. m	13. q
4. a	14. e
5. j	15. g
6. c	16. k
7. p	17. l
8. d	18. n
9. t	19. h
10. b	20. s

39.

A	B
1. Anhydrous glycerine	a. To stain fresh roots
2. Glycerine jelly	b. To retain green colour of the plant material
3. Euparal, Isobutanol, glycerine jelly	c. To measure nematodes
4. Lactophenol /FAA technique	d. Does not need heating and destaining
5. Copper sulphate and sulphur dioxide	e. Hardening of water rich tissues by freezing
6. Ethylene glycerol and liquid nitrogen	f. Tissue dehydration
7. McBryde technique	g. Ribbon mounting
8. Sodium hypochlorite-acetic acid technique	h. To measure image brightness
9. Bleaching solution	i. For permanent mounts
10. Camera lucida	**j. Sequential fixative for electron microscopic observation**
11. Cryosection	k. To enhance nematode electrical conductivity
12. Formalin-propiono-propanol (FPP)	l. To preserve nematode infested plant tissues
13. Tertiary-butyl-alcohol (TBA)	**m. Infiltration of embedding resin in electron microscopy**
14. Haupt's adhesive	n. To preserve juveniles of root-knot nematdes
15. Johansen's Quadruple stain	o. To mount the vulval cones of cyst nematodes
16. Photoelectric cell technique	p. To prepare living specimen
17. Sodium cacodylate buffer	q. To stain nematodes in forest litter
18. 1,2-epoxypropane	r. Fixative used in histopathological studies
19. Metal coating	s. To mount end-on view of nematode head/tail
20. Osmium tetroxide	t. Stain used in histopathological studies

ANSWERS 39

1. i	11. e
2. s	12. r
3. o	13. f
4. l	14. g
5. b	15. t
6. n	16. h
7. d	17. j
8. a	18. m
9. q	19. k
10. c	20. p

40.

A	B
1. Esterase	a. To detect mRNA
2. MDH	b. To estimate total phenols in nematode infested tissue
3 Phosphate glucose isomerase	c. To differentiate and to identify Steinernema carpocapsae
4. Alkaline phosphatase	d. To estimate total sugars in nematode infested tissue
5. Nitrocellulose	e. To extract cyst DNA
6. Southern blotting technique	f. To estimate total insoluble polysaccharides in nematode infested tissue
7. Northern blotting technique	g. To estimate L-phenylalanine ammonia lyase in nematode infested tissue
8. Western blotting technique	h. To estimate total nitrogen in nematode infested tissue
9. Phenol/chloroform technique	i. To differentiate and to identify Radopholus
10. Folin-Ciocalteau reagent	j. To estimate total amino acids in nematode infested tissue
11. Sodium potassium tartrate	k. To estimate total potash in nematode infested tissue
12. Schiff's reagent	l. To differentiate and to identify Meloidogyne incognita
13. Ninhydrin solution	m. To estimate polyphenol oxidase in nematode infested tissue
14. Arsenomolybdate colour reagent	n. To estimate total proteins in nematode infested tissue
15. Anthrone reagent	o. Used in Blotting technique
16. L-phenylalanine solution	p. To estimate reducing sugars in nematode infested tissue
17. Phenolphthalein indicator	q. To detect DNA
18. Modified Kjeldahl method	r. To estimate phosphorus in nematode infested tissue
19. Cobaltinitrite method	s. To differentiate and to identify Bursaphelenchus
20. Pyrogallol	t. To detect protein

ANSWERS 40

1. l	11. n
2. s	12. f
3. l	13. j
4. c	14. p
5. o	15. d
6. q	16. g
7. a	17. r
8. t	18. h
9. e	19. k
10. b	20. m

41.

A	B
1. King's B medium	a. To locate carbohydrates in *Anguina tritici*
2. Tryptic soy agar	b. To examine vulva area of mature cysts
3. Kennight's agar	c. Used in aerial infrared photography technique
4. Soil sprinkling technique	d. Used in mass culturing of *Ditylenchus myceliophagus*
5. White traps	e. Used in mass culturing of *Ditylenchus dipsaci*
6. LOTEK technique	f. Used in mass production of ***Pseudomonas fluorescens***
7.Alginate gel	g. Used to monitor fumigant residues in soil
8.Fluorescein / rhodamine isothiocyanate	h. To measure nematode respiration
9.Totomount water agar technique	i. To isolate entomopathogenic nematodes
10. Elchochrome infrared film 8443	j. Used in axenic culturing of *Aphelenchoides* spp.
11. Density gradient centrifugation	k. Used in long term storage of viable *Ditylenchus dipsaci*
12. Wheat grain spawn	l. To isolate nematode trapping fungi
13. Citrus leaf callus tissue technique	m. Used to culture *Radopholus similis*
14.White's standard nutrient solution	n. Used in mass production of endophytic bacteria
15. Bacto Soytone oligidic medium	o. Used in scanning electron microscopy
16.Aqueous solution of DMSO- 7.5% v/v	p. For rearing migratory endoparasitic nematodes
17.Coconut mesocarp	q. *in vivo* culturing of entomopathogenic nematodes
18. Gas sensors & Di-Trapex	r. To purify the eggs of *Meloidogyne* spp.
19. Critical point drying technique	s. For delivery of entomopathogenic nematodes
20.Biological oxygen monitor	t. Used in preparation of avermectins

ANSWERS 41

1. f	11. r
2. n	12. d
3. t	13. p
4. l	14. e
5. i	15. j
6. q	16. k
7. s	17. m
8. a	18. g
9. b	19. o
10. c	20. h

42.

A	B
1. *Heterodera avenae*	a. Excretory pore near stylet knobs
2. *Heterodera graminis*	b. Bullae branched
3. *Heterodera cajani*	c. Internal colour of female is golden
4. *Heterodera sorghi*	d. Dorsal arch rounded
5. *Meloidogyne graminicola*	e. Cyst ambifenestrate
6. *M.incognita*	**f. Tail long and filiform**
7. *M. arenaria*	g. Oesophageal glands enclosed in basal bulb
8. *M.hapla*	h. Underbridge absent
9. *M.javanica*	i. Juveniles with scales or spines on cuticle
10. *Globodera rostochiensis*	j. Stylet with backwardly sloping knobs
11. *G.pallida*	k. Dorsal arch low with smooth striae
12. *Heterodera oryzicola*	l. Tail cylindrical or conoid
13. ***H.filipzevi***	m. Body slim and spiral
14. *Tylenchus*	n. Bullae elongated
15. *Pratylenchus*	o. Tail with mucro
16. *Ditylenchus*	p. Internal colour of female is white or cream
17. *Criconema*	q. Vulval knobs present
18. *Hemicycliophora*	**r. Dorsal arch usually flattened**
19. *Helicotylenchus*	s. Cyst bifenestrate
20. *Hirschmaniella*	t. Head cap low, sloping posteriorly

ANSWERS 42

1. h	11. p
2. q	12. e
3. b	13. s
4. n	14. f
5. k	15. l
6. a	16. g
7. t	17. i
8. r	18. j
9. d	19. m
10. c	20. o

43.

A	B
1. Gnotobiotic	a. One known species
2. Axenic	b. Pea early browning
3. Synxenic	c. Brome grass mosaic virus
4. Monoxenic	d. "Yellows" in black pepper
5. Dixenic	e. Tobacco rattle virus
6. Xenic	f. 'Spreading decline' of citrus
7. *Xiphinema coxi*	g. Unknown species
8. *Xiphinema diversicaudatum*	h. 'Slow decline" of citrus
9. *Longidorus macrosoma*	i. "Summer crimp" of chrysanthemum
10. *Trichodorus viruliferus*	j. "Peach tree short life" disease
11. *Trichodorus allius*	k. Known species only
12. *Radopholus similis*	l. 'Onion bloat"
13. *Pratylenchus coffeae*	m. "Crinkle leaf disorder" of coffee
14. *Radopholus citrophilus*	n. "Slump" disease of citrus
15. *Tylenchulus semipenetrans*	o. Cherry leaf roll virus
16. *Aphelenchoides besseyi*	p. One or more known species
17. *Macroposthonia xenoplax*	q. "Dry rot" of potato
18. *Hemicriconemoides spp.*	r. Two known species
19. *Ditylenchus dipsaci*	s. Raspberry ring spot virus
20. *Ditylenchus destructor*	t. No other species

ANSWERS 43

1. k	11. e
2. t	12. d
3. p	13. n
4. a	14. f
5. r	15. h
6. g	16. i
7. o	17. j
8. c	18. m
9. s	19. l
10. b	20. q

44.

A	B
1. *Radopholus similis + Fusarium oxysporum f.sp. cubense*	**a. Cauliflower**
2. *Meloidogyne incognita + Fusarium oxysporum f.sp. lycopersici*	b. Carnation
3. *Meloidogyne incognita + Fusarium oxysporum f.sp. pisi*	c. Tomato
4. *Meloidogyne javanica + Fusarium oxysporum f.sp. phaseoli*	d. Strawberry
5. *Meloidogyne incognita + Fusarium oxysporum f.sp. cucumerinum*	e. Chilli
6. *Meloidogyne incognita + Fusarium oxysporum f.sp.conglutinans*	f. French bean
7. *Meloidogyne incognita + Fusarium oxysporum f.sp. solani*	g. Bean
8. *Meloidogyne incognita + Fusarium oxysporum f.sp. dianthi*	h. Gerbera
9. *Meloidogyne incognita + Fusarium oxysporum f.sp. coffeae*	i. Pea
10. *Pratylenchus penetrans +Verticillium dahliae*	j. Ginger
11. *Meloidogyne incognita + Verticillium alboatrum*	k. Cucumber
12. *Meloidogyne incognita + Rhizoctonia solani*	l. Areca nut
13. *Meloidogyne incognita + Pythium aphanidermatum*	m.Okra
14. *Tylenchorhynchus brassicae +Rhizoctonia solani*	n. Banana
15. *Meloidogyne incognita + Macrophomina phaseolina*	o. Barley
16. *Meloidogyne arenaria + Phytophthora cryptoge*	p. Crossandra
17. *Radopholus similis + Cylindrocarpon obtussisporum*	q. Brinjal
18. *Meloidogyne incognita + Rhizoctonia solani*	r. Coffee
19. *Anguina tritici +Corynebacterium tritici*	**s. Cauliflower**
20. *Rotylenchulus reniformis +Rhizoctonia solani*	t. Tomato

ANSWERS 44

1. n	11. c
2. t	12. q
3. i	13. e
4. g	14. s
5. k	15. f
6. a	16. h
7. p	17. l
8. b	18. j
9. r	19. o
10. d	20. m

45.

A	B
1. *Rotylenchulus reniformis+Agrobacterium tumefaciens*	a. Brinjal
2. *Meloidogyne incognita +Agrobacterium radiobacter var. tumefaciens*	b. Garlic
3. *Aphelenchoides fragariae +Rhodococcus (Corynebacterium) faciens*	c. Gladiolus
4. *Meloidogyne incognita +Clavibacter michiganense*	d. Rhubarb
5. *Meloidogyne incognita + Ralstonia solanacearum*	e. Peach
6. *Globodera pallida + Ralstonia solanacearum*	f. Pea
7. ***Meloidogyne incognita +Corynrbacterium flaccumfaciens***	g. Rose
8. ***Ditylenchus dipsaci + Pseudomonas fluorescens***	h. Capsicum
9. *Meloidogyne incognita + Pseudomonas marginata*	i. Grapevine
10. *Meloidogyne incognita +Pseudomonas caryophilli*	j. Ginger
11. *Pratylenchus vulnus + Agrobacterium rhozogenes*	k. Chrysanthemum
12. *Ditylenchus destructor + Clavibacter michiganense sub sp. insidiosum*	l. Cardamom
13. *Meloidogyne incognita + Ralstonia solanaceraum*	m. Potato
14. *Meloidogyne javanica + Ralstonia solanaceraum*	n. Dioscorea
15. *Hoplolaimus uniformis+ Fusarium oxysporum f.sp. pisi*	o. Strawberry
16. *Pratylenchus coffeae + Pythium aphanidermatum*	p. Coleus
17. *Meloidogyne incognita +Fusarium oxysporum f.sp. dioscoria*	q. Beans
18. *Meloidogyne incognita + Macrophomina phaseolina*	r. Mint
19. *Meloidogyne incognita + Rhizoctoina solani*	s. Carnation
20. *Pratylenchus penetrans +Verticillium alboatrum*	t. Tomato

ANSWERS 45

1. i	11. g
2. e	12. d
3. o	13. j
4. t	14. h
5. a	15. f
6. m	16. k
7. q	17. n
8. b	18. p
9. c	19. l
10. s	20. r

46.

A	B
1. Dorsal oesophageal gland opening in procorpus	a. Mononchida
2. Male with rays in delicate caudal alae	b. Dolichodoridae
3. Movable hollow spear	c. Tylenchulidae
4. Amphids always post-labial	d. Diplogastrida
5. *Xiphinema*	e. *Tylenchus*
6. Stoma narrow, cardia small, without gland cells	f. Chromadorea
7. Lateral field and striations distinct, caudal alae distinct	g. Hoplolaimidae
8. Stoma wide, cardia with with 3 gland cells	h. Criconematidae
9. Dorsal oesophageal gland opening in metacorpus	i. Tylenchina
10. Male with prominent rays in lobed caudal alae	j. Aphelenchoididae
11. Strong, cuticularized stoma, predators	k. Rhabditida
12. *Belonolaimus*	l. Adenophorea
13. Individuals connect at tail tip in water to form ball of nematodes	m. Longidoridae
14. *Globodera*	n. Dorylaimida
15. *Hemicycliophora*	o. Pratylenchidae
16. Amphids apertures on lips	p. Dorylaimina
17. *Seinura*	q. Aphelenchina
18. *Helicotylenchus*	r. Nygolaimina
19. *Paratylenchus*	s. Strongylida
20. *Nacobbus*	t. Heteroderidae

ANSWERS 46

1. i	11. a
2. k	12. b
3. n	13. d
4. l	14. t
5. m	15. h
6. p	16. f
7. e	17. j
8. r	18.
9. q	19. c
10. s	20. o

47.

A	B
1. n	a. Distance from base of stylet to median bulb valve/stylet length.
2. V	b. Stylet length/body length
3. T	c. Body length / tail length
4. P	d. Number of body annules between vulva and anus
5. G1	e. % length of male gonad relative to body length
6. a	f. % distance of phasmid from anus in relation to length of tail
7. b	g. % length of anterior female gonad in relation to body length
8. c	h. % length of posterior female gonad in relation to body length
9. o	i. Body length / greatest body diameter
10. $P_p\|\|$	j. Conus of stomatostyle/total stomatosyle length
11. SE	k. % distance of anterior phasmid from anterior of nematode in relation to body length
12. P^a	l. Total number of body annules
13. SV/s	m. % distance of posterior phasmid from anterior of nematode in relation to body length
14. m	n. Distance from anterior end to excretory pore in *μm*
15. R	o. Number of specimens on which measurements are based
16. RV	p. Stylet or spear length
17. RVan	q. % distance of vulva from anterior
18. s/L	r. Number of body annules between vulva and tail tip
19. G_2	s. % distance of dorsal esophageal gland opening from stylet knobs
20. s	t. Body length / distance from anterior to esophago-intestinal valve

ANSWERS 47

1. o	11. n
2. q	12. k
3. e	13. a
4. f	14. j
5. g	15. l
6. i	16. r
7. t	17. d
8. c	18. b
9. s	19. h
10. m	20. p

48.

A	B
1. Triterpenoids	a. *Calotropis*
2. Pyrocatchol	b. Organophosphate
3. Pyrrolizidine	c. Sugarcane molasses
4. Glycoside	d. Geranium extract
5. Volatile fatty acids	e. Tomato resistant variety
6. Sucrose	f. Sewage sludge
7. Allyl isothiocyanate	g. Cowpea resistant variety
8. Thionemone	h. Pangola grass
9. Organic N	i. *Catharanthus roseus*
10. Non-legume green manure crop	j. Mustard oil
11. Non-legume green leaf manure crop	k. Onion seeds
12. Tear gas	l. *Gloriosa superuba*
13. Ethoprop	m. Cucumber
14. Methomyl	n. Organocarbamate
15. Citronellol	o. Organic amendments
16. Methionine	p. Halogenated hydrocarbon
17. Serpentine	q. Neem
18 . Colchicine	r. Sunhemp
19. Kievitone phytoalexin	s. Buck wheat
20. Orthodihydroxy phenolics	t. *Asparagus*

ANSWERS 48

1. m	11. a
2. h	12. p
3. r	13. b
4. t	14. n
5. o	15. d
6. c	16. k
7. j	17. i
8. q	18. l
9. f	19. g
10. s	20. e

49.

A	B
1. Banana var. Gold Finger	a. *Meloidogyne incognita*
2. Citrus var. Swingle Citrumello	b. *Pratylenchus penetrans*
3. Grapevine var.Dakshi	c. *Rotylenchulus reniformis*
4. Passion Fruit var. Kaveri	d. *Heterodera cajani*
5. Tomato var.Kalyanpur selection	e. *Rotylenchulus reniformis*
6. Brinjal var.Black Beauty	f. *Rotylenchulus reniformis*
7. Chilli var.Teja	g. *Meloidogyne javanica*
8. Capsicum var. California Wonder	h. *Meloidogyne incognita*
9. Potato var.Superior	i. *Radopholus similis*
10. Onion var.Evergreen	j. *Meloidogyne incognita*
11. Sweet potato var. Gold Rush	k. *Meloidogyne incognita*
12. Cowpea var.CO 6	l. *Rotylenchulus reniformis*
13. French bean var. Blue Lake	m. *Meloidogyne incognita*
14. Okra var. Parbhani Kranthi	n. *Rotylenchulus reniformis*
15. Musk melon var. S-445	o. *Tylenchulus semipenetrans*
16. *Coffeae robusta*	p. *Meloidogyne incognita*
17. Black pepper var. Pournami	q. *Rotylenchulus reniformis*
18. Tuberose var. Shringar	r. *Rotylenchulus reniformis*
19. Potato var. Kufri Dewa	s. *Meloidogyne incognita*
20. Potato var. Red Aboda	t. *Meloidogyne javanica*

ANSWERS 49

1. i	11. r
2. o	12. d
3. e	13. p
4. m	14. k
5. q	15. g
6. a	16. f
7. t	17. h
8. n	18. s
9. b	19. j
10. c	20. l

50.

A	B
1. Hypodermis	a. Unicellular, uninucleate cells found embedded in pharyngeal tissue.
2. Cephalic framework	b. Only in female nematodes
3. Amphids	c. The stoma is provided with armatures such as tooth, teeth, denticles, etc.
4. Deirids	d. In Dorylaimida, the intestine posteriorly connects with this organ
5. Phasmids	e. Below the cuticle
6. Stoma	f. Only in male nematodes
7. Oesophagus	g. Cylindrical or barrel-shaped or slightly wide feeding apparatus
8. Oesophageal glands	h. Consists of nematodes feeding on other soil micro-arthropods, nematodes etc.,
9. Cardia	i. Ring or basket-like cuticularized structure present around the stoma.
10. Intestine	j. A nonfunctional gonad which is also reduced in size
11. Rectum	k. Semipermeable
12. Anus	l. Located around the level of excretory pore.
13. Cloaca	m. The reduced, nonfunctional branch in reproductive system
14. Prerectum	n. Located posterior to anus or cloaca
15. Pseudo-prodelphic	o. Provided with sphincter
16. Omnivore-predators	p. The longest part in the digestive system
17. Bacterial-feeding	q. Paired chemoreceptor organs
18 . Mononchida	r. Tripartite in Tylenchida
19. Prevulval uterine sac	s. Esophago-intestinal junction
20. Cuticle	t. The anterior most part of the digestive tract

ANSWERS 50

1. e	11. o
2. i	12. b
3. q	13. f
4. l	14. d
5. n	15. j
6. t	16. c
7. r	17. g
8. a	18. h
9. s	19. m
10. p	20. k

51.

A	B
1. Resembling genus of *Tylenchorhynchus*	a. *Helicotylenchus*
2. Resembling genus of *Ditylenchus*	b. *Longidorus*
3. Resembling genus of *Belonolaimus*	c. *Heterodera*
4. Resembling genus of *Hoplolaimus*	d. *Subanguina*
5. Resembling genus of *Rotylenchulus*	e. *Scutellonema*
6. Resembling genus of *Pratylenchus*	f. *Criconema*
7. Resembling genus of *Radopholus*	g. *Diptenchus*
8. Resembling genus of *Meloidogyne*	h. *Trichodorus*
9. Resembling genus of *Heterodera*	i. *Neosteinernema*
10. Resembling genus of *Rotylenchus*	j. *Mermis nigrescens*
11. Resembling genus of *Criconemoides*	k. *Tylenchus*
12. Resembling genus of *Hemicriconemoides*	l. *Monochamus alternatus*
13 . Resembling genus of *Anguina*	m. *Radopholus*
14. Resembling genus of *Xiphinema*	n. *Hirschmaniella*
15. Resembling genus of *Longidorus*	o. *Romanomermis culicivorax*
16. Resembling genus of *Paratrichodorus*	p. *Dolichodorus*
17. Resembling genus of *Steinernema*	q. *Meloidodera*
18. Nematode infecting grass hopper	r. *Hemicycliophora*
19. Nematode infecting mosquito	s. *Tylenchulus*
20. Vector of *Bursaphelenchus xylophilus*	t. *Paralongidorus*

ANSWERS 51

1. k	11. f
2. g	12. r
3. p	13. d
4. a	14. b
5. s	15. t
6. m	16. h
7. n	17. i
8. c	18. j
9. q	19. o
10. e	20. l

TRUE OR FALSE

I. Mention whether the following statements are true or false.

1. Tylenchoidea and Mermithoidea have nematodes attacking insects
2. *Harpssporium anguilulae* is the most familiar endozoic fungus that parasitizes predaceous nematodes
3. *Xiphienma diversicaudatum* transmits viruses and viroids
4. Nematode egg size is not constant but varies according to the species
5. In general, small nematodes respire at a slower rate than large ones
6. Predacious nematodes are characterized by a small open stoma armed with teeth
7. Bursa is present in *Trichodorus*
8. *Meloidogyne* may be either monarchic or diorchic
9. *Seinura* is a predacious nematode with odantostyle
10. Dorylaims are present in less population in soil than *Mononchus*
11. **Turbellarians are minute flatworms, predominantly herbivores that feed on nematodes**
12. *Longidorus* transmits NEPO viruses only
13. Amphids are funnel shaped with narrow apertures in *Paralongidorus*
14. *Xiphinema indica* transmits Arabis mosaic virus
15. In some nematodes, number of moults may be six or seven
16. Female intersexuality is more common than male intersexuality
17. As per De Man's formula, "L" stands for tail length of nematode
18. Small amounts of fat and glycogen are stored in the hypodermis
19. In *Hemicycliophora*, spicules are extended inside the body at all times
20. Nematode eggs of different species do not divide in the same way to form the embryo
21. **Nematodes do have flame cells in their body**
22. Nematode borne viruses mostly are soil and pollen borne
23. No special respiratory organs are known in nematodes, hence they do not respire
24. Most Adenophorea are diorchic
25. Members of Criconematidae, generally , are fast-moving
26. The gubernaculum is present in all nematodes
27. Potato is a good host for *Meloidogyne naasi*
28. The majority of species of *Meloidogyne* reproduce by parthenogenesis
29. All nematode mt DNAs are circular, single-stranded DNA molecules
30. Dagger nematodes prefer vascular tissues while stubby root nematodes feed on epidermal
31. Viruses do not pass through the process ecdysis in nematodes
32. **Bacillary bands are the modified mesophyl gland cells opening** through a complex cuticular pore
33. Most of the food of phytonematodes is digested intracorporeally
34. Nematodes of the class Secernentea characteristically have a single testis with one exception
35. In female nematodes which are swollen to a spherical shape, the vulval location is always terminal
36. The canal type of excretory system is not common in tylenchids
37. Some dorylaimids are known to have two nerve rings
38. The shape of the tail tip in nematodes is of less taxonomic value

39. Stomatostylet is the characteristic feature of all orders
40. Most nematodes moult four times in their life cycle
41. 'Vas efferens', a specialized tubular part of the seminal vesicle, is present in some male nematodes
42. *Ditylenchus* has a conoid tail
43. Both hypodermal and caudal glands are present in the hypodermis
44. Second stage of *Anguina tritici* can never undergo dormancy
45. **The centrifugal flotation method is based on the differences in specific gravity**
46. **Non-vermiform female nematodes can be extracted by using fluidizing column technique**
47. Amphids may not be present in all groups of nematodes
48. Longitudinally nematode body is divided into two regions
49. **In Dorylaimida, first molting occurs inside the egg after hatching**
50. Mustard shaped cysts are formed by *Heterodera*

I. ANSWERS

1. T	11. F	21. F	31. T	41. T
2. F	12. T	22. F	32. F	42. T
3. F	13. F	23. F	33. F	43. T
4. T	14. F	24. T	34. T	44. F
5. F	15. F	25. F	35. T	45. T
6. F	16. T	26. F	36. F	46. T
7. F	17. F	27. F	37. T	47. F
8. T	18. F	28. T	38. F	48. F
9. F	19. F	29. F	39. F	49. F
10. F	20. T	30. T	40. T	50. F

II

1. *Bacillus radiobactor* is effective against *Globodera pallida*
2. *Galleria mellonella* and *Helicoverpa armigera* **are efficient hosts to mass culture entomopathogenic** nematodes
3. Secernentea has an advanced tubular type of excretory system.
4. Oxygen is a major factor in the hatching of eggs of phytonematodes
5. Chitin is synthesized by the egg and not by the female nematode reproductive system
6. Shaft is a part of the pharynx
7. *Radopholus similis* was brought under State Quarantine Act in Florida during 1963
8. The Baunacke method used for cyst extraction is also known as 'white bowl method'
9. The only hexose found in nematodes is ascarylose
10. **Auxin deficiency leads to smaller syncytial formation by cyst ne**matodes
11. Five pathotypes of *Heterodera avenae* was reported in by Mathur *et.al.*
12. *Xiphinema cobbi* was reported from USA after 1980
13. Bacterivorous nematodes are present in a dung heap
14. The common name of *Meloidodera* is cystoidal nematode
15. Lesion nematode is also termed as non-meadow nematode
16. The potato genes Gpa2 and Gro1-4 confer resistance to a narrow range of pathotypes of a single potato cyst nematode species
17. Fenestration is the arrangement and form of the translucent cuticular areas around the vulva in cyst nematodes
18. The gonads of *Globodera* and *Heterodera* can be clearly distinguished in iv juvenile stage
19. Delay in egg hatching may be due to the accumulation of oxygen
20. Ritzemaboss reported the occurrence of physiological races in phytonematodes in 1888
21. Lipids are the major energy reserve source in phytonematodes
22. Root-knot nematodes in general can survive without moisture for 6-36 months
23. Metabolically, nematodes are typical poikilotherms
24. *Aphelenchoides composticola* feeds on spores and mycelia of *Agaricus*.
25. Nerve ring is present around the anus
26. The chromosome number of *Meloidogyne incognita* is n= 32–38
27. High temperatures cause sex reversal in males of root-knot nematodes
28. Earlier name of the genus Hemi*criconemoides* was *Bakernema*
29. *Xiphinema italiae* also transmits grapevine fan leaf virus
30. Intestinal cells always constant in their number and shape
31. Reducing agents like glutathione and sodium dithionite can attract few species of nematodes
32. Haemoglobin is present in several animal parasitic nematodes
33. The nematode cuticle is always a three layered structure
34. Few nematode species may obtain carbon dioxide from plant roots also

35. Among several isozyme systems, carboxylesterase/esterase EST proved to be the most useful in discriminating *Radopholus* species
36. The muscle cells of nematode body are innervated by processes which pass from the muscle to the nerve and not, as in other animals, by nerves sending fibres to the muscle.
37. **'Z-organ' is of taxonomic significance in** *Xiphinema*
38. **Surface tension of water in soil has least influence on its availability to nematodes**
39. **The scientific name of spine nematode is** *Criconemoides*
40. *Hirschmaniella* has a dorsal overlapping
41. Parthenogensis is seen in *Radopholus*
42. Quiescence due to desiccation has been observed in *Aphelenchoides ritzemabosi*
43. Filipjev in 1946 gave the name *Anguina tritici* to wheat gall nematode
44. Nematodes are basically aquatic animals but have adopted terrestrial habitat
45. For extraction of cysts from soil, dried sample is preferred than wet.
46. Amphids play a major role in the orientation of juveniles to the chemical source
47. Plant-parasitic nematodes produce ethanol as a major end-product under aerobic conditions
48. In *Longidorus*, guiding ring is located at the tip of the tail
49. Water can move freely into and out of the nematode body through the cuticle.
50. Coconut red ring nematode is not mycophagous

II. ANSWERS

1. T	11. T	21. T	31. T	41. T
2. T	12. F	22. F	32. T	42. T
3. T	13. T	23. T	33. T	43. F
4. T	14. F	24. T	34. F	44. T
5. T	15. F	25. F	35. F	45. T
6. F	16. T	26. T	36. T	46. T
7. F	17. T	27. F	37. T	47. F
8. T	18. F	28. F	38. F	48. F
9. T	19. F	29. T	39. F	49. T
10. F	20. T	30. F	40. F	50. F

III

1. Pinimbin is responsible for trap formation in *Arthrobotrys*
2. *Meloidogyne arenaria* does not infect chickpea
3. Acid phosphatase is secreted by phytonematodes which is involved in parasitism
4. **Extraction efficiency diminishes with increasing sample size**
5. *Ditylenchus* comes under Anguilliidae
6. **Acetic acid neutralizes the shrinking effect caused by formalin during nematode fixing**
7. Cyst nematodes feed from a syncytium, which is composed of numerous cells after fusion of the endoplasts
8. High population density in nematodes does not affect the sex ration towards maleness
9. **The pseudocoelomic fluid does not contains hematin**
10. High salinity, generally repels nematodes
11. Common name of *Hemicriconemoides* is sheathoid nematode
12. The giant cells by root-knot nematodes are formed through repeated ectomitosis without cytokinesis
13. *Hemicycliophora* is a sedentary ectoparasite
14. **Zuckerman and Cobb were the first to determine the host finding mechanism in nematodes**
15. Amphids are actually the paired lateral cephalic sense organs
16. Nematodes grow continuously, except during molting
17. Urea is not excreted by many nematodes
18. All phytonematodes may not inhibit soil at some time
19. Stylet knobs need not be of different shapes in nematodes
20. Coiling and clumping are two behavioural responses by some species of nematodes to removal of water
21. Nematode eggs are not plasmolyzed in hypertonic solutions.
22. Males of citrus nematode develop outside the roots
23. *Hoplolaimus* has a tulip-shaped stylet knob
24. Chlorogenic acid contracts muscles in nematodes and is a mechanical transmitter
25. The root incubation technique is not a better option to extract endoparasites only
26. In the class Secernentea, excretory canal opens through the excretory pore
27. In some nematodes new cuticle may develop without shedding of the older external cuticle
28. Generally egg to egg life cycle duration in nematodes ranges between 5-7 weeks
29. The tomato genes Mi-1 and Hero A confer broad-spectrum resistance against several root knot nematode species
30. Lectins are a class of putative anti-nematode lipids
31. **Enzyme activity in nematodes is always influenced by their parasitic habits**
32. Males of Adenophorea are provided with well developed pre-anal supplementary organs
33. The sensory organs in nematodes are made up of cuticle and hypodermis
34. **Nurse cells are the modification of cortex-pericycle cells**

35. Fleming's reagent used to stain the roots mainly consists of chromic acid and jasmonic acid
35. Transcription factors are the proteins that activate expression of genes through interaction with promoters.
37. Nematode proteins are mainly assayed by 'electrophoresis'
38. "Nematode wool' is formed by all species of Ditylenchus except *Ditylenchus myceliophagus*
39. *Ditylenchus* is an amphidelphic nematode
40. *Heterodera carotae* and *H. cajani* have small egg sacs compared to their small cyst size
41. *Steinernema carpocapsae* can control *Plutella xylostella*
42. Adult stage in *Aphelenchoides besseyi* does not undergo dormancy
43. The dorsal gland has a long extension that terminates in an ampulla in which the secretory granules are collected
44. Flame cells are absent in all nematodes
45. In Japan, *Rhadinaphelenchus cocophilus* is of quarantine importance on palms
46. Seinhorst extraction apparatus cannot be used for cyst extraction
47. **Reniform nematode modifies pericycle cells into nurse cells**
48. RNA interference (RNAi) technology is novel and successful for managing nematodes.
49. Ampullae, the pouch-like sense organs are combinations of nerve endings and pseudo cells
50. Soil rich in organic matter will have high phytonematode population

III. ANSWERS

1. T	11. T	21. T	31. T	41. T
2. F	12. F	22. T	32. T	42. F
3. T	13. T	23. T	33. T	43. T
4. T	14. F	24. F	34. F	44. T
5. F	15. T	25. F	35. F	45. T
6. T	16. T	26. T	36. T	46. F
7. F	17. F	27. T	37. T	47. T
8. F	18. F	28. F	38. T	48. T
9. F	19. F	29. T	39. F	49. F
10. T	20. T	30. F	40. F	50. F

IV

1. Nematode RNA is assayed by PCR tests followed by gel electrophoresis
2. Phytonematodes generally ingest cytoplasm and cell sap including intact nucleus
3. Majority of didelphic tylenchids have prodelphic arrangement
4. The nerve ring is present around the cloaca
5. Chemicals should pass through intercellular barriers like cytoplasmic membrane to kill nematodes
6. Nematode infection, in general, leads to increase in the respiration rate of the host at later stages
7. cAMP acts as an attractant to *Rotylenchulus reniformis*
8. *Xiphinema* induces terminal galls on roots
9. Eggshell is the only part in nematodes that contain chitin
10. The coelomyarian muscle cell is a storage organ for glycogen
11. Among all types, marine nematodes are known for their fast development
12. Stylet knobs are 'sloping backwards' in *Hemicycliophora*
13. The mitochondrial DNA in nematodes have relatively shorter DNA sequence compared to chromosomal DNA
14. The jelly-like mass containing the eggs in *Heterodera* is secreted by the walls of the spermatheca
15. Excretory pore in nematodes is usually located in the anterior one third of the body
16. The word fenestra refers to the thin-walled area on the vulval cone or perineal area of mature cysts
17. Plant root exudates always increases the phytonematode population
18. Phasmid is a caudal sense organ in all nematodes
19. Most phytonematodes possess an asymmetrical tubular type of excretory system
20. *Meloidogyne* males may either be mono-or diorchic based on the period of ovary development
21. Nematodes eggs can withstand -4^{0}C
22. *Hemicycliophora* is a sedentary ectoparasite
23. Necrosis of host cells caused by migratory endoparasites is mainly due to the release of ammonia
24. **Nanoparticles that can be manipulated by the application of magnetic field are known as radar** nanoparticles
25. The majority of nematodes including tylenchids are telogenic
26. In *Meloidogyne* **sex chromosomes are present and the sex ratio may be influenced by environmental** factors
27. The Longidoridae and Trichodoridae are sedentary root ectoparasites
28. A PCR machine can work in the range of 4-99^0 C temperature.
29. Agarose gels have a higher resolving power than polyachrylamide gels
30. Dextrose at 1000 ppm can kill nematodes
31. The excretory pore in Tylenchida is generally located in the intestine region
32. Maltase is a commonly occurring enzyme in root-knot nematode body
33. Tylenchids always has two pairs of spicule
34. An oxygen supply of 5-21% is best suited to phytonematodes, in general

35. A buffer also adds colour to the sample
36. Sandy loam soils are generally less suitable for phytonematodes
37. **Mi-1 also confers resistance against specific isolates of the potato aphid,** *Macrosiphum euphorbiae* **and against the white fly, Bemisia tabaci**
38. Migratory nematodes usually feed on cortical cells
39. Very few pulse crops can be used as trap crops for nematode management
40. **Primer is a single-stranded synthetic oligonucleotide designed to hybridize to a known sequence of DNA as a starting point for DNA amplification.**
41. Hisar Lalit cultivar of tomato is resistant to root-knot nematodes
42. Most phytonematodes are monoecious
43. ***Belonolaimus*** **is a sedentary ectoparasite, while** ***Hemicycliophora*** **is a migratory ectoparasite**
44. Many taxa under Adenophorea possess excretory system
45. **Centrifugation floatation technique is more suited to extract** nematodes from soil and organic debris
46. Generally high oxygen concentration is injurious to nematodes
47. J_2 of *Meloidogyne javanica* are less aggressive on tobacco sometimes than *M. incognita*
48. Lips of plant feeding nematodes are immovable
49. The entire cuticle of *Globodera rostochiensis* is not covered in small subsurface punctations
50. Caudalids are usually absent in Hoplolaiminae

IV. ANSWERS

1. F	11. T	21. T	31. F	41. T
2. F	12. T	22. T	32. F	42. F
3. F	13. T	23. F	33. F	43. F
4. F	14. F	24. F	34. T	44. F
5. F	15. T	25. T	35. T	45. T
6. T	16. T	26. F	36. F	46. T
7. T	17. F	27. F	37. T	47. F
8. T	18. F	28. T	38. T	48. T
9. T	19. T	29. F	39. F	49. F
10. T	20. F	30. T	40. T	50. F

V.

1. In nematodes, the major nitrogenous waste product is ammonia
2. Polyachrylamide gels are highly effective for separating small fragments of DNA
3. **The excretory system in nematodes lacks flame cells and cilia**
4. 'Isocytous' refers to the equal height of all the intestinal cells
5. Amphids are normally of complex type in tylenchids and are present close to the oral aperture
6. Egg sac matrix does not contain glycoproteins
7. Spermatheca is always found in female nematodes
8. Hypodermis secretes and maintains the cuticle.
9. Circomyarian cells are rare, which are found among specialized muscles of vulva and spicule
10. Fine striae on cuticle are absent in *Tylenchus*
11. The area on the cuticle overlapping epidermal chords is termed **'lateral field'**
12. **The pseudocoelom is the principal extracellular fluid compartment, which acts as a primitive circulatory** system.
13. *Heterodera avenae* can survive at 65^{0}C
14. Lateral lines on the cuticle are taxonomically important, which range in number from 2-6
15. In *Rhadinaphelenchus* intestine overlaps oesophageal glands in a short lobe
16. In *Hemicycliophora* all the lips are somewhat circular in shape
17. Engineering nematode resistance in crop plants through host-derived RNAi is largely based on the selection of target gene.
18. Life cycle of pin nematode may take one year or more
19. **Juveniles of some nematode genera emerge from the egg before the first molt**
20. Normally males are lesser than the females or may be absent in many nematode species
21. In the family Trichodoridae, the stylet is solid type.
22. Phytonematodes contain 50-80 % lipid per dry weight that is stored in the hypodermis
23. Along with auxin and cytokinin, ethylene is also found in the galled root tissues caused by *Heterodera*
24. Intestine is the largest part of the alimentary canal of nematodes
25. In tylenchids, females have three and males have six rectal glands
26. Criconematidae lack phasmids
27. The sensory organs in Tylenchida are relatively poorly developed compared to Dorylaimida
28. Pesudocoelom is not responsible for the turgescent state of the nematode body which is responsible for the functioning of somatic musculature
29. Nematodes hardly move 0.1 or 0.2 meters in a year
30. **Most nematodes are killed by flooding or drying of soil**
31. *Ditylenchus* species induces 'spikkels' in narcissus
32. Phytonematodes normally possess 16 sensory structures in the cephalic region
33. Deirids are essential for terrestrial life as majority of nematodes possess it
34. Caudalids are usually present in the family Hoplolaiminae

35. Coarse textured soils are most favourable to *Ditylenchus dipsaci* than clayey soils
36. Single pair of deirids is present in Tylenchoidea
37. *Longidorus martini* transmits mulberry ring spot virus
38. Compared to plant pathogens, marine nematodes survive better in anaerobic state.
39. Glycogen is not the only major storage carbohydrate in nematodes
40. In phytonematodes, amino acids are major source of carbohydrate
41. *Pratylenchus coffeae* is a bisexual nematode with prometric reproduction
42. In *Radopholus*, oesophageal glands overlap the dorsal side of the intestine
43. *Bursaphelenchus xylophilus* needs insect vectors for transmission from diseased to healthy plants
44. Sex pheromones are generally secreted by adult females, but sometimes males also secrete.
45. A syncytium is characterized by small secondary vacuoles, dense cytoplasm, numerous organelles and small amoeboid nucleus
46. During hatching, the water content in nematode body may increase from 10-20% to 40-50%
47. *Meloidogyne hapla* is a native of California as reported by Raski
48. *Radopholus similis* can survive well under dry conditions
49. *Aphelenchoides avenae* is fungivorous
50. The growth curve in nematodes is not always sigmoidal in shape

V. ANSWERS

1. T	11. T	21. T	31. T	41. F
2. T	12. T	22. F	32. T	42. T
3. T	13. F	23. F	33. F	43. T
4. T	14. T	24. T	34. T	44. T
5. F	15. F	25. T	35. F	45. F
6. F	16. T	26. T	36. F	46. T
7. T	17. T	27. T	37. T	47. T
8. T	18. F	28. F	38. T	48. F
9. T	19. T	29. F	39. F	49. T
10. F	20. T	30. T	40. F	50. F

VI.

1. *Meloidogyne incognita* causes more damage to plants in saline soils
2. Telescopic movement is more common in nematodes than undulatory movement
3. Chitin, a polymer of N-acetylglucosamine is a major component of nematode eggshells
4. Deirids are present in Criconematidae
5. Functionally phasmids are chemoreceptors, but can also serve as excretory organs
6. High concentration of CO_2 encourages respiration in nematodes
7. *Aphelenchoides* is known as leaf and bud nematode
8. Neutral pH is favourable for hatching eggs of root-knot nematodes
9. Nematodes interact with other microorganisms only as synergistic or neutral types.
10. *Xenorhabdus bvienii* is not a symbiotic bacteria for *Steinernema kraussiae*
11. *Helicotylenchus* has a short stylet
12. *Dolichodera* is a genus of cyst nematodes
13. *Bacillus licheniformis* is not a potential bio agent against *Meloidogyne incognita*
14. Nematodes relay on their skin for respiration
15. In phytonematodes, sperm is tailess
16. ß-1,4-endoglucanases and pectate lyases are cell wall degrading proteins
17. The duration of development from the single cell stage to the juvenile varies in different nematode genera
18. Oxygen is an important factor for the hatching of eggs in phytonematodes
19. Plant-feeding nematodes that have long bodies have lower fecundity than short-bodied nematodes.
20. Both banana race and citrus race of *Radopholus similis* **were first identified in India**
21. Necrosis and discoloration of infested tissues are mainly due to alkaloid compounds
22. *Globodera pallida* is a cool climate preferring nematode
23. *Radopholus similis* causes tunnels and cavities in roots of banana
24. **Striae are the shallow or superficial transverse grooves that penetrate to the deeper layers of the cuticle.**
25. *Tylenchulus semipenetrans* is a major pest on lemon
26. Photoreception is common in most nematodes
27. Lips are hexaradiate in Tylenchoidea
28. The common name of *Hoplolaimus* is lance nematode
29. *Anguina tritici* cannot also infect rye crop
30. Green leaf manure can limit the population of nematodes, in general
31. *Meloidogyne arenaria* is commonly associated with cotton
32. *Trichodorus* can transmit plant viruses
33. Solanaceous vegetable crops are highly preferred by *Meloidogyne incognita*
34. *Glomus fasciculatum* can be used to minimize nematode populations in soil
35. RNA interference (RNAi) is a gene-doubling phenomenon

36. Resistance to nematodes in plants is generally characterized by failure of the nematodes to produce functional feeding sites in the host after invasion and to develop subsequently as reproducing females
37. Root-knot nematodes feed from multinucleate giant cells developed by the expansion of stele cells within the differentiating vascular cylinder
38. The Erlenmeyer bottle method is used for extraction of active nematodes from soil
39. Secretions injected through the stylet are non-granular and often form a viscous drop at the stylet tip.
40. A single multinucleate giant cell is produced by *Bellodera utahensis*
41. *Ditylenchus dipsaci* contains less of cholesterol than phytosterol
42. *Heterorhabditis bacteriophora* is effective against root weevils
43. Nematodes with long stylets and those that use the stylet actively to penetrate cells have a well-developed headframe
44. The fourth stage juveniles hatch from the eggs in response to host-plant root exudates and invade the root just behind the apex, preferentially in the differentiation and elongation zone
46. Addition of organic matter to soil can increase available water and so decrease irrigation requirements
46. **In a fallow field, there is plenty of stimulus to hatch juvenile nematodes and many of them may remain** viable in the soil for the next crop cycle
47. Sudan grass is an excellent example of a cover crop to manage nematodes that is also useful as a green manure
48. Phytonematodes usually contain less lipid than do animal parasites
49. Tea is also a host for *Radopholus similis*
50. Ammonia is a rare common and less studied by-product of decomposition of organic materials

VI. ANSWERS

1. F	11. T	21. F	31. F	41. F
2. F	12. T	22. T	32. T	42. T
3. T	13. F	23. T	33. T	43. T
4. T	14. T	24. F	34. T	44. F
5. T	15. T	25. T	35. F	45. T
6. F	16. F	26. F	36. T	46. F
7. T	17. T	27. T	37. F	47. T
8. T	18. T	28. T	38. T	48. F
9. F	19. T	29. F	39. T	49. T
10. F	20. F	30. T	40. F	50. F

VII

1. **In general, efficacy against nematodes increases as %N in amendments increases and as C:N ratio** decreases
2. Ammonia, NH_3, is much more toxic to nematodes than ammonium ion, NH_4^+
3. **In nematode phylogenetic tree, based on 18s RNA gene sequences, five different clades (I-V) have been defined**
4. The predominant adult form of *Caenorhabditis elegans* is bisexual, with each individual producing both eggs and sperm
5. *Helicotylenchus* may be semi- endoparasite or ectoparasite
6. Only females of *Ditylenchus dipsaci* can produce pheromones
7. Fenestration refers to the arrangement and form of the translucent cuticular areas around the vulva in cyst nematodes
8. Major energy reserve in parasitic nematodes is carbohydrate
9. Measures of the degree of genetic differences between populations or species based on differences in allele frequencies is the genetic distance
10. In rice root-knot nematode, males occur rarely compared to other species of *Meloidogyne*
11. Nematodes with long stylets feed several cell layers deep and do not move frequently to new feeding sites; hence they are relatively sedentary.
12. In moist nematodes, the median zone may be absent.
13. Micro-arrays are the glass slides bearing short stretches of oligonucleotide probes or cDNA clones used to measure expression levels of many thousands of genes simultaneously
14. Multiple sequence alignment refers to a matrix of sequence data, in which the individual columns represent homologous characters.
15. Ovejector is a common region of the two uteri and connected to the vagina
16. The evolutionary history of a single taxon is studied in the Phylogeny
17. *Aphelenchoides besseyi* is not seed borne
18 .Several *Bursaphelenchus* species are associated with longhorn beetles
19. O-glycans present in amphidial secretions J_2 of *Heterodera schachtii* is important in chemoreception
20. *Meloidogyne incognita* is attracted by ammonia and several nitrogenous salts released by decaying material
21. Inhibition of acetylcholinesterase is the target site for the control of phytonematodes by fumigant nematicides
22. Nematodes possess a wide variety of sterols and can biosynthesize them de novo
23. Stylet knobs are absent in Aphelenchina
24. Rainfall affects the movement of nematodes in the soil and on the soil surface
25. Simulation is the process of conducting experiments on a model of a system
26. The Mi gene confers resistance in tomato against jassids also
27. Chitin is present in the eggshells of various PPNs but absent in plant cells.
28. Galactose is a major carbohydrate detected on the cuticle of *Globodera pallid*
29. DNA-based detection tools make excellent methods of nematode diagnosis since they are slow and steady

30. Nematodes in axenic culture have greater longevities than nematodes cultured on live bacteria
31. The age pigment, lipofuscin in nematodes, is confined to the intestinal epithelium
32. Simulation models are useful for research and analysis of crop damage through the partitioning of energy flow
33. The surface coat of phytonematodes is characterized by a net positive charge
34. Of all the phytonematodes, in most countries, quarantine measures have been strictly followed in potato cyst nematodes
35. Microorganisms have been regarded as 'biofactories' for the production of metallic nanoparticles.
36. All nematodes are harmful in nature
37. Nematodes with long stylets are ectoparasites; those with short stylets may be ectoparasites or endoparasites.
38. The free-living nematodes like rhabtidis can be cultured on agar plates with great difficulty
39. ELISA is a protein based molecular technique employed in Nematology
40. Semiautomatic elutriator technique is useful to extract cysts
41. Males of *Tylenchulus semipenetrans* always develop inside the roots
42. The penetration inhibition test is used in bioagent testing techniques
43. Trehalose is present in many nematode species where its concentration often exceeds that of glucose but is usually lower than that of glycogen
44. Lactic acid is not necessarily the end-product of carbohydrate catabolism in nematodes
45. Urea is excreted by some nematodes
46. In very dry soils, the osmatic pressure is low
47. In general, phytonematodes have an efficient osmoregulatory mechanism.
48. Plant parasitic nematodes excrete organic acids and peptides also
49. Nematodes grow between moults and the mature worm may continue to grow after the final moult
50. The outer layer of the nematode egg is present in all nematode species

VII. ANSWERS

1. T	11. T	21. F	31. T	41. F
2. T	12. F	22. F	32. T	42. F
3. T	13. T	23. T	33. F	43. T
4. F	14. T	24. T	34. T	44. T
5. T	15. T	25. T	35. T	45. T
6. F	16. F	26. F	36. F	46. F
7. T	17. F	27. T	37. T	47. T
8. T	18. T	28. T	38. F	48. F
9. T	19. T	29. F	39. T	49. T
10. F	20. F	30. T	40. T	50. F

VIII

1. The jelly-like mass containing the eggs in citrus nematode is not secreted by the excretory system
2. Morphologically species of *Xiphinema* and *Longidorus* are almost similar
3. Small nematodes respire at a faster rate than large ones
4. Nematode cuticle is semimetabolically active structure
5. Hatching of eggs of *Meloidogyne javanica* is maximum at a temperature of 30^0C
6. Plant-parasitic nematodes with postcorpus (digestive glands) bounded by a membrane are all ectoparasites
7. All genera that feed or that can feed endoparasitically have overlapping esophageal glands with dorsal gland usually more developed.
8. Hypodermal cells become very active during hatching of the cuticle
9. *Paratrichodorus minor* transmits pepper ring spot virus
10. The avermectins have an indirect action on neurotransmission in nematodes
11. In unfavorable conditions, nematodes can switch their food preference, a condition known as omnivory
12. Nematode community structure can be used as a bioindicator in environmental monitoring
13. Lip or cephalic region in nematodes is always set off from the body.
14. Lateral guiding pieces are present only in male nematodes
15. Surface tension of water controls its availability to nematodes in plant tissues
16. The head of *Hirschmaniella* generally, is not set-off
17. Lac plant (*Flemingia semialata L.*) as a new host of *Meloidogyne incognita* **was reported for the first** time in 2016 from Karnataka.
18. Immature females of *Rotylenchulus* are slender, in the form of the letter "C'
19. The xylem and phloem cells are affected by the reniform nematode invasion into the host
20. Grape is not a host for citrus nematode, in several regions
21. *Nacobbus serendipiticus* does not attack tomato
22. *Aphelenchoides bessyii* is a seed borne nematode
23. *Meloidogyne exigua* infects coffee but not M. incognita
24. Sweet potato variety 'Gold rush' is susceptible to the reniform nematode
25 *Bursaphelenchus* never infects seeds
26. Eggplant is also a host for *Globodera rostochiensis* in several parts of the world
27. Dagger nematodes also induce giant cells in host tissues
28. 'Ovijector' is a small muscular organ at the vulval opening present in some nematode species
29. *Hemicriconemoides* is also known as true sheath nematode
30. Current studies indicate that nematodes are actually related to the arthropods and priyapulids in a newly recognized group, Ecdysozoa
31. Adenophorea include a wide range of marine, freshwater and soil nematodes but few plant parasites
32. Flat worms are not the representative members of Platyhelminthes
33. A cladistic representation of a phylogeny, whereby only the branching order is displayed is termed 'cladogram'

34. Segmentation is never the main difference between Annelida and Nematoda
35. In nematodes, muscle cells branch toward the nerve cells as in other animals
36. **Pseudocoel aids nematodes to gain or lose rigidity by way of fluid pressure**
37. **Unlike most animals, nematodes do not depend on cilia or flagella** for excretion
38. The juveniles, for the most part, have the same number of cells as adults
39. The presence of setae is the most advanced character in nematode forms
40. *Meloidogyne hapla* can reproduce by both cross-fertilization and meiotic parthenogenesis
41. In nematodes, generally, males are slightly larger than females, which usually display a bent tail
42. *Caenorhabditis elegans* **was the first multicellular organism that had its genome sequenced in 1998**
43. All the mitotic parthenogenetic nematode forms are evolved from meiotic parthenogenetic ancestors or less likely from amphimictic ones during maturation of the oocytes
44. **All nematodes do have a complex, but illdefined digestive tract**
45. Nematodes are suitable objects to study evolution of development because species from all branches of the phylogenetic tree can be analyzed
46. The basic principle of phylogeny is that all are related and evolved from a common ancestor
47. Citrus nematode does not infect grape
48. Pseudocoelom is also present in Rotifera
49. Nematoda is the only phylum of pseudocoelomates that includes a few amount of species
50. Nematodes are subdivided into basal 'Enoplea', that includes clades 1 and 4

VIII. ANSWERS

1. F	11. T	21. F	31. T	41. F
2. T	12. T	22. T	32. F	42. T
3. T	13. F	23. T	33. T	43. T
4. F	14. T	24. F	34. F	44. F
5. T	15. F	25. F	35. F	45. T
6. T	16. T	26. T	36. T	46. T
7. T	17. T	27. T	37. T	47. F
8. F	18. T	28. T	38. T	48. T
9. T	19. F	29. F	39. F	49. F
10. F	20. F	30. T	40. T	50. F

IX

1. Most nematodes have genomes ranging from 50 to 250 Mb.
2. Intersexes are usually functional females with some male somatic characters like spicules
3. The rate of intron loss is very high in nematodes
4. Usually stunt nematodes are found in moist to wet, in low areas **of fields and near irrigation ditches.**
5. *Hemicycliophora thienemanni* is a commonly occurring species
6. A considerable part of host injury by nematodes is due to the plant tissue reaction to the secretion injected into them during feeding
7. Cells away from the giant cells caused by root-knot nematodes, undergo hypertrophy and proliferation to form a gall
8. **'Necrotic crown' or 'white flagging' symptom is caused by rice ufra nematode**
9. The surface cut of the tissue infested by red ting nematode may sometimes appear light pink or cream to dark brown also, in all oil palms
10. Spreading decline of citrus is caused by *Radopholus citriphilus*
11. Minerals and amino acids are the major electrolytes detected on root-gall exudates
12. *Hemicycliophora arenaria* causes minute galls on lemon roots.
13. Most epidemiological elements of nematode-plant relationships may not be dependent on density.
14. Nematodes which grow large and have long life cycles with low rates of population increase are known as 'k strategists'
15. Total energy consumption during the life cycle of a female root-knot nematode is 1 cal.
16. 'Black head' disease in banana is due to the lesion nematode
17. Nematodes are often considered as stress-related pathogens
18. Lesion nematodes are a major problem on apple than *Meloidogyne incognita*
19. *Anguina tritici* carries the fungus *Dilophospora alopecuri* on its cuticle and takes with it into host tissue during its penetration
20. Giant cells are insensitive to a translocatable physiological factors produced by soil fungi
21. The haploid chromosome numbers of most nematodes is n=4 to 12
22. *Hoplolaimus indicus* is commonly associated with the rhizosphere of vegetable crops
23. In cultivated land, nematode injured spots are often spherical in the direction of cultivation, because nematodes are moved by machinery.
24. Post feeding by a nematode juvenile when the stylet is withdrawn, the opening in the cell wall is sealed with an electron-dense feeding plug.
25. The response of potato to *Globodera rostochiensis* is due to stomatal closure through water stress that leads to reduced photosynthesis.
26. Occurrence of clonal induction of proteinase inhibitor proteins is a common molecular event in hosts infested by nematodes
27. 'Bearded or whiskered' roots are due to the infestation of *Heterodera schachtii*
28. Chrysanthemum variety 'Orange Beauty' is resistant to *Aphelenchoides ritzemabosi*
29. Predaceous nematodes are predominant in soil compared to phytoparasitic forms
30. Small, yellowish, slightly raised, pimple-like spots on the foliage of narcissus caused by root-knot nematodes are known as 'spikkels'

31. HOX genes are transcription factors that are closely clustered in the genomes of most animals
32. Secondary soil fungi, usually get attracted towards the giant cells or syncytia
33. *Tylenchorhynchus* feeds on root epidermal cells, leaving behind simple micropuncture type wounds.
34. Woody plants in advanced stages of decline incited by nematodes will have profused foliage.
35. Among all species of spiral nematode infesting banana, *Helicotylenchus multicinctus* is most predominant
36. *Dolichodorus* is relatively a small nematode
37. The main obvious effect on the host root tip feeding by *Trichodorus* is the increased of cell division in the apical meristem
38. The levels of electrolytes leaking from nematode-infested roots increase progressively following infection
39. The feeding plug secreted by cyst nematodes once the feeding site is established, may originate from the amphids
40. The population of actinomycetes is normally increased in the rhizosphere soil in tomato infested with root-knot nematodes
41. New root initials generally are killed by heavy infestations of the sting nematode, a symptom **reminiscent of Zn deficiency**
42. Root-knot nematode infested plants produce elevated levels of 14C metabolites that attracts *Rhizoctonia solani*
43. Of late, *Meloidogyne javanica* is also a major problem on banana in several regions
44. **'Curly root tip or fish hook' may also be caused by cyst nematodes**
45. The presence of a lectin on the traps of *Arthrobotrys oligospora* binds to a carbohydrate on the nematode cuticle
46. The quality and quantity of rhizosphere exudation depends on the soil organisms including nematodes
47. Small galls on the roots of oats, barley and other grasses are also caused by *Ditylenchus angusuts*
48. *Xiphinema* spp. cannot cause surface necrosis of roots
49. The genome of phytonematodes of the order Tylenchida includes genes that encode for endoglucanases
50. The total biomass of a female *Meloidogyne* is 200 microgram, including the egg mass

IX. ANSWERS

1. T	11. T	21. T	31. T	41. F
2. T	12. T	22. T	32. T	42. T
3. T	13. F	23. F	33. T	43. T
4. F	14. T	24. T	34. F	44. F
5. T	15. T	25. T	35. T	45. T
6. T	16. F	26. F	36. F	46. T
7. T	17. T	27. T	37. F	47. F
8. F	18. T	28. T	38. T	48. F
9. F	19. F	29. F	39. T	49. T
10. T	20. F	30. F	40. F	50. T

X

1. 'Black Beauty', a grape variety is resistant to *Meloidogyne incognita*
2. Radish is a good trap crop for nematodes
3. *Seinura*, a free living nematode is commonly found associated with mushroom cropping beds
4. Hot water treatment of ginger rhizome at 50-55^0 C for 10 minutes is effective against *Radopholus similis*
5. The famous quote "nature does not work with pure cultures' was by Fawcett in 1931.
6. The alterations induced in host by one nematode species may not always predispose all fungal species
7. **One of the first examples of a lectin-mediated interaction between a fungus and its host was reported** in nematode-trapping fungus *Pochonia chlamydosporia*
8. Nematodes posses genes for parasitism but not genes for virulence.
9. Several phytonematodes also transmit spores of many soil fungi
10. Nematode surfaces have a coat which contains different lipids probably in the form of glycoproteins
11. Root-knot and cyst nematode feeding sites differ in ontogeny and appearance
12. Pathotypes are differentiated based on the different breeding lines of the different plant species.
13. **The first members of a parasitome to be cloned from phytonematodes** were beta-2, 6-endoglucanases
14. Sampling at the time of planting theoretically will give a better estimate of the initial nematode problem, where population levels are low enough for detection.
15. *Longidorus* is always a subterranean nematode
16. Top 2.5 to 5 cm of soil needs to be discarded while sampling, as nematodes usually do not live sue to extreme environmental conditions
17. Using microplots, types of barriers that may be constructed, methods of installations with various nematodes cannot be determined
18. The *Mi* gene of tomato encodes a member of the plant resistance protein family characterized by the presence of a putative nucleotide-binding site and a leusine-rich repeat.
19. Cellulases in phytonematodes are important in the initial steps of pathogenesis.
20. Among phytonematodes, the only sterols detected in *Ditylenchus triformis* and *D. dipsaci* are phytosterol and sitosterol
21. In a complex association, nematodes may favor all stages of fungal infection and development
22. **Root-knot nematode eggs do not contain free and esterified sterols**
23. ISPM is the standard abbreviation for International Standards for Phytosanitary Measures
24. Azasteroids and other related non- sterol amines dramatically affect sterol metabolism in nematodes.
25. Mitochondrial genomes of metazoa are typically composed of a single, circular molecule that varies in size from 55-85 kilobases
26. It is suggested that gonochoristic nematodes did not evolve into hermaphrodites and that some of those became parthenogenetic.
27. Several highly evolved nematode forms may have only 2 lips
28. Polygenic resistance is comparatively less stable than monogenic resistance
29. Lectins are plant proteins possessing at least one non-catalytic domain, which binds irreversibly to a **specific mono –or oligopolysaccharide**
30. The presence or absence of a bulb in oesophagus is of less evloutionary importance
31. The antibiotic DAPG (2,4-diacetylphloroglucinol) produced by some bacteria are toxic to several nematodes
32. The photoecdysteroid 20-hydroxyecdysone is a major hatching hormone in nematodes

33. Proteinase inhibitors of various plants have been used to control the proteolysis of root-knot nematodes by inhibiting gut serine proteinases
34. Novel nematode management approaches never act either on nematode targets at the nematode-plant interface, or in the plant response.
35. The nematode distribution is always uniform or constant and changes may not occur rapidly
36. It is not advisable to collect soil samples for nematode population estimation, when field is very wet
37. Phytonematodes may directly affect the release of exudates in a quantitative way by rupturing the root cell membrane during feeding, penetration and migration within the root.
38. Abamectin is a mixture of avermectins containing more than 20% avermectins B1a and less than 80 avermectin B1b.
39. A predaceous nematode may inject a toxin into the nematode prey, causing immobilization within seconds
40. Sorghum produces a cyanogenic glucoside compound called 'dhurrin' that breaks down to release toxic phenol when the plant tissue is damaged
41. Some sterols do not contain a double bond at C-5 and they are referred to as 'stanols'
42. Burrowing nematode on banana is of quarantine importance in India
43. A biotype refers to a group of genetically identical individuals sharing a common biological feature.
44. In damage models developed for nematodes, population density is designated as 'Pd' and tolerance limit as '*T*'
45. Engineering resistance can be achieved by transforming plants with protease inhibitors, lectins, toxic proteins and dsRNA targeted against the phytonematodes
46. Several nematode R genes are members of a family non-characterized by a nucleotide-binding site and leucine-rich repeats
47. Nuclei of syncytial cells undergo endoreduplication of their RNA content but do not divide
48. The amphidial glands are the smallest and most simple of the anterior sensory organs of nematodes
49. Parasitism genes enable nematodes to overcome the effects of genes for resistance
50. Nematode-induced loss of resistance to *Fusarium udum* in pigeon pea is associated with reduced levels of the isoflavonoid phytoalexin 'cajanol'.

X. ANSWERS

1. F	11. T	21. T	31. T	41. T
2. T	12. F	22. F	32. F	42. T
3. F	13. F	23. T	33. T	43. T
4. T	14. F	24. T	34. F	44. F
5. T	15. T	25. F	35. F	45. T
6. T	16. T	26. F	36. T	46. F
7. F	17. F	27. T	37. T	47. F
8. T	18. T	28. T	38. F	48. F
9. T	19. T	29. F	39. T	49. T
10. F	20. F	30. F	40. F	50. T

ESSAY TYPE

1. What are the factors which decide antagonistic nematode interactions?
2. List ten major phyla under subkingdom Metazoa.
3. Differentiate Secernentea from Adenophorea.
4. Discuss ectoparasitic nematodes and their different types with suitable examples and diagrams.
5. Discuss in brief the interaction among two sedentary endoparasites.
6. **List four major factors which influence the occurrence of a particular species in an environment.**
7. Mention the unique characters of Nematoda.
8. **What is perineal pattern technique? Discuss its significance.**
9. Differentiate a giant cell from a syncytium.
10. List four ultrastructural changes induced by phytonematodes in host plants.
11. Mention major roles of hormones in nematode metabolism.
12. Narrate in brief the role of soil moisture in the activity of nematodes.
13. **Discuss briefly on dormancy in plant parasitic nematodes.**
14. **Explain in brief the significance of pheromones in nematode.**
15. **Write the technical definition of a nematode.**
16. What are avermectins? Mention their two major modes of action.
17. **Mention in brief the significance of nematode secretions.**
18. Explain the importance of sensory stimuli in plant parasitic nematodes. How do you exploit them for nematode management?
19. List four major advantages and two disadvantages of entomopathogenic nematodes.
20. **Briefly narrate the mechanisms of interaction of nematodes with fungi/bacteria.**
21. Explain in detail about Numerical taxonomy.
22. **Narrate the significance of Biochemical taxonomy in Nematology.**
23. Explain the importance of radio isotope in nematode taxonomy.
24. Write in detail about electrophoresis and enzyme phenotype techniques.
25. With the use of latest keys how do you identify Pratylenchidae, Paratyenchidae and Nacobbidae.
26. Explain the principles of Staining in Nematology with suitable examples.
27. Explain in detail about *in vitro* culturing of plant parasitic nematodes.
28. **Narrate the important biochemical techniques for identification** of species and races in nematodes.
29. Distinguish taxonomic characters of Hoplolaimidae, Criconematoidae, and Tylenchulidae.
30. Signify the importance of Type specimen and how do you designate it?
31. Explain in detail about Alpha taxonomy, Beta and Gamma taxonomy using latest keys how do you distinguish Monochidae, Rhabditidae and Diplogasteridae.
32. Explain in detail International code of zoological nomencalture.
33. Mention various signals produced in nematodes. Explain any one of them.
34. **Define pheromone. How do you regulate pheromone production artificially?**
35. Mention various histopathological changes initiated by nematodes in plants.
36. Mention types of interaction between nematodes & other microorganisms.
37. Narrate major mechanisms of interaction between nematodes and AM fungi.

38. Mention major roles of hormones in nematode metabolism.
39. List four most important soil factors that affect the activity of nematodes.
40. Mention major types of habitats of nematodes.
41. Mention two genera of predacious nematodes.
42. Mention major genera of free living nematodes.
43. Furnish common names for *Hoplolaimus* and *Criconema.*
44. Mention important enzymes which digest fats in nematodes.
45. Mention major sense organs in nematodes.
46. **Provide scientific names for rice 'ufra' nematode and vinegar nematode.**
47. **What is the significance of the term 'sheath' in sheath nematode?**
48. What is the meaning of 'stubby root'?
49. Mention the primary sources of inoculum for *Anguina* and *Heterodera.*
50. **Discuss briefly on dormancy in plant parasitic nematodes.**
51. **Explain in brief the significance of pheromones in nematode.**
52. What are avermectins? Mention their two major modes of action.
53. **Mention in brief the significance of nematode secretions.**
54. Explain the importance of sensory stimuli in plant parasitic nematodes. How do you exploit them for nematode management?
55. List four major advantages of entomopathogenic nematodes.
56. **Mention the significance of nematode secretions along with the major types of secretions.**
57. **What are the various inherent factors that influence the nematode survival?**
58. Substantiate the statement "all nematodes cannot transmit viruses'.
59. List the major requirements for evolution of nematode genera / species.
60. Substantiate with examples 'highly specialized phytonematodes'.
61. Discuss the major features of foliar phytonematodes with examples.
62. **Briefly narrate the mechanisms of interaction of nematodes with fungi/bacteria.**
63. **Briefly explain the role of soil organic matter in the activity of nematodes.**
64. **Briefly narrate the significance of cuticular secretion sin nematodes.**
65. Mention various signals produced in nematodes. Explain any one of them.
66. **Define pheromone. How do you regulate pheromone production artificially?**
67. Mention various histopathological changes initiated by nematodes in plants.
68. Mention types of interaction between nematodes & other microorganisms.
69. Narrate major mechanisms of interaction between nematodes and AM fungi.
70. **What is the significance of enzymes in nematodes? Mention examples of enzymes produced by two** major phytonematodes.
71. How do you differentiate giant cells from syncytia?
72. **Briefly discuss host modification by nematodes.**
73. Narrate the mechanism involved in virus transmission by nematodes.
74. Mention major features of the phylum Nematoda.

75. Provide the characteristic features of two major classes of the phylum Nematoda.
76. Explain in brief the novel concepts in nematode management.
77. Discuss the genetics of nematode parasitism.
78. Signify the role of molecular and cytogenetical approaches in systematics.
79. Explain nematode phylogenetic and evolutionary concepts.
80. Furnish the importance of avermectins in nematode management.
81. Discuss in detail the establishment and use of nematode threshold level in crop losses.
82. **Briefly explain the importance of azasteroids.**
83. **Explain the significance of Surface receptors.**
84. **Mention the significance of nematode secretions along with the major types of secretions.**
85. **What are the various inherent factors that influence the nematode survival?**
86. **Briefly explain the role of soil organic matter in the activity of nematodes.**
87. Mention major types of habitats of nematodes, in general.
88. What are avermectins? Mention their two major modes of action.
89. Explain the importance of sensory stimuli in plant parasitic nematodes. How do you exploit them for nematode management?
90. **Briefly narrate the mechanisms of interaction of nematodes with fungi/bacteria.**
91. Explain in brief the novel concepts in nematode management.
92. **List four economically important nematode diseases in India along with scientific names of causal agents.**
93. Classify nematodes based on their mode of feeding with examples.
94. Explain in brief general life cycle of a plant parasitic nematode.
95. Write a labeled neat diagram of a female plant parasitic nematode.
96. Explain in brief symptoms caused by nematodes in crop plants.
97. Write a labeled neat diagram of a male plant parasitic nematode.
98. Explain in brief, general life cycle of a plant parasitic nematode.
99. Substantiate the statement 'sedentary endoparasites are highly specialized pathogens.
100. **Mention the significance of cuticular secretions in nematodes.**
101. What are different types of associations nematodes establish with other soil microorganisms?
102. What do you know about the mechanisms involved in nematodes-AMF interaction?
103. Narrate the mechanisms involved in the formation of knot/gall on roots by root-knot nematodes.
104. **Define competition exclusion. Mention the factors influencing it.**
105. Explain in brief the hatching process in nematodes.
106. In your opinion, what are the major roles of a Nematologist in research, teaching and extension?
107. How do you differentiate and diagnose Meloidogyne incognita and M. hapla morphologically?
108. List four ultra-structural /histopathological variations in nematode-infected host tissues.
109. **Briefly discuss the digestion of fats by nematode enzymes.**
110. List the factors which decide antagonistic interactions.
111. List ten major phyla under subkingdom Metazoa.

112. Differentiate Secernentea from Adenophorea.
113. Discuss ectoparasitic nematodes and their different types.
114. Explain in brief the biology and life cycle of Radopholus similis on banana.
115. Mention various types of reproduction in plant parasitic nematodes.
116. Explain any one type with example.
117. Explain in brief the process of excretion and respiration in nematodes.
118. Explain in detail the digestive system in nematodes with a neat diagram.
119. **Mention any five major historical milestones in Nematology providing the names of the scientist with the year.**
120. Narrate in brief the economic importance of nematodes.
121. Draw a neat diagram of a male plant parasite nematode and label major parts.
122. Draw a neat diagram of a body wall of nematode and label major parts.
123. Mention the taxonomic position for the genera *Globodera, Pratylenchus, Ditylenchus, Xiphinema and Criconema.*
124. Differentiate : Coelom and pseudocoelom ; Females of *Meloidogyne* and *Globodera*; Prodelphic and amphidelphic ; Cuticle and hypodermis.
125. Explain in brief the biology and life cycle of Radopholus similis on banana.
126. Write a neat, labeled diagram of a female phytonemaotde.
127. Mention the taxonomic position for *Meloidogyne, Radopholus, Anguina, Aphelenchoides* and *Heterodera.*
128. **Explain the significance of Sensory organs, Intestine, Stoma,. Muscle cells and Gubernaculum.**
129. Explain in brief nervous system in nematode.
130. **Mention any five major historical milestones in Nematology providing the names of the scientist with the year.**
131. **Briefly discuss on the modification of stoma in nematodes.**
132. Explain in brief the feeding process in phytonematodes.
133. List the major plant parasitic nematodes associated with black pepper. Provide a suitable management schedule for the most predominant nematode.
134. List the major plant parasitic nematodes associated with rose and carnation. Provide a suitable management schedule for the most predominant nematode.
135. Explain in brief the symptomatology of *Meloidogyne graminicola* on rice.
136. Explain in brief the symptomatology of *Globodera* rostochiensis on potato.
137. How do you manage root-knot nematodes on papaya by cultural practices?
138. **Define Integrated Nematode management. Suggest a suitable integrated management schedule for** *Radopholus similis* on banana.
139. **Define Integrated Nematode management. Suggest a suitable integrated management schedule for** *Ditylenchus angustus* on rice.
140. **Briefly narrate the biology of** *Heterodera avenae* on wheat.
141. **Briefly narrate the biology of** *Radopholus similis* on banana.
142. Explain the life cycle and the modes of spread of *Aphelenchoides besseyi* in rice.

143. Explain the life cycle and the modes of spread of *Meloidogyne javanica* in tomato.
144. **Define biological management. Provide the schedule for the biological management of root-knot** nematode on pumpkin.
145. **Define biological management. Provide the schedule for the biological management of nematode** complex in banana.
146. Mention various principles of nematode management with a suitable example under each.
147. How do you classify nematodes based on their feeding habit? Provide an example under each category. List ten most economically important plant parasitic nematodes of India. Explain the symptomatology of one predominant nematode on a fruit crop.
148. List the major plant parasitic nematodes associated with wheat. Provide a suitable management schedule for one most predominant nematode.
149. List the major plant parasitic nematodes associated with citrus. Provide a suitable management schedule for one most predominant nematode.
150. Explain in brief the symptomatology of reniform nematode.
151. Explain in brief the symptomatology of lesion nematode.
152. Furnish a management schedule for a major nematode on rice.
153. Furnish a management schedule for a major nematode on chrysanthemum.
154. **Narrate briefly the life cycle, spread and management of** *Aphelenchoides besseyi* in rice.
155. **Narrate briefly the life cycle, spread and management of** *Xiphinema index* in grape.
156. How do you manage root-knot nematodes in brinjal by non-chemical methods?
157. **Define quarantine. What is its significance in plant nematology? Explain with an example.**
158. **Explain the significance, symptomatology and life cycle of cereal** cyst nematode on oats.
159. What is the role of resistant varieties in nematode management? List at least two resistant/tolerant varieties of tomato / potato against major nematodes.
160. Compare between biological and chemical methods of nematode management. List merits and demerits under each category with suitable examples.
161. Compare the biology of root-knot nematode with that of golden cyst nematode of potato.
162. Discuss the use of various organic amendments in nematode management with examples.
163. **Discuss the benefits of using mycorrhiza in nematode management.**
164. **List the major plant parasitic nematodes associated with ornamental crops. Suggest specific** management schedule for the most common nematode.
165. What are the important phytonematodes associated with papaya. Provide a suitable management schedule for the predominant nematode on papaya.
166. Explain integrated management of rice ufra or white tip disease in rice.
167. Explain in brief the symptomatology of *Anguina tritici* on wheat.
168. Explain in brief the symptomatology of *Tylenchulus semipenetrans* on citrus.
169. How do you manage root-knot nematode on rice by cultural practices?
170. **Define Integrated Nematode management. Suggest a suitable integrated management schedule for** *Meloidogyne incognita* on tomato.
171. **Define Integrated Nematode management. Suggest a suitable integrated management schedule for** Aphelenchoides besseyi on rice.

172. **Briefly narrate the biology of** *Meloidogyne graminicola* **on rice.**
173. **Briefly narrate the biology of** *Radopholus similis* **on banana.**
174. Explain mode of spread and management of *Meloidogyne incognita* in carnation.
175. Explain the life cycle and the symptomatology of *Meloidogyne incognita* in brinjal
176. **Define biological management. Provide the schedule for the biological management of root-knot nematode in cucurbits.**
177. **Define biological management. Provide the schedule for the biological management of burrowing nematode in banana.**
178. Mention various principles of nematode management with a suitable example under each.
179. Explain the feeding process in phytonematodes.
180. **List five potential phytonematodes of India. Furnish the biology of one predominant nematode.**
181. What are the major nematodes associated with rice? Provide symptomatology of one important nematode.
182. List the major plant parasitic nematodes associated with potato. Provide a suitable management schedule for one predominant nematode.
183. Explain in brief the symptomatology of burrowing nematode.
184. Explain in brief the symptomatology of reniform nematode.
185. Furnish a management schedule for a major nematode on wheat.
186. Furnish a management schedule for a major nematode on tuberose.
187. **Narrate briefly the life cycle, spread and management of** *Meloidogyne graminicola* **in rice.**
188. **Narrate briefly the life cycle, spread and management of** *Pratylenchus* **in banana.**
189. Explain non-chemical methods to manage root-knot nematodes in cucumber?
190. **Define quarantine. What is its significance in plant nematology?**
191. Explain the symptomatology and life cycle of cereal cyst nematode on barley.
192. Narrate the role of resistant varieties in nematode management. List at least two resistant / tolerant varieties of tomato against nematodes.
193. Compare between biological and cultural methods of nematode management. List merits and demerits under each.
194. Compare the biology of burrowing nematode with that of cereal cyst nematode of wheat.
195. Discuss the use of various organic amendments in nematode management with examples.
196. **Discuss the benefits of using AM mycorrhiza in nematode management.**
197. List major nematode pests of rice. Brief the symptoms caused by any one nematode.
198. Explain in brief the major management practices for 'toppling-over' disease of banana.
199. Explain in brief the major management practices for 'ear-cockle' disease of wheat.
200. Brief the biology of root-knot nematode on tomato.
201. Explain in brief the symptoms caused by citrus nematode in lemon.
202. Explain in brief the symptoms caused by white tip nematode in rice.
203. How do you classify plant parasitic nematodes based on their feeding habits? Explain with examples.
204. Explain the relationship between the rhizosphere and various activities of phytonematodes.
205. Explain in brief the symptoms and management practices for golden cyst nematode of potato.

206. Explain the steps involved in the feeding process of phytonematodes.
207. List four major nematode pests of fruit crops.
208. Classify the symptoms caused by plant parasitic nematodes.
209. How do you group plant parasitic nematodes based on their mode of feeding? Explain in brief the symptoms, biology and major management practices for wheat gall nematode.
210. Explain in brief the feeding process in nematodes.
211. Explain in brief the symptoms, spread and management of root-knot nematode in crossandra or tube rose.
212. **Define integrated management in nematodes. Explain in brief integrated management of burrowing** nematode attacking banana.
213. Explain in brief the symptoms, biology, spread and management for ufra disease in rice.
214. Explain in brief the symptoms, biology, spread and management for citrus nematode in orange.
215. Mention major roles of hormones in nematode metabolism.
216. List four most important soil factors that affect the activity of nematodes.
217. Mention major types of habitats of nematodes.
218. Write a short note on the carbohydrates as surface receptors on nematode cuticle.
219. Discuss aspects of the permeability nature of the nematode eggshell.
220. Mention two genera of predacious nematodes.
221. Mention major genera of free living nematodes.
222. Furnish common names for *Hoplolaimus* and *Criconema.*
223. Mention important enzymes which digest fats in nematodes.
224. Mention major sense organs in nematodes. Explain one important sense organ in detail.
225. **List major natural and artificial egg hatching factors.**
226. **Provide scientific names for rice 'ufra' nematode and vinegar nematode.**
227. **What is the significance of the term 'sheath' in sheath nematode?**
228. How do you differentiate 'stubby root' and 'coarse root'?
229. Mention the primary sources of inoculum for *Anguina* and *Heterodera.*
230. **Discuss briefly on dormancy in plant parasitic nematodes.**
231. **Explain in brief the significance of pheromones in nematode.**
232. Explain in detail the adaptation aspects of nematodes to different climatic conditions.
233. What are avermectins? Mention their two major modes of action.
234. **Mention in brief the significance of nematode secretions.**
235. **Explain the significance of sensory stimuli in nematode behaviuor.**
236. List four major advantages of entomopathogenic nematodes.
237. **Mention the significance of nematode secretions along with the major** types of secretions.
238. **What are the various inherent factors that influence the nematode** survival?
239. Substantiate the statement "all nematodes cannot transmit viruses'.
240. List the major requirements for evolution of nematode genera / species.
241. Substantiate with examples 'highly specialized phytonematodes'.

242. Discuss the major features of foliar phytonematodes with examples.
243. **Briefly narrate the mechanisms of interaction of nematodes with fungi/bacteria.**
244. **Briefly explain the role of soil organic matter in the activity of nematodes.**
245. **Briefly narrate the significance of cuticular secretion sin nematodes.**
246. Mention various signals produced in nematodes. Explain any one of them.
247. **Define pheromone. How do you regulate pheromone production artificially?**
248. Mention various histopathological changes initiated by nematodes in plants.
249. Mention types of interaction between nematodes & other microorganisms.
250. Narrate major mechanisms of interaction between nematodes and AM fungi.
251. **What is the significance of enzymes in nematodes with suitable examples.**
252. How do you differentiate giant cells from syncytia?
253. **Briefly discuss host modification by nematodes.**
254. Narrate the mechanism involved in virus transmission by nematodes.
255. Mention major features of the phylum Nematoda.
256. Provide the characteristic features of two major classes of the phylum Nematoda.
257. Explain in brief the novel concepts in nematode management.
258. Discuss the genetics of nematode parasitism.
259. Signify the role of molecular and cytogenetical approaches in systematics.
260. Explain nematode phylogenetic and evolutionary concepts.
261. Furnish the importance of avermectins in nematode management.
262. Discuss in detail the establishment and use of nematode threshold level in crop losses.
263. Discuss sex ratio and its importance in nematodes.
264. **Briefly explain the importance of azasteroids.**
265. **Explain the significance of Surface receptors.**
266. Explain the interaction between migratory endo- and sedentary endoparasitc nematodes?
267. Explain in brief the ecological interactions among ectoparasitic nematodes.
268. **Define hormone. Discuss in brief juvenile hormones.**
269. Write about 'receptors' in nematodes.
270. Classify nematodes based on their mode of feeding with suitable examples.
271. Mention the diagnostic characters of the phylum Nematoda.
272. Compare and correlate the surgical strategies of Heterodera avene and Pratylenchus coffeae in soil and host tissues.
273. **Briefly explain 'giant cell' and 'syncytium'.**
274. Mention four major ultrastructural variations induced by plant parasitic nematodes in host tissues.
275. **Briefly discuss the interaction among migratory endoparasitc nematodes.**
276. Mention various ultra structural changes that normally occur in nematode-infected plants.
277. **Discuss about pheromone diffusion and flow in soil.**
278. Discuss in brief on the nematode receptors and recognition phenomenon.
279. Discuss in brief 'highly specialized plant parasitic nematodes'.

280. Narrate and compare the life cycles of *Trichodorus, Longidorus* and *Xiphinema.*
281. Explain various histopathological variations brought about by phloem feeders and cortical feeders.
282. Narrate the role played by major enzymes produced in nematode body.
283. Discuss and correlate the nematode physiology with nematode management.
284. **What is the significance of Sex Attractants? Explain in detail with specific examples.**
285. Critically discuss physiological specialization in nematodes in relation to nematode management.
286. What is chemotaxis in nematodes?
287. **Define Oocytes amd explain oogenesis in nematodes.**
288. Mention the main function of intestine in nematodes.
289. **What are pathotypes? Narrate their significance in the pathogenicity.**
290. What is the average oxygen consumption by a nematode?
291. **Define 'Neoplastic Change' in nematodes.**
292. What are transfer cells? Signify their role.
293. **Define microvilli and discuss their importance.**
294. Explain the roles of oesophageal glands.
295. What are ecotypes? Discuss in detail with examples
296. **Define aerobic starvation and emphasize its significance.**
297. Mention any two adaptive changes brought about by nematodes in plants.
298. Mention two important nematode parasites of woody tissues.
299. Critically illustrate major steps in the life cycles of *Heterodera* and *Radopholus.*
300. Discuss various biochemical changes brought about by nematodes in plants.
301. Explain the organization of the nematode body.
302. Mention major types of nematode locomotion. Explain any two in detail with diagrams and examples
303. Explain how does excretion of nitrogenous and other waste substances occur in nematodes.
304. Compare and discuss types of plant injuries brought about by endo- and semi endo-parasites with **specific examples.**
305. Mention three major contributions in Nematological research in India with the year and the Scientist.
306. List major soil factors in nematode-soil environment and explain any one.
307. Explain the association Root-knot nematodes with fungi.
308. **Define sensory stimuli. List important stimuli for nematodes.** Explain any one in brief.
309. Explain modes of attack of insect pests by EPN's, with suitable examples.
310. Write in brief on current developments on EPN's in India.
311. **Define Avermectins. Mention the modes of action of Avermectins.**
312. Explain four major control strategies by exploring sensory stimuli in nematodes.
313. List out major EPN species effective against pests of cabbage, apple and tomato.
314. List two fungal and two bacterial bioagents that are effective against nematodes
315. Mention modes of action of oil cakes and green manuring against nematodes
316. Classify nematodes based on the particle shapes of the viruses they transmit, at least with one example.
317. **Mention the nematode and other pathogens (Scientific names only) involved in complex diseases on** banana,. tomato and wheat.

318. Substantiate with four points, why *Meloidogyne* is predominant compared to other genera.
319. List out the various organisms associated with nematodes in soil.
320. Mention four commercially feasible, potential bioagents of plant parasitic nematodes
321. Discuss in brief community pattern in a rhizosphere.
322. **Define community and diversity.**
323. **Explain in brief about the factors influencing sex ration in nematodes.**
324. **Define community analysis. Mention its various parameters.**
325. **Define trophic group. How nematodes are classified based on this?**
326. **How do you induce shift in sex ration in nematodes by artificial** means?
327. Explain the factors affecting population change of a nematodes in a community
328. Narrate the role of water in the migration of nematodes in soil (any four points).
329. Discuss in brief the effect of pH on survival of nematodes.
330. Discuss dormancy in nematodes.
331. Mention various soil fauna in rhizosphere and in your opinion which combination with nematodes is **more significant from the point of view of plant parasitism?**
332. Discuss the effect of organic material in the survival capacity of nematodes.
333. Explain in brief the importance of nematophagous fungi.
334. Write the composition of humus.
335. **Define Community. Explain in brief the diversity factor in nematodes.**
336. Discuss evolution of Tylenchida.
337. Brief future trends in nematode taxonomy.
338. **Explain enzyme phenotypic identification of** *Meloidogyne.*
339. How do you designate type specimen.
340. Discuss cyto taxonomy.
341. Explain in detail about Numerical taxonomy.
342. **Narrate the significance of** *Biochemical taxonomy* in Nematology.
343. Explain the importance of radio isotope in nematode taxonomy.
344. Write in detail about electrophoresis and enzyme phenotype techniques.
345. With the use of latest keys how do you identify Pratylenchidae, Paratyenchidae and Nacobbidae?
346. Explain the principles of Staining in Nematology with suitable examples.
347. Explain major taxonomic keys and their application.
348. **Briefly explain the nematode neurosecretions.**
349. List major microtechniques used I Nematological research.
350. **What is the significance of histopathological staining.**
351. How do you isolate the nematophagus fungi.
352. Explain in detail about *in vitro* culturing of plant parasitic nematodes.
353. **Narrate the important biochemical techniques for identification** of species and races in nematodes.
354. Distinguish taxonomic characters of Hoplolaimidae, Criconematoidae, and Tylenchulidae.
355. Signify the importance of Type specimen and how do you designate it?

356. Explain in detail about Alpha taxonomy, Beta and Gamma taxonomy using latest keys.
357. How do you distinguish Monochidae, Rhabditidae and Diplogasteridae?
358. Explain in detail International code of zoological nomencalture.
359. **Write briefly on vinous habitats of nematodes.**
360. Narrate the relationship of soil type and nematode survival.
361. What is the role of soil moisture in nematode survival?
362. List important dormant structures in plant parasitic nematodes.
363. List any four nematode phyla which are closely related.
364. Mention two plant parasitic nematodes that can survive well in high moisture habitat.
365. Differentiate osmobiosis with anoxybiosis.
366. Write short notes on nematode wool, signaling in nematodes and soil temperature v/s nematodes.
367. Mention the factors that affect nematode survival
368. Explain dormancy in nematodes and related terms
369. List out the various organisms associated with nematodes in soil.
370. Mention four commercially feasible, potential bioagents of plant parasitic nematodes
371. Discuss in brief community pattern in a rhizosphere.
372. **Define community and diversity.**
373. Write short notes on endoparasitic fungi and trapping devised in nematode trapping fungi.
374. **Explain in brief about the factors influencing sex ration in nematodes.**
375. How do you differentiate 'Frequency and density'?
376. What do you know about 'Quiescence and Cryptobiosis'?
377. **Define community analysis. Mention its various parameters**
378. **Define trophic group. How nematodes are classified based on this?**
379. List the major nematode problems in plantation crops. Explain the symptomatology and management practices for any one nematode problem in this crop.
380. **How do you induce shift in sex ration in nematodes by artificial** means?
381. Explain the factors affecting population change of a nematodes in a community.
382. Narrate the role of water in the migration of nematodes in soil (any four points).
383. Write short notes on clustering and Index of similarity.
384. **How do you induce shift in sex ration in nematodes by artificial** means?
385. Narrate the role of water in the migration of nematodes in soil (any four points).
386. **Write in brief the significance of Nematodes in Agriculture.**
387. Mention four major historical milestones in Nematology.
388. Differentiate the terms 'Fecundity and Fertility' with examples.
389. List four economically important nematode diseases in India with casual agents.
390. Write a labeled neat diagram of a female plant parasitic nematode.
391. How do you classify nematodes based on their feeding habit?
392. Explain in brief symptoms caused by nematodes in crop plants
393. Explain integrated nematode management with an example.

394. Differentiate Parthenogenesis and Hermophroditism, Monorchic and Didelphic, Trap crop and antagonistic crop.
395. Write short notes on general life cycle of a plant parasitic nematode, migratory endoparasitic nematode, pseudocoelom.
396. **Define sensory stimuli. List important stimuli for nematodes. Explain any one in brief.**
397. What are the major types of mathematical models for assessment of crop damage?
398. Explain modes of attack of insect pests by EPN's, with suitable examples.
399. Write in brief on current developments on EPN's in India.
400. **Define Avermectins. Mention the modes of action of Avermectins.**
401. Explain four major control strategies by exploring sensory stimuli in nematodes.
402. Write in brief about various mechanisms of interactions between nematodes and fungi with suitable examples.
403. **Mention the significance of nematode secretions. Explain secretions from various sensory organs of** nematode body.
404. **Define avermectins. Explain the modes of action of avermectins.**
405. **Narrate the significance of hormone metabolism in nematodes.**
406. Mention any four families of entomopathogenic nematodes
407. **Write short notes on poison baits for nematode control, cauliflower disease complex, juvenile** hormones.
408. **Briefly explain the general distribution of nematodes in nature.**
409. Narrate the nematode relationship with other animals.
410. **Briefly explain the general life cycle of a plant parasitic nematode.**
411. Classify nematodes based into trophic groups.
412. List the factors that determine rate of increase and size of a nematode population.
413. **List the factors which influence sex ratio in nematodes & explain** any one.
414. Explain the role of soil moisture in nematode ecology.
415. **Write then significance of organic matter in soil environment.**
416. Differentiate Blastocele and Schizocele with suitable diagrams.
417. Discuss in brief major genes involved in development and reproduction.
418. Explain major transgenic strategies for enhancement of nematode resistance in plants.
419. Discuss the important foliar nematodes under the genus *Aphelenchoides*.
420. List eight major points to substantiate that '*Meloidogyne* is ubiquitous' in nature.
421. What do you know about proteinase inhibitor coding genes?
422. **Discuss in brief the significance of nematicidal proteins** with suitable examples.
423. Discuss the major morphological features of *Bursaphelenchus xylophilus*.
424. Explain in brief the chemo-disruptive peptides.
425. Discuss in brief, plant quarantine structure and policy in India.
426. Write a short notes on the nematode effectors.
427. How do you differentiate *Tylenchorhynchus* from *Tylenchus*?
428. **Briefly explain the syncytia specific promoters**

429. List major steps involved in developing a mathematical model for nematode crop damage assessment
430. Highlight the importance of nematode resistance genes.
431. **Mention five characteristic features of the genus *Pratylenchus***
432. Why do nematodes have cylindrical body?
433. Write a detailed procedure to screen chemicals for their nematicidal activity.
434. **While flooding is a means to manage most nematodes, standing water does not affect rice root nematode in rice fields. Give reasons.**
435. **Discuss briefly molecular Koch's postulates pertaining to nematodes.**
436. **Briefly explain the use of nematodes as indicators of toxic environmental contaminants.**
437. List major differences between Coelenterata and Nematoda.
438. **What do you know about the association and the final effect of interaction between phytonematodes** and Mycorrhizal fungi?
439. **Mention significant features of the subclass Diplogasteria.**
440. List two major characteristics each for Mononchida, Dorylaimida and Rhabditida.
441. **Explain the technique to study the efficacy of nematicidal plants** against burrowing nematode.
442. What procedure you adapt to study the effects of nematicidal plants against fungi associated with the rhizosphere?
443. Discuss anti-nematode genes in detail.
444. What are the advantages of adopting molecular-based detection tools for the diagnosis of phytonematodes?
445. Discuss on the association of phytonematodes with *Rhizobium* in cowpea.
446. **Briefly discuss DNA-based markers used for diagnosis and description** of new species nematodes.
447. Differerntate the terms 'Habitat and Niche', 'Triploblastic and Pseudocoelomic'.
448. What do you know about PCR-based detection methods for diagnosis of nematodes?
449. What are the major considerations before sampling for nematodes?
450. Discuss in detail on the host races of root-knot nematode in India.
451. How do you isolate and characterize the active principles of nematicidal plants?
452. List major regulatory measures to manage phytonematodes. Explain any one in detail.
453. **Discuss the significance of crop rotation in nematode management.**
454. Explain in detail nematode trap crops and antagonistic crops with examples.
455. Suggest the most important non-chemical methods of management of the burrowing nematode in banana.
456. **Briefly explain the methodology for soil fumigation to manage** nematodes.
457. What are the different methods of nematicide application?
458. Discuss the modes of action of fumigant nematicides.
459. Differentiate contact and systemic nematicides with respect to their mode of action with merits and demerits.
460. **Briefly discuss the movement of nematicides in soil**
461. List major principles of Integrated Nematode Management.
462. What do you know about the application of recombinant DNA technology in Nematology?

463. Explain the roles of lectins in nematode management.
464. What are the major desirable characteristics of a typical biocontrol agent?
465. List potential bioagents of phytonematodes.
466. Write detailed note on nematophagous fungi with examples.
467. Discuss suppressive soils in relation to nematodes.
468. Explain the role of AM fungi in nematode management with suitable examples.
469. How do you study the effects of major physic-chemical factors on the nematicidal properties on botanicals.
470. Correlate nematode community with its ecosystem ecology.
471. **Discuss polyspecific nature of phytonematodes.**
472. Signify the role of toxins and polysaccharides in pathogenesis.
473. Narrate the process of nematode feeding.
474. Comment on sex ratio in nematodes.
475. Suggest a suitable procedure to evaluate nematicidal effects on hatching and emergence of juveniles from cysts.
476. What do you know about secretory–excretory products in nematodes?
477. List major ten enzymes present in nematodes.
478. Discuss the details of oesophageal gland secretions in parasitism.
479. Explain in detail about nematodes of quarantine importance.
480. **Briefly explain molecular targets of nematode effectors in host cells.**
481. Narrate the step-wise symptom development by *Anguina tritici* in wheat.
482. **Define effectors and signify their importance in phytonematodes.**
483. Explain in detail the interaction effects of the burrowing nematode with panama wilt fungus in banana.
484. Signify the role of soil organic amendments in nematode management.
485. How do you estimate infestation levels of *Meloidogyne* spp. in roots by using a rating chart?
486. Differentiate the terms 'Natality and Mortality'.
487. What are the important factors that can affect the performance of organic amendments?
488. **What factors influence virus retention, specificity and dissociation in nematode vectors of viruses?**
489. Why are systemic nematicides preferred for nematode management?
490. What are the major enzymes secreted by phytonematodes during penetration and formation of feeding sites?
491. What are North Carolina host differentials and list out host differentials for *Meloidogyne* species and races?
492. Substantiate that nematode community is more diverse and stable in a forest bed compared to an **agricultural field.**
493. Discuss various methods of nematicide application.
494. What are the major modes of action of fumigant nematicides?
495. How do you extract *Creconemoides* species from soil? Which is the best technique?
496. **Briefly explain the application of recombinant DNA technology in nematode management**
497. What do you know about endotoxins of *Bacillus thuringiensis*?

498. What are the modes of action of rhizobacteria?
499. What are the mechanisms of nematode management by *Trichoderma viride*?
500. Discuss the role of opportunistic fungi in nematode management
501. List major desirable characteristics of an ideal biological control agent.
502. **Briefly explain the role of phenols in nematode resistance.**
503. **Discuss the significance of crop rotation in nematode management** with suitable examples.
504. How do you obtain large quantities of living, adult females of root-knot nematodes? Explain the technique.
505. **Briefly list major steps in nematode microtechnique studies. Explain ribbon mounting and embedding** techniques.
506. Discuss in detail the procedure for culturing nematodes on carrot discs.
507. What are the major steps involved in the preparation of nematode mounts?. Explain each step in detail.
508. Explain the process of cyst formation and development of *Globodera pallida* in potato.
509. What are the fundamental components of nematode sampling?
510. What are the advantages of using entomopathogenic nematodes in agriculture?
511. **Briefly explain major properties of entomopathogenic nematodes.**
512. **Mention five potential examples of entomopathogenic nematodes** used in biological control.
513. **Briefly discuss the application of molecular markers** in identification of phytonematodes.
514. **Define gene cloning and plantibodies.**
515. Narrate the life cycles for reniform nematode, citrus nematode and red ring nematode of coconut.
516. Explain the fate of nematicides applied to the soil.
517. Discuss in brief on the biosafety policies in India.
518. **What is the significance of DNA bar-coding ? Discuss in detail.**
519. How do you mass culture *Ditylenchus dipsaci*?
520. Discuss, in general, occurrence of physiological races in nematodes.
521. Explain in detail the life cycle of *Rhadinaphelenchus cocophilus*.
522. Write a note on the ectoparsitic nematodes and discuss their modes of feeding.
523. **List top five economically important phytonematodes in India. Explain the life cycle, symptomatology** and management for any one nematode.
524. Discuss the major aspects of interaction between root-knot nematodes and *Rhizobium*.
525. Write in detail major symptoms caused by nematode on below ground plant parts, with suitable examples.
526. How does a host plant's physiology is affected by a nematodes?
527. **What is the meaning of Post Entry Quarantine? Mention is significance.**
528. How does a host responds to a nematode infection? Discuss different types of the responses.
529. Explain oogenesis and spermatogenesis in nematodes.
530. List major extraction techniques for nematodes from soil. Explain the methodology for any one method.
531. Write in detail major punctuations, transverse markings and longitudinal markings on a nematode cuticle.

532. **Briefly discuss the muscle layer in nematodes.**
533. What do you know about the stomodeum in nematodes?
534. Explain with suitable diagrams major types of excretory systems present in nematodes, in general.
535. Give a brief description of a nematode egg with a suitable labeled diagram.
536. **What are the major points to be considered during field evaluation of a nematicide?**
537. **List major techniques of artificial culturing of nematodes in laboratory conditions. Explain one method in detail.**
538. **Briefly discuss the vertical distribution of nematodes in soil.**
539. What do you know about the osmoregulatory mechanisms being operated in nematodes?
540. Differentiate the water agar assay technique of testing nematicides from the microtiter plate assay technique.
541. Explain the nematode cuticle with its different layers.
542. What do you know about hypodermis and its glands?
543. Explain in detail various methods of nematicide application.
544. Compare the symptomatology of ufra and white tip diseases in rice.
545. Explain in detail the management strategy for wheat gall nematode.
546. List major six families under Tylenchoidea. Explain the features for any two families.
547. Provide major constituents of Johansen's quadruple stain, Sass safranin /fast green stain and Triple quadruole stain.
548. Discuss in detail about major peripheral nervous organs in nematodes.
549. **Briefly explain longitudinal ridges and its types in nematodes.**
550. Write a brief note on the digestive enzymes of a phytonematode.
551. How do you broadly classify nanoparticles?
552. Discuss in detail, the nematode distribution in nature.
553. Write short note on the osmoregulation and excretion in nematodes.
554. Discuss growth of nematodes.
555. **Briefly explain the type of locomotion in nematodes.**
556. How do you mass culture golden cyst nematode of potato?
557. Discuss in detail the mass production of ***Pseudomonas fluorescens***.
558. How do you isolate nematode-trapping fungi?
559. **Furnish the protocol in the form of a flow chart for mass production of** *Trichoderma viride*.
560. Detail the procedure to mass culture entomopathogenic nematodes, in general.
561. List four important text books in Plant Nematology with titles and authors.
562. Furnish three major periodicals in Nematology.
563. **Define the terms Transcriptomics, Proteomics and Metabolomics.**
564. How do you test the phytotoxicity of a nematicide? Discuss the technique.
565. Compare the mechanisms of actions of organocarbamates with that of organophosphates.
566. **Explain in detail the significance of the study of nematode population dynamics.**
567. **Mention the significance of sampling for nematodes. What are the criteria to be considered while soil** sampling?

568. Discuss on the ecology based nematode management systems.
569. What do you know about engineering nematode resistance in crop species?
570. Explain in detail molecular markers for nematode-resistance genes.
571. Discuss on natural resistance genes pertaining to nematodes.
572. Distinguish the basic differences among RFLP, RAPD, AFLP techniques.
573. What do you know about nematode virulence and parasitism genes?
574. **Briefly narrate the** transgenic strategies for enhancement of nematode resistance in plants.
575. **What is nematicide detoxification? what are its implications** in nematode management?
576. Correlate the interaction effects of root-knot nematodes with major soil fungal pathogens in vegetable crops.
577. Narrate the role of nanotechnology in nematode management.
578. Discuss the aspect on nematodes as model systems for nutritional studies.
579. How do you adapt a technique for critical-point drying of nematodes?
580. What do you know about nematode effector proteins?
581. Explain how nematodes act as the indicators of pollution, in general.
582. Substantiate the statement "phytonematodes are threat to crop production".
583. **Briefly discuss the salient features of the anatomy of a male burrowing nematode with a neat diagram.**
584. **How do you employ biochemistry as a tool in identification and its usefulness in understanding the** nature of nematode parasitism?
585. Discuss the major steps in the formation and anatomy of giant cells induced by *Meloidogyne species.*
586. Furnish the information on the various mechanisms of resistance to root-knot nematodes.
587. Discuss in detail the preliminary and advanced evaluation of nematicides.
588. Explain the major modes of action of nematicides, in general.
589. **What is the significance of the cropping systems in nematode management?**
590. Furnish the detailed procedure to study the penetration of host roots by endoparasitic nematodes.
591. How do you use 'enemy plants' to manage phytonematodes? Substantiate with suitable examples.
592. Discuss in brief about crop loss assessment due to nematodes.
593. What do you know about the forecasting of nematode diseases?
594. List major taxonomic characters which have been employed in nematode **identification.**
595. Write a detailed note on the hypodermis in a nematode body.
596. Describe the general process of embryogenesis in nematodes.
597. How do plants respond to a nematode infestation at different growth stages?
598. How do you inhibit steroid / hormone metabolism in nematodes?
599. **Define ecdysteroids. Briefly explain their significance.**
600. **What are juvenile hormones? Briefly discuss their importance** in nematodes.
601. **Define sterols. Discuss about the sterol metabolism inhibition** in nematodes.
602. How do you use density gradient technique to purify the eggs of root-knot nematodes?
603. List major nematode stimuli in nature. How do you exploit nematode sensory stimuli in nematode management?

604. Discuss in detail major modes of action of avermectins.
605. **Define pheromone. Explain in detail pheromone metabolism in nematodes.**
606. List major ecological factors affecting nematode distribution and survival in soil and explain any one in detail.
607. Differentiate syncytia and giant cells.
608. Mention precautions to be taken while spraying a chemical to manage nematodes in field.
609. What do you know about nematodes being used as models to study biological aging?
610. Discuss major criteria to be considered to develop a mathematical model for the crop damage assessment by nematodes.
611. List major accomplishments in Nematology in the 'omic' era.
612. What is the procedure to detect nematode infestation by using aerial infrared photography technique?
613. What do you know about Pest Risk Analysis? Discuss its importance
614. Furnish the relationship between viruses and nematode vectors with respect to the transmission, acquisition and inoculation.
615. **Define Plant Quarantine. Discuss different types of plant quarantine pertaining to nematodes.**
616. What do you know about biosafety and biosecurity?
617. **Define Intellectual Property Rights and mention its various forms.**
618. List major nematode problems in tuber crops. Explain symptomatology for one nematode.
619. **Briefly discuss on innovations in teaching, research and extension in Nematology.**
620. Summarize the effects of climate change on nematodes under natural conditions.
621. What are the strategies to be considered before crops loss assessment by nematodes?
622. Differentiate single-point models and multiple-point models.
623. Provide the 'Cobb's formula' used for measuring nematodes.
624. **List major factors influencing the choice of nematode bioassay** techniques
625. **Explain the technique to study the effect of cell free filtrates of fungi on mortality of second stage** juveniles of nematodes.
626. **What do you know about nematode extraction efficiency and its relativity?**
627. List four major cyst extraction techniques and explain any one in detail.
628. What do you know about microplot techniques in Nematological research?
629. **How do you employ remote sensing techniques in Nematology? Mention its significance along with** major techniques.
630. What are the major techniques for testing nematicides? Explain one technique in detail.
631. **Explain the significance of botanicals in nematode management. List four major botanicals proved** effective against phytonematodes.
632. What are the most important sampling procedures for phytonematodes in horticultural orchards? Explain one pattern in detail.
633. Discuss in detail on host plant resistance in nematodes.
634. Explain the technique of *in vitro* culturing of entomopathogenic nematodes?
635. **Discuss in brief about nematode specific resistance genes.**
636. **Briefly explain major aspects of Transgenic proteinase inhibitor -based management of phytonematodes.**

637. List major criteria to be considered for nematode drawings for photographic reproduction.
638. **Define hatching factor. Give major five hatching factors specific to nematode genus with examples**
639. Discuss in detail various types of locomotion in nematodes and also in particular to phytonematodes.
640. Discuss in detail on the interaction of root-knot nematode with bacterial pathogens.
641. What are the causes of variability in nematodes?
642. Discuss the role of nematode cuticle with a neat labeled diagram.
643. What are the major specialized morphological and physiological adaptations of nematodes to plant parasitism?
644. Furnish the technique to evaluate potential biocontrol agents of *Globodera rostochiensis*.
645. **How do you correlate the influence of environment on nematode sex** determination?
646. Explain oogenesis and spermatogenesis in nematodes.
647. How do phytonematodes respond to the environmental stress?
648. List major digestive and oxidoreductive enzymes produced by nematodes
649. **Briefly explain the formation processes of syncytia and giant cells.**
650. **Briefly discuss the nematode penetration and migration into host** tissues.
651. Explain in general, the interaction process of nematodes with soil fungal pathogens.
652. How do you culture axenic nematodes?
653. Summarize the latest status on virus transmission by nematodes.
654. Write a short notes on the chemical composition of nematodes.
655. Substantiate the statement that 'cuticle is a 'metabolically active organ' in nematodes.
656. Discuss in brief, the respiration in nematodes.
657. **What are the major factors that influence the egg hatching in nematodes? Explain any one factor in** detail.
658. **Discuss in brief the effect of phytonematodes on the nodules and N-fixation in legumes.**
659. What are the major rules pertaining to the nomenclature of nematodes?
660. Explain in detail the technique to study the *in vitro* **effect of cell free filtrates of fungi on nematode** hatching?
661. Discuss in detail 'soil as environment ' for nematodes.
662. **In your opinion, what are the five most distinguishing characters** of nematodes?
663. List four economically important nematode diseases in India. Explain the characteristic symptoms and practical management practices for one disease.
664. **Explain the technique to study the effect of cell free filtrates of bacteria on mortality of second stage** juveniles of nematodes.
665. What are the various types of survey? Which one is most commonly used for nematode survey?
666. **What are the benefits of the combined Cobb's sieving and Baermann funnel technique?**
667. List two major diagnostic features each for *Anguina, Pratylenchus, Heterodera, Helicotylenchus* and *Hoplolaimus.*
668. Discuss the position of nematodes in the animal kingdom by comparing with other phyla.
669. **Briefly explain the major nematode morphometric parameters, discussing** De Man formulae.
670. Discuss muscle layers in nematodes with examples and diagrams.

671. Narrate the processes of virus acquisition and transmission by major nematode species.
672. What do you know about the physical methods of nematode management? Explain with suitable examples.
673. Provide integrated approach for the management of cereal cyst nematode on wheat.
674. Narrate the technique to isolate bioagents from egg- mass of nematodes.
675. How do you manage a complex diseases on banana involving *Radopholus similis* and *Fusarium oxysporum* f.sp. cubense?
676. List major techniques for staining of nematodes in plant tissues. Furnish the detailed procedure for a most commonly used technique.
677. How do you classify nematicides? Mention each with suitable examples
678. Explain merits and demerits of biological mode of nematode management.
679. Differentiate the use of 'radar' and 'sodar' techniques of remote sensing in Nematology.
680. Discuss in detail the regulatory method of nematode management with examples.
681. Why hot water treatment has gained importance to manage nematodes?
682. **Substantiate the efficacy of green manure cropping in nematode management with suitable examples .**
683. Explain in detail various mechanism by which organic amendments manage nematodes
684. **What are the benefits of nematicides over other methods of nematode management? Explain with** examples.
685. Diagrammatically illustrate different types of excretory systems in nematodes.
686. Which is the most common type of reproduction in root-knot nematodes?. Discuss its details.
687. How do you diagnose the infestation by reniform nematodes in crops?
688. Mention the most common type of nematode –bacterial complex disease in vegetable crops. Explain the symptomatology.
689. Correlate the effect of climate change on nematode populations in crops.
690. What is cinemicrography? Explain the detailed procedure.
691. What are amphids? What are its different types? Explain with diagrams.
692. In what manner cultural nematode management practices are better than chemicals? Substantiate with examples.
693. List major techniques of preserving nematode infested plant tissues. Explain the detailed procedure of a commonly used technique.
694. Draw a neat diagram of nematode body wall and label the parts.
695. Explain the mode of formation of a giant cell in host, with suitable diagrams.
696. List major phytonematodes infesting rice. Explain the survival and symptomatology for one nematode.
697. Discuss on the host range of root-knot nematodes.
698. Explain the race occurrence in the burrowing nematode with examples.
699. Discuss the survival and distribution of cyst nematodes in potato during winter and summer seasons.
700. Explain in detail the life cycle of *Tylenchulus semipenetrans* in citrus, with suitable diagrams.
701. Explain the symptomatology of lesion nematode in crop plants with diagrams.
702. Write the detailed procedure for the technique to estimate total amino acids in nematode infested plant material.

703. List three major recently introduced, commercially available bio-agents against nematodes.
704. Discuss in detail the habitats of nematodes in general, with examples under each.
705. What are the major characters of the phylum Rotifera?
706. Explain inner body cuticle of a nematode in detail with labeled diagram.
707. Explain the procedure for staining faint surface structures in **nematodes, with its significance.**
708. Explain the plant parasitism in Adenophorea.
709. **According to you, what are the five major historical developments** in Nematology in India?
710. Explain the detailed procedure for the most commonly used technique to mount the vulval cones of cyst nematodes.
711. Compare and discuss the potentiality of bacterial and fungal bioagents against nematodes.
712. Mention major six super families in the class Adenophorea.
713. Differentiate epicuticle and exocuticle in nematodes.
714. **List major punctuations in nematodes and mention their significance.**
715. **Explain the significance of lips in phytonematodes in plant parasitism.**
716. Mention different types of pharynx in nematodes, with labeled diagrams.
717. List major post embryonic changes that occur in nematodes.
718. What are the major orders in Secernentea?
719. **What do you know about hypodermis? Mention its significance in** nematodes.
719. Explain in detail the symptomatology of *Anguina tritici* in barley.
720. Compare the life cycles of *Rhadinaphelenchus cocophilus* and *Criconema.*
721. Explain the survival pattern and symptoms caused by cereral cyst nematode in oats.
722. Discuss major characters of the order Aphelenchida.
723. Compare the characteristic features of the phyla Nemata and Nematomorpha
724. **Write a neat diagram of pseudocoelom in nematodes and mention its significance.**
725. What are the major steps involved in nursery bed fumigation in polyhouse crops to manage nematodes? Explain with neat diagrams
726. **What is bare-root dip treatment? Mention its significance in nematode** management.
727. Explain major types of stoma in nematode with examples and diagrams.
728. **What are ganglia? Mention the significance in nematodes.**
729. Differentiate the structures alae, striae and ray in nematodes, with neat diagrams.
730. What do you know about somato-intestinal muscles in nematodes?
731. **How do you lay out a field demonstration for the efficacy of a treatment over control? Explain in** detail.
732. How do you differentiate deirid from hemizonid?
733. **Mention the significance of ovary in nematodes. Explain various** types with diagrams and examples.
734. What are the major nematode problems in tube rose? Explain the symptomatology for one nematode.
735. Mention the major pre-and post-embryonic changes in nematodes.
736. Explain in detail the sexual dimorphism in nematodes.
737. Write a short note on the 'telamon' structure in nematodes.

738. Discuss the spermatogenesis phenomenon in Tylenchida.
739. **List five major stains used in nematode staining techniques. Explain the detailed procedure for one** staining technique.
740. Mention the role of rectal glands in nematodes.
741. Write a neat diagram of oesophagus in Tylenchida and label the parts.
742. **Explain the significance of pseudocoelomic fluid in nematodes.**
743. What are papillary nerves? Mention various types and their significance.
744. In spite of the severe economic losses caused by nematodes, why they are not very popular?
745. Summarize the role of nematodes in agriculture with examples and illustrations.
746. How do you distinguish root-knots and bacterial nodules in pulse crops?
747. Explain fumigants with examples and modes of application.
748. **List most important fixatives and preservatives used for nematode fixing. Mention the roles of these.**
749. Explain in detail biochemical mechanism of nematode resistance, with special reference to *Radopholus similis* infesting banana.
750. Discuss in brief major problems associated with the use of chemical nematicides
751. How do you economize the chemical nematicide usage?
752. Expalin nature of resistance in tomato and brinjal against Meloidogyne javanica.
753. Write short notes on the embryo rescue technique and protoplast fusion technique.
754. **Discuss in detail the significance of Recombinant DNA technology in nematode management.**
755. What do you know about the compatibility of bioagents used to manage nematodes and fungicides?
756. **Discuss in detail the efficacy of** *Pochonia chlamydosporia* against root-knot nematodes.
757. **Compare the bioefficacy of Trichoderma harzianum and Purpureocillium lilacinum with respect to** nematode management and application methods.
758. Furnish a detailed management schedule for root-knot nematode infesting capsicum in polyhouse using *Trichoderma viride*.
759. Explain in detail various reproductory organs in a male nematode, with labeled diagrams.
760. Explain in detail different types of oesophagus in Rhabditid and Diplogastrids, with labeled diagrams.
761. **Discuss the methods and tools currently used for the identification** of plant parasitic nematodes.
762. Explain the concept of DNA barcode for nematode taxonomy.
763. **What is Real time PCR technique? Explain the procedure and its significance in relation to nematodes.**
764. Discuss in detail, microarrays.
765. **Briefly discuss the influence of the host crop on the population** of a phytonematode.
766. How do you correlate the effect of variations in the environmental factors on nematode penetration into the host plant?
767. **Discuss the role and significance of traditional breeding for** resistance to nematodes.
768. **What are the factors that influence the movement of nematodes in soil?** Provide examples.
769. **Write a short note on the phytonematodes and their host specificity**
770. Which are the most commonly occurring fungal parasites in the rhizosphere of tomato plant that are antagonistic to root-knot nematodes? Explain their modes of action.

771. **List important techniques used for killing and fixing nematodes. Explain one most commonly used** technique.
772. List major differences between the orders Tylenchida and Dorylaimida.
773. Write a brief note on the super family Criconematoidea with major families and their characteristics.
774. **List major five species under Heterodera. Explain the symptomatology, life cycle and management** aspects for an economically important species.
775. Describe the life cycle, spread and feasible management options for the red ring nematode in coconut.
776. Explain the symptomatology, life cycle and management schedule for white tip nematode in rice.
777. Signify the roles of 'feeding sites' in diseases caused by nematodes.
778. How do you categorize ectoparsitic nematodes based on their feeding habits?
779. What are 'meadow' nematodes? Explain the damage caused by them with management choices.
780. Write a brief note on the foliar nematode diseases of ornamentals with symptoms and life cycle.
781. What do you know about nematodes attacking mushrooms? Signify the loss with appropriate management schedule.
782. **Briefly explain the major bioassay techniques employed in Nematology.**
783. Discuss the nematode population distribution in space and time.
784. **Correlate the influence of seasons on the nematode populations,** in general.
785. Provide a suitable, practically feasible integrated management approach for root-knot nematodes in gerbera grown under protected cultivation.
786. How do you differentiate 'slow decline' from 'spreading decline' in citrus? Explain with reference to the etiology and symptoms.
787. Illustrate pictorially, major types of stoma in nematodes.
788. Diagramatically represent the cross section of esophagus with suitable labelling.
789. **How do you determine the presence of a nematode problem in a field?** Discuss in detail.
790. **List five major nematicides and provide their common names, modes** of action and formulation.
791. What are the basic requirements to establish a Nematology laboratory? List essential minimum equipments required.
792. Discuss in brief recent innovations in teaching, research and extension Nematology.
793. Write the step-wise procedure for screening for resistance in vegetable crops against root-knot nematodes.
794. Furnish the taxonomy details of the genus *Meloidogyne*, **with specific reference to the adult female** characters.
795. Discuss in detail the potentiality of phytonematodes as predisposing factors for secondary plant pathogens in soil.
796. **What do you know about 'anaesthetization' of nematodes? What is its significance?**
797. Furnish detailed procedure for mass culturing of *Ditylenchus destructor* on potato tubers.
798. What is the purpose of surface sterilization in Nematology? Provide a detailed procedure for surface sterilization and inoculation of eggs of *Meloidogyne javanica.*
799. What do you know about the role of endophytic bacteria in nematode management?
800. Mention major techniques for axenizing nematodes. Explain anyone method in detail.

801. Provide the systematic position of the genus *Heterorhabditis*.
802. Write the symptomatology, survival and practically feasible management measures for *Ditylenchus destructor.*
803. Furnish the detailed information on the mode of infection and management of *Aphelenchoides composticola*.
804. Write a brief note on cyst forming nematodes with special reference to the symptoms and host range.
805. Discuss in detail the life cycle, survival and symptoms caused by the wheat gall nematode.
806. **List five practically feasible management strategies for** *Radopholus similis* in black pepper.
807. What do you know about the nematode problems in ornamentals? List economically important nematodes on these crops.
808. Nematodes are a major menace in vegetables. Substantiate this statement.
809. **In spite of their significance, nematode diseases are not as popular as diseases caused by fungi, viruses** and bacteria. How do you defend this statement?
810. List major nematodes infesting ginger. Provide suitable management measures.
811. What extension activities you suggest to improve the awareness among the farming community about nematode diseases?
812. Morphologically how do you distinguish between *Xiphinema diversicaudatum* and *X. americanum*?
813. **Discuss in detail about the 'modified host cells' by different nematodes with suitable examples and** labeled diagrams.
814. Discuss on the nematode problems in onion. Furnish the management options for one major pest.
815. What are the economically important nematode problems in grass? Provide suitable management schedule.
816. What do you know about 'Kalahasthi melody'? Discuss in detail.
817. **List five major seed borne nematode diseases and provide suitable** management measures.
818. Discuss in brief the major agnotobiotic or xenic culture techniques employed for nematodes.
819. How do you correlate the 'Pest Risk Analysis' with nematode diseases?
820. What do you know about the 'Remote sensing' in particular reference to nematode diseases?
821. Write short notes on the 'nematode excretion'.
822. Discuss in detail on the interrelationship between the nematode population dynamics and economic thresholds.
823. **Briefly narrate the general sequences during the development of** reproductive system in nematodes
824. List ten major nematode morphological characters of taxonomic importance.
825. Discuss the major aspects related to the community structure of nematode parasites of higher plants.
826. How do you culture microbivorous and predatory nematodes? Explain major techniques.
827. **How do you lay out a field demonstration on the management of burrowing** nematode in banana?
828. Explain the 'Hatching test technique' of *Heterodera* and *Globodera* cysts
829. Discuss in detail on 'Polyhouse Nematology' with suitable examples.
830. Differentiate Globodera rostochiensis and G. pallida with special reference to the cyst morphology, spread and practical management options.
831. **Furnish field applicable integrated management measures for Pratylenchus coffeae.**

832. **Briefly discuss the role of biochemical marker for identification** of phytonematodes.
833. Summarize the major molecular aspects of nematode parasitism.
834. **What are the major abnormalities resulting from nematode injury to buds, growing points and flower** primordia?
835. List major nematode pests on tobacco. Provide suitable management measures to one economically important pest.
836. What are the major nematode diseases of oil seed crops? Describe the characteristic symptoms.
837. Discuss on the nematode problems in ginger or turmeric with special reference to the symptoms, life cycle and management.
838. Explain in brief the nematode -fungal and or bacterial complex **diseases in flower crops.**
839. List advantages and disadvantages of using entomopathogenic nematodes.
840. Write a short note on the nematode resistance genes.
841. **Briefly discuss the influence of soil pore size and soil texture on nematode population with examples.**
842. What do you know about 'undulatory propulsion' movement in nematodes?
843. Describe major steps involved in the orientation of nematodes toward plant roots
844. Correlate the effects of essential climatic factors on the population of R*adopholus similis* in banana.
845. **Summarize briefly the economic importance of nematodes in agriculture**
846. Discuss in detail the morphology of second stage juveniles and females of root-knot nematodes.
847. **Briefly explain the phylogeny of** *Meloidogyne.*
848. Explain the interaction among concomitant populations of nematodes in plant rhizosphere.
849. What is the role of a Nematologist in developing resistant cultivars?
850. How do you estimate the egg content of cysts in Heterodera species? Outline the technique.
851. List various sense organs located on the cuticle of a nematode.
852. Write a short note on the cephalic and caudal sensory organs in nematodes, with labeled diagrams.
853. Write the events that occur during attraction and penetration into the host tissues by the nematodes.
854. How do you differentiate the plant symptoms caused by nematodes and nutritional disorders? Discuss with suitable examples.
855. Explain in brief the effect of nematode infestation on the respiration of host plants.
856. Substantiate the statement that "nematode infestation apart from reducing crop yields, also lead to abnormal end products" with suitable illustrations and examples.
857. **Briefly discuss the significance of root dip treatment to manage** phytonematodes.
858. List major nematode problems in medicinal crops. Explain the management strategies for one major nematode.
859. **Mention five nematode-bacterial complex diseases with causal agents.** Provide practical management measures for any one complex disease.
860. Explain the symptomatology of root-knot nematodes and fungal complex diseases in polyhouse grown crops.
861. **Discuss the benefits and drawbacks of using mycorrhizal fungi in** the management of nematodes.
862. Provide a suitable integrated management schedule for root-knot nematodes infesting capsicum grown under protected cultivation.

863. What are the characteristic symptoms of mulberry infested by root-knot nematodes? Explain the life cycle and major management options.
864. Compare and discuss the use of bioagents and chemicals for nematode management.
865. **What do you know about beneficial nematodes? Discuss in brief.**
866. Provide an integrated management schedule for mushroom nematodes.
867. List economically important nematode problems in aromatic plants. Explain the symptomatology for any one crop against one major nematode.
868. Morphologically illustrate various stages of *Radopholus similis*.
869. **Define a bio-agent. List major merits and demerits of using a bio-agent against nematodes.**
870. List economically important nematodes on black pepper. Provide practicable management measures for one nematode.
871. **Compare a 'cyst' with a 'gall' with respect to the origin, morphology and significance.**
872. **Briefly explain the nature of nematode secretions**
873. What are the major roles of Nematologists in the successful implementation of 'Integrated management' of plant diseases?
874. Explain the symptomatology of two major nematode problems in turmeric, with diagrams
875. **What is the significance of nanotechnology in nematode management?**
876. Discuss on nematode damage models.
877. Discuss in brief the plant parasitism in Secernentea with examples and diagrams.
878. Write a brief note on the alimentary canal of nematodes, with diagrams of major organs.
879. Discuss in detail the virus transmission mechanism in *Xiphinema.*
880. **Explain the procedure for Floatation-modified Fenwick can technique for the extraction of cyst** nematodes.
881. Discuss in brief 'alginate capsules' formulations related to entomopathogenic nematodes.
882. Compare and explain the variations in oesophagus in mermithoid, rhabditoid and diplogasteroid groups of nematodes.
883. Discuss various functions of the central nervous system in nematodes.
884. Explain the roles of avermectins in nematode management.
885. Discuss the interaction and its effect between sedentary endo-and migratory ectoparasitic nematodes with examples.
886. What is the meaning of 'ecological interaction' among phytonematodes? Explain with examples.
887. Explain the variations in amphidelphic female reproductive system, with examples and diagrams.
888. Write a short note on the storage of infective entomopathogenic nematode juveniles
889. What do you know about 'incompatibility' between a nematode and a host? Discuss its consequences
890. Write a neat diagram of an adult female of *Rotylenchulus reniformis*
891. Discuss the latest developments in seed treatment nematicides.
892. List major techniques to extract eggs from soil. Explain any one method in detail.
893. What is the exact mode of action of carbofuran 3G on phytonematodes?
894. Write a detailed note on the nematodes infesting shoot portion of plants with special reference to the symptoms and life cycle.

895. Explain the role of temperature on the disease caused by a nematode on a susceptible host.
896. Explain the life cycle of *Ditylenchus angustus* on rice and suggest suitable management practices.
897. **List specific responsibilities of a Nematologist while collaborating with a breeder in developing a** nematode-resistant variety.
898. Explain 'three part oesophagus' in tylenchids with a neat labeled diagram
899. What do you know about acetylcholine esterase? Discuss its sig**nificance in nematodes.**
900. What is the protocol for the mass production of entomopathogeni**c nematodes on artificial media?**
901. List four major fumigants along with their common names, which were very popular before their ban.
902. Compare and correlate the mechanism of transmission of viruses by Longidorus and Trichodorus.
903. **How do you diagnose the phytotoxic effects of a chemical in field ? Discuss in detail.**
904. Furnish the detailed procedure for the technique to estimate polyphenol oxidase in nematode infested plant material.
905. Discuss the interaction between *Radopholus similis* and *Pratylenchus coffeae* in a banana rhizosphere.
906. List major precautions to be taken before preparing a chemical **spray to manage nematodes in field.**
907. Differentiate amphidial nerve from somatic nerve with special reference to their functions.
908. **What are plantibodies? Furnish their significance in the development** of nematode resistance.
909. Discuss the symptomatology of *Bursaphelenchus xylophilus* and provide practically feasible management schedule.
910. How do you extract sluggish nematodes from soil? Provide a detailed procedure.
911. What do you know about 'Prominence Value and Importance value'?
912. **Narrate the significance of mesenteron in nematodes with a neat** labeled diagram.
913. **What are the factors that influence resistance in a host plant** against nematodes?
914. List major techniques employed to preserve biocontrol agents of nematodes. Explain the detailed procedure for one method.
915. Discuss in detail the environmental effects on nematicidal activity in soil.
916. Suggest a suitable cropping sequence to manage *Radopholus similis* in banana.
917. How does hermophroditism occurs in nematodes? Discuss its signi**ficance on pathogenicity.**
918. Explain the effect of mulching on nematode populations in soil with suitable examples.
919. Discuss the role of moisture on nematode orientation, migration and penetration into the host.
920. Provide the detailed technique to isolate bacteria associated with entomopathogenic nematodes.
921. Explain the biochemical mechanism of incompatibility between a host and phytonematode.
922. **Discuss major phytonematodes attacking floral organs of crop plants and suggest suitable management** measures.
923. Discuss isozymes techniques used to differentiate plant parasitic nematodes.
924. Menton three major techniques to prepare internal organs of nematodes for observation in scanning electron microscope. Explain any one technique in detail.
925. Explain the major steps involved in the formulation of a simulation nematode model.
926. What is the type of interaction that exists between a migratory endo-and a sedentary endoparasitic nematode? Explain with suitable examples
927. **List major types of nematicide application at field level. Mention merits and demerits of most** commonly used methods in India.

928. Write short notes on the uterus, oviduct and vagina in a female nematode
929. Why free living nematodes make good bioindicators? Discuss in detail.
930. Mention major molecular detection techniques employed in Nematology.
931. What do you know about the regulation and expression of transferred genes against nematode?
932. Explain the technique to measure nematode respiration with the biological oxygen monitor.
933. **Discuss briefly molecular Koch's postulates pertaining to nematodes.**
934. Substantiate why *Globodera* is considered as the highly specialized nematode.
935. Write a neat labeled diagram of an adult female nematode of Globodera rostochiensis.
936. What do you know about the phenol /Chloroform technique for extraction of cyst DNA? Explain the procedure.
937. Narrate in detail major molecular responses expressed by the host against nematode infestation.
938. **Discuss the role of bioinformatics in identification and analysis of effector proteins from** phytonematodes.
939. List major techniques to isolate nematode trapping fungi. Explain the procedure for one technique.
940. Discuss in detail the role of sense organs in nematode locomotion.
941. Narrate different ways of engineering resistance genes in crop plants to suppress the nematode infection and populations in the soil below the threshold level.
942. How do you differerntiate 'Epitheliocele from Gymnocele'?
943. What is 'etiological interaction" among nematodes? Discuss different types of etiological interactions with examples.
944. Furnish the detailed technique for the extraction of total DNA from nematodes.
945. Diagrammatically illustrate and discuss the perineal pattern for four most common species of *Meloidogyne.*
946. Explain in brief strategies for the development of resistant transgenics.
947. What is the procedure to collect the endospores of *Pasteuria penetrans* in studies related to the bio-management of nematode?
948. Explain major steps in the embryonic and post embryonic development of any phytonematode.
949. Explain the roles of proteinase inhibitors in nematode management.
950. **Discuss in detail the significance of PGPR in nematode management.**
951. Explain in detail the step-wise procedure for culturing Radopholus similis within mesocarp of coconut.
952. List major mechanisms involved in nematode management by green manuring.
953. How do you prepare the nematode infested root material for the biochemical estimations?
954. **Define the terms 'Phytosanitary Requirements, Phytosanitary Certification and International Standard** for Phytosanitary Measures'.
955. Discuss the mechanism of viruses transmission by nematodes.
956. Discuss the movement of nematicides in soil.
957. **Briefly explain the nematodes of quarantine importance in India.**
958. **List most commonly used fixatives and buffers in for nematode specimen in electron microscopic** techniques.
959. Discuss the position of Nematoda within metazoans.
960. **Discuss in brief the general guidelines for field application of** entomopathogenic nematodes.

961. **Discuss in detail the significance of modeling and computer simul**ation techniques in Nematology.
962. Discuss the major destructive and adaptive cellular changes caused by nematodes in plants.
963. Explain the most appropriate technique for the mass production of endophytic bacteria, antagonistic to phytonematodes.
964. Discuss in brief, the embryology of nematodes.
965. What do you know about the application of microwave energy for nematode management in soil?
966. Explain in brief the signal transduction in plant–nematode interactions.
967. Discuss in detail the procedure for the technique to mass culture *Purpureocillium lilacinum* against phytonematodes?
968. Signify the need of 'Extension Nematology" in the present scenario.
969. **What are photomicrographic techniques? Mention its significance** along with major types.
970. **What are azasteroids? Explain their significance in nematodes.**
971. Explain in detail the technique to study the *in vitro* **effect of cell free filtrates of bacteria on nematode** hatching?
972. **Define RNA interference. How do you exploit it in nematode management?**
973. **Explain in detail the significance of multispectral sensing in** Nematological research.
974. Explain the role of nematode enzymes in pathogenesis.
975. Discuss in brief the cytogenetic aspects of nematode evolution.
976. **Explain in brief about the ecological classification and distri**bution of nematodes.
977. Explain the modes of nematicidal action of *Purpueocillium lilacinum* and ***Pseudomonas fluorescens***
978. How do you manage phytonematodes in organic farming system?
979. Discuss in detail the bioassay technique of nematode populations.
980. What is Biointensive Integrated Nematode management? Discuss with illustrations
981. Narrate the improved technique to enumerate and observe parasites and predators of soil nematodes.
982. **What are the factors that influence the choice of nematode bioass**ay techniques?
983. List major types of installation of micro plots and discuss major limitations of microplot technique sin Nematological research.
984. Discuss in detail major approaches of estimating crop yield losses due to nematodes.
985. **What are nematode altered plants? Discuss the significance**
986. What are the most commonly used techniques employed for drawing and measuring nematodes? Provide the detailed procedure for one major technique.
987. How do you isolate nematode resistance genes? Provide the stepwise procedure to isolate *Mi-1* from tomato.
988. List important types of sections used in histopathology and histochemical studies in Nematology
989. Discuss the distribution of nematodes in soil during different seasons
990. Mention general requirements for nematode material preparation for observation in electron microscope.
991. **What is the significance of non-phytoparasitic nematodes in nature**? Explain with examples.
992. In what way the phylla Kinorhyncha and Gastrotricha are related to each other? Discuss.
993. Which is the most important sense organ in nematodes? Defend your answer.

994. Discuss the importance of somatic musculature in nematodes and its relation to plant parasitism
995. Explain in detail North Carolina Differential Host test for the identification of *Meloidogyne* species and races
996. How do you assess the reaction of banana plant to migratory endoparasitic nematodes under green house conditions? Discuss the most appropriate technique.
997. Write a brief note on the growth and development of a phytonematode under normal soil conditions.
998. How does change in planting time can prevent nematode infestation?
999. Discuss the embryology in nematodes with labeled diagrams
1000. List five major modes of actions of botanicals in nematode management.
1001. Briefly discuss on the phytonematodes which come under Global Quarantine Act.
1002. What do you know about the 'Nematode Advisory Service'? Discuss in detail.
1003. Discuss in detail the 'Bar Codes' for identification of strains and isolates of nematode specimens.

SOME REFERENCE BOOKS

SOME REFERENCE BOOKS

Ayoub, S. M. 1981. Plant Nematology. An Agricultural Training Aid. NemaAid Publications. Sacramento, CA. U.S.A. 195 pp.

Bajaj, H.K., R.S. Kanwar, D.C. Gupta (eds) 2011..Handbook of Practical Nematology , **Scientific Publishers,** Jodhpur, Rajasthan, India, 150 pp..

Barker, K. R., C. C. Carter, and J. N. Sasser. 1985. An Advanced Treatise on Meloidogyne. Vol. II. Methodology. N. C. State Univ., Raleigh, 223 pp.

Bird, A. F. and J. Bird. 1991. The Structures of Nematodes (2nd ed). Academic Press, NY. 316 pp.

Croll, N. A. and Bernard E. Matthews. 1977. Biology of Nematodes. Halsted Press, John Wiley & Sons, Inc, New York, 117 pp.

Dasgupta, M.K.. 1998. Phytonematology. Calcutta, Naya Prokash., 846 pp.

Dropkin, Victor H. 1980. Introduction to Plant Nematology. John Wiley & Sons, Inc. 336 pp.

Goodey, J. B. 1963. Laboratory Methods for Work with Plant and Soil Nematodes. Great Britain Ministry of Agriculture, Fisheries and Food. Tech. Bull. No. 2. H.M.S.O., London, 72 pp.

Kirjanova, E. S. and E. I. Krall. 1980. Plant-Parasitic Nematodes and Their Control. Amerind Publ. Co. Pvt. Ltd., New Delhi, 748 pp.

Lamberti, F. and C. E. Taylor (eds.). 1979. Root knot Nematodes (Meloidogyne species), Systematics, Biology and Control.Academic Press Inc., London, 477 pp.

Lee, D. L. 1965. The Physiology of Nematodes. Oliver and Boyd, Edinburgh.154 pp.

Maggenti, A. R. 1981. General Nematology. Springer-Verlag, New York, 385 pp.

Mai, W. F., O. G. Mullin, H. H. Lyon and K. Loeffler. 1996. Plant-Parasitic Nematodes: A Pictorial Key to Genera. Comstock, New York. 277 pp.

Mohilal, N. and Gambhir, R.K., (Eds.) 2006. Plant Nematology in India: Commemoration volume in honour of late Professor Ch.Dhanachand.RB Printing House, Nagamphal, Imphal, India, 178 pp.

Norton, D. C. 1978. Ecology of Plant-Parasitic Nematodes. John Wiley & Sons, New York, 268 pp.

Parvatha R.P. 2008. Integrated nematode management in horticultural crops. Scientific Publishers, Jodhpur Rajasthan, India, 300 pp.

Parvatha Reddy, P., 1987. A Treatise on Phytonematology. Agricole Publishing Academy, New Delhi, 381 pp.

Perry, R.N. and D. A. Wharton (Eds). 2011. Molecular and physiological basis of nematode survival. Wallingford, Oxfordshire ; Cambridge, MA : CABI, 320 pp.

Ravichandra, N.G., 2008. Plant Nematology. I.K. International Publishers, New Delhi, 710 pp

Ravichandra,N.G., 2010.Methods and Techniques in Plant Nematology, PHI Publishers, New Delhi, 696 pp

Ravichandra, N.G., 2014. Horticultural Nematology, Springer Publications, Netherlands, 565 pp.

Sasser J. N. and W. R. Jenkins (eds.). 1960. Nematology Fundamentals and Recent Advances with Emphasis on Plant-Parasitic and Soil Forms. Univ. North Carolina Press, 480 pp.

Sasser, J. N. and C. C. Carter. 1985. An Advanced Treatise on Meloidogyne. Vol. I. Biology and Control. N. C. State Univ.,Raleigh. 422 pp.

Swarup, G. and D.R. Dasgupta. 1986. Plant-parasitic Nematodes of India. Indian Agricultural Research Institute, New Delhi, 497 pp.

Taylor, A. L. 1967. Introduction to Research on Plant Nematology: An FAO Guide to the Study and Control of Plant-parasitic nematodes. FAO, Rome, 139 pp.

Taylor, A. L. and J. N. Sasser. 1978. Biology, Identification and Control of Root knot Nematodes. N.C. State Univ. Press, 111 pp..

Thorne, Gerald. 1961. Principles of Nematology. McGraw-Hill, New York, 553 pp.

Trivedi, P.C. (Ed.) 2003. Nematode management in plants. Scientific Publishers, Jodhpur, Rajasthan, India, 317 pp.

Veech, J. A. and D. W. Dickson (Eds.). 1987. Vistas on Nematology. Society of Nematologists. 509 pp.

Walia, R. K. and Khan, M. R. 2018. A Compendium of Nematode Diseases of Crop Plants. ICAR-All India Coordinated Research Project on Nematode in Cropping Systems, New Delhi, pp. 106.

Walia, R.K. 2014. Text book of Introductory Plant Nematology. Second Revised Edition, Indian Council of Agricultural Research New Delhi, 240 pp

Wallace, H. R. 1973. Nematode Ecology and Plant Disease. Edward Arnold Publishers, Lt., London, 228 pp.

Webster, J. M. (ed.). 1972. Economic Nematology. Academic Press, New York, 563 pp.

Zuckerman, B. M. and R. A. Rohde (Eds.). 1981. Plant-parasitic nematodes. Vol. III. Academic Press, New York, 508 pp.

Zuckerman, B. M., W. F. Mai, and R. A. Rohde (Eds.). Plant-parasitic Nematodes. 2 volumes. Academic Press, New York. Vol. I, 345 pp., Vol II, 347 pp.

GLOSSARY

GLOSSARY

A

abd Anal body diameter, used as a measure of distances between structures, or size of structures, in posterior region.

Abduct To draw away from a position near or parallel to the median axis. See adduct.

Abductor Muscle Any muscle that draws away from the main axis or extends or draws parts from the body. See adductor muscle.

Aberrant Deviating from the usual type or form. Abnormal.

Abiogenesis The theoretic arising of living from nonliving matter. Spontaneous generation. See biogenesis.

Abiotic Pertaining to or characterized by non- living, inanimate phenomenon or objects.

Abnormal Deviating from the usual type or form. Aberrant.

Aboral Remote from or opposite to the mouth.

Aborigines The indigenous fauna and flora of a geographical region.

Abort To arrest embryonic development.

Aborted Unsuccessful development rendering a part or organ unfit for normal function.

Abortion The unsuccessful or imperfect development of any entity or normally present part or organ.

Abortive Unsuccessful embryonic development. Rudimentary.

Abraded Scraped, rubbed or worn away.

Abrupt A sudden transition. Without gradation.

Abscissa (pl. **Abscissas, Abscissae**) The transverse line perpendicular to the vertical line (ordinate) to show by graphic design the relations of two series of facts.

Absent Buccal Capsule When an embedded buccal capsule has walls of the same consistency as the lining of the esophagus the buccal capsule is said to be absent.

Absorb To take something within the body. See adsorb.

Abullate Lacking bullae. See bullae.

Abundance A measure of the frequency of entities such as species in a community or guild measured in terms of numbers of individuals, biomass of individuals, contribution to ecosystem function or metabolic footprint of the entities.

Abyssal Fauna Organisms dwelling at oceanic depths below 6,000 feet, quiet water, complete darkness. See benthonic fauna, pelagic fauna.

Acanthiform-shaped like a thorn. See corniform.

Acaudal Lacking a tail.

Accentuation Increased distinctness.

Accessory Certain subordinates that are auxiliary in function, course, etc., to the principal.

Accessory Piece The gubernaculum.

Accessory Sexual Structures The spicules and gubernaculum of the male nematode.

Acclimate To cause or the process of becoming accustomed to a new environment. To acclimatize.

Acclimatize To acclimate.

Acclivous A gentle rise. An upward slope.

Accrescent A gradual increase in thickness toward the apex.

Accretion The manner by which crystalline and certain organic forms increase by growth of external addition, Also, the joining of parts normally separate.

Acerate Needle-shaped. See acerose, acicular, pointed.

Acerose Needle-shaped. See acerate, acicular, pointed.

Acephalous Lacking a head.

Acescence The process of becoming sour

Acicular Needle-like. Having a long, slender point.

Aciculum (pl. **Aciculums, Acicula**) A seta.

Acidobiotic Living in an acid environment.

Acidophilic A successful mode of life in an acid environment.

Acidophobic Intolerant of an acid environment.

Aciform-shaped like a needle. See pointed.

Aciniform Formed like a cluster of grapes.

Acoelomate The condition of lacking a true coelom. That is, the body cavity, if any, is not entirely surrounded by mesoderm.

Acquired Character A character that originates during the life of an organism due to environmental or functional cause.

Acre-foot One foot of water over an acre; 43,560 cubic feet; or 325,851 U.S. gallons.

Acre-Inch One-twelfth of an acre-foot.

Acrocentric Having a subterminal centromere.

Acriniform Having a radiated form.

Active See alacrious.

Acuate Pointed. See pointed.

Acute Pointed. With a sharp terminus. See pointed.

Acuminate Tapering to a slender point. See pointed.

Acuminose Subacuminate.

Adcloacal Situated in the proximity of the cloaca.

Adanal Situated in the proximity of the anus.

Adanal Bursa A bursa which does not enclose the tail terminus. See Leptoderan.

Adanal Copulatory Papillae The adanal supplements.

Adanal Supplements Organs of secretion and attachment adjacent to the anus of some male nematodes,

Adaptation Modification of an organism or its parts or organs to make it more fit for existence under the conditions of its environment.

Addendum (pl. **Addenda**) An addition. A supplement. An appendix.

Adduct To draw towards the median axis or one part toward another. See abduct.

Adduction A drawing toward. A drawing together. See abduction.

Adductor Muscle Any muscle that draws toward the main axis or brings parts into opposition. See abductor muscle.

Adeniforn Having the shape of a gland. Gland-like.

Adhesion The attractive force between two dissimilar bodies that are in contact. See cohesion.

Adhesion Tubes Hollow tube-like setae by which some nematodes ambulate on a surface. See ambulatory setae.

Adhesive Bristles Adhesion tubes.

Adhesive Tube The spinneret.

Aditus A passage or opening as an entrance.

Adnate Grown together especially of unlike parts.

Adsorb To take something upon the surface of a body. See absorb.

Adult A sexually mature individual. A condition achieved after the fourth molt.

Adunc Inwardly curved. Hooked. See hamate.

Adventitious Not arising from the usual place. Arising sporadically.

Adventive An introduced species.

Aerate To imbue with air.

Aeration The exchange of air in soil with air from the atmosphere. The composition of the air in a well aerated soil is similar to that in the atmosphere; in a poorly aerated soil, the air is considerably higher in carbon dioxide and lower in oxygen.

Aerial Inhabiting the air. Growing or existing above ground.

Aeriform Of the nature or form of air. Gaseous.

Aerobic Requiring the presence of oxygen to live. See anaerobic, facultative aerobic, facultative anaerobic, obligatory aerobic, obligatory anaerobic.

Aerolated A situation where the transverse striae enter the lateral fields

Aestivation Dormancy during periods of unfavorable environment. See dormant.

Afferent Bearing or conducting toward an organ or position. See efferent.

Affinity Relationship. Attraction. A relation among organisms where plan of organization and structure indicate position in the overall system.

Agar A gelatin-like substance extracted from seaweed. Agar is used in culture media to cause it to set.

Aggregated See agminate.

Agminate Aggregated. Grouped. Clustered.

Agnobiotic Culture A nematode population with one or more kinds of organisms present. See pure culture, aseptic culture, monoxenic culture, axenic culture, xenic, synxenic, monxenic, dixenic, trixenic, polyxenic.

Alacrious Active. Lively.

Alae (sing. **Ala**) Longitudinal thickening of the cuticle forming wing-like expansions. See longitudinal alae, cervical alae, caudal alae, bursa, lateral alae.

Alar Relating to or belong to the alae.

Alate Having alae. Possessing lateral fields.

Albumin A class of proteins. One segregate of the cuticle.

Albuminoid Concretion Food reserve deposits found in modified intestinal cells of some nematodes.

Aliform Wing-shaped.

Alimentary Canal The alimentary tract.

Alimentary Tract The tubular food-carrying passage extending from the mouth to the anus. Its parts are oral aperture, stoma, esophagus, intestine, prerectum, rectum, anus.

Aliquot A part of a whole that divides the whole without a remainder,

Alleyway A path or separation between experimental plots.

Alloplasm Functionally-specialized living matter differentiated from and less active than protoplasm, as that of cilia.

Allotype A paratype of the opposite sex of the specimen designated as the holotype. See type.

Alternation of Generations See metagenesis.

Alveolate Having pits or cells like a honeycomb.

Alveolus (pl. **Alveoli)** A small cell, cavity or pit.

Amalgamated Lips Lips fused giving a smooth or nearly smooth contour.

Ambifenestrate In some species of the genus Heterodera where the vulval bridge is slender and the hatching pore is typically an hourglass shape. See fenestrate, semifenestrate, bifenestrate, circumfenestrate .

Ambulacral Spines Ambulatory setae.

Ambulatory Setae Setae, sometimes hollow and tube-like, used in locomotion. See adhesion tubes, stilt bristles, setae.

Amendment Any material, such as lime, gypsum, sawdust, or synthetic conditioners that is worked into the soil to make it more productive. Strictly, a fertilizer is also an amendment but the term â€œamendmentâ€□ is used most commonly for added materials other than fertilizer.

Amphicoelous Biconcave

Amphicyrtic Biconvex

Amphid (pl. **Amphids)** Paired lateral sense organs which generally open to the exterior on or near the lip region. .

Amphid Aperture The opening leading to the pouch of the amphid. .

Amphidelphic Having two ovaries, generally one extending anteriad and the other posteriad of the vulva. See monodelphic, prodelphic, didelphic. .

Amphidial Duct The connecting passage between the amphidial opening and the amphidial pouch.

Amphidial Gland An organ located posterior to the nerve ring a.id connects with the central nervous system through the lateral ganglion.

Amphidial Nerve The nerve extending anteriad from the nerve ring to the amphid.

Amphidial Opening The amphid aperture.

Amphidial Orifice The amphid aperture.

Amphidial Pocket The amphidial pouch

Amphidial Pore The amphid aperture.

Amphidial Pouch The cavity or chamber of the amphid which contains the sensilla. A dilation of the amphidial gland. Sensilla pouch

Amphidial Tubes The passages containing the amphidial nerves which connect the fibrillar terminals and the sensilla. An extension of the amphidial gland.

Amphimixis The union of germplasm of two organisms in sexual reproduction.

Ampliation An enlargement.

Ampulla A membranous sac or vesicle.

Ampullate Flask-shaped.

Ampuliform Flask-shaped. Dilated.

Anabiosis Act of being revived, especially after a period of desiccation.

Anabiotic Anabiosis

Anabolism The constructive processes iii metabolism, from digestion to assimilation. See metabolism, catabolism.

Anaerobic Living in the absence of atmospheric oxygen. See aerobic, facultative aerobic, facultative anaerobic, obligatory aerobic, obligatory anaerobic.

Anal Aperture The anal opening.

Anal Body Diameter (Width) The body diameter at the level of the anus; sometimes abbreviated as abd or abw. See Anal Diameter

Anal Cleft The anal opening

Anal Diameter The body diameter at the level of the anus.

Anal Fenestra In the genus *Heterodera*, the opening resulting from the breakdown of the thin walled, transparent cuticular region of the anus.

Anal Ganglion Receives the ventral cord and from which the anolumbar connectives extend to the lumbar ganglia.

Anal Muscles Muscle cells/tissue which function to make the anus operative.

Anal Opening The orifice to the exterior at the terminus of the rectum and delimited by the anus.

Analogue A part or organ similar in function to a part or organ of another individual but differing in structure and origin. See homologue.

Analogous Similar in function but differing in origin and structure. See homologous.

Analogy Similarity in function between parts or organs of a different origin. See homology.

Analytical Key The resolving of the characteristics of nematodes into their constituent parts resulting in the selection of a particular specimen.

Anaphase A state in mitosis or meiosis in which daughter chromosomes migrate toward the spindle poles. See prophase, metaphase, telophase.

Anastomosis (pl. Anastomoses) The union or intercommunication of the lumen of a system or network.

Anceps Sword-shaped.

Androdioecious Descriptive of a population expressing the characteristics of androdioecy

Androdioecy The condition in which a population consists of both males and hermaphrodites

Androgynous Combining the characters of both sexes. Hermaphroditic. See hermaphrodite, reproduction.

Androtype A male type. See type.

Aneuploid Having a chromosome number that is not a multiple of the haploid number. See monoploid, euploid, heteroploid, hyperploid.

Aneuploidy The process or history of becoming polyploid.

Angulate Having angles. Angled.

Anion An ion bearing a negative charge. See ion, cation.

Anisoglottid Having the metarhabdions of the glottid apparatus at different levels. See isoglottid.

Anisomorphic Of dissimilar form, especially pertaining to the metarhabdions of the glottid apparatus. See isomorphic.

Annulate Comprising or furnished with rings. Ringed.

Annulations Deep transverse striae which occur at regular intervals in the cuticle.

Annule The interval between the annulations. A transverse ornamentation of the cuticle surface formed by deep striae at regular intervals. Formed of or giving the shape of rings.

Annulet A small ring.

Annulose Ringed.

Annulus A ring-like structure or marking.

Anolumbar Connectives Connections joining the anal or preanal ganglia with the lumbar ganglia.

Anorectal Connectives Commissures extending from the anal ganglia to the dorsal surface of the rectum joining the dorsorectal ganglion from which a median nerve extends posteriad to the tail.

Antagonistic Symbiosis A symbiotic association which is disadvantageous or destructive to one of the symbionts.

Anteapical Situated just before the apex. Subapical.

Antemortem Before death. See necropsy, postmortem.

Anteriad Directed forward. Opposed to posteriad.

Anterior In front. Before. The front position as opposed to the posterior.

Anterior Cephalic Crown A crown of six papillae or setae located immediately posterior to the labial crown of papillae. See labial crown of papillae, posterior cephalic crown.

Anterior Cephalid The anteriad cephalid at which the two lateral cords arise. See cephalid.

Anterodorsal Toward the front and dorsum.

Anteroposterior Axis The long axis from head tail. Longitudinal axis.

Anteroventral Toward the front and the venter.

Anthelminthic A compound fatal to worms or causing the expulsion of worms, especially intestinal worms.

Antibiosis An antagonism between two or more organisms, especially microorganisms in soil, to the detriment of one of them.

Antipathetic Symbiosis A symbiotic association which is advantageous or necessary to one or both symbionts.

Antrorse In a forward or upward direction. See postrorse, retrorse.

Antrum A cavity or hollow space. A sinus.

Anus The extremity of the rectum. The posterior opening of the alimentary canal.

Aperture An opening. Hole. Orifice. The diameter of the opening.

Apex (pl. **Apices**) A proximal continuation of the aphelenchoid spicule shaft. The tip. The point of culmination.

Aphelenchoid Esophagus Having a narrow procorpus with a strongly formed median bulb followed by a narrow tube which extends to the intestine. A basal swelling is lacking and the three esophageal glands lie outside the esophagus proper.

Apically Near, toward or directed toward the apex.

Apicad Toward the apex.

Apical Cell A cell of the epithelium that forms the gonoduct wall and is situated at the distal end of the ovary and to be differentiated from the terminal cell.

Apical Lip Notches Indented lip margins at the confluence of the lips.

Apiculate Abruptly terminated by a fine, distinct point.

Apneuistic Respiration through the cuticle.

Apomixis, Apomictic Parthenogensis Reproduction without fertilization and by mitotic division of egg cells. See Parthenogenesis, Mitotic Parthenogenesis.

Automixis, Automictic Parthenogenesis Reproduction without fertilization and with a meiotic reduction division of unfertilized egg cell. Diploid number may be re-established by fusion of nucleus with polar body. See Parthenogenesis, Meiotic Parthenogenesis.

Apophysate Bearing an apophysis.

Apophysis (pl. **Apophyses**) A process. A prominence.

Apotype A specimen, other than the type, upon which a subsequent or amended description or figure is based. Hypotype. Plesiotype. See type.

Apparition Appearance, Visible.

Appendage An external organ or limb.

Appendix An outgrowth or process, especially an addition to what is relatively complete in itself.

Appendicle A small appendage.

Appendicule A large, single, ventral preanal supplementary organ which is extensible.

Applied Nematology Economic nematology. The practical use of pure nematology for control of nematode pests.

Appose To place surfaces opposite each other.

Apposition The condition of having surfaces against each other. See Fornent, juxtaposition.

Aquatic Living in a water habitat. See arenicolous, edaphic, hypogaeic, marine, rheophile, terrestrial, terricolous, xerophilous.

Arakaoderan bursa Caudal alae completely surrounding the cloacal area. See leptoderan, peloderan.

Arc A bow-like curvature.

Arcade A circular band of tissue which usually surrounds the esophagus at the base of the lips.

Arcade Cells Cells situated along the esophagus, two opposite each esophageal lumen radius and one opposite each sector for a total of nine. The name is derived from the arch-shaped connections at the base of the lips.

Arcadial Tissue A mass of cells surrounding the stoma.

Arch A structure or form usually curved. In perineal patterns of Meloidogyne females the curved formation of cuticular striae above the phasmids.

Arcuate Curved like a bow. Arched.

Are A measurement of area equal to 100 square meters or 119.6 square yards.

Area Application Broadcast application of soil nematicides.

Arenicolous Inhabiting sand. See hypogaeic, aquatic, edaphic, marine, rheophile, terrestrial, terricolous, xerophilous, fossorial.

Areolate A small area, especially of cuticle, delimited by transverse and longitudinal markings.

Arista A bristle like appendage.

Aristate Having a slender or spiny terminus.

Arrested Growth A stopping of the growth processes.

Arrowhead-shaped See sagittate.

Ascarocollagen The non-hydrolyzed form of ascarogelatin. It corresponds to the fiber, internocortical and fibril layers of the cuticle.

Ascarogelatin A segregate of the cuticle of the collagen group.

Ascarlyose The sugar 3,6-dideoxy-L-arabinohexose.

Asepsis The state or condition of being aseptic. See sepsis.

Aseptic Free of microorganisms. See septic.

Aseptic Culture A nematode population containing a single species only and free of contamination by all other organisms. Aseptic population. See pure culture, pure population, agnotobiotic, axenic, xenic, synxenic, monoxenic, dixenic, trixenic, polyxenic.

Aseptic Population An aseptic culture.

Asexual Reproduction Reproduction where the union of gametes is not involved. See sexual reproduction.

Association Ecologically, uniform groups of strata over a large area.

Asteriform Star-shaped. Like a star.

Atavism Reversion to a more primitive type.

Athrocyte A coelomocyte capable of absorbing foreign material from body fluids and storing it in crystalline form. See coelomocyte.

Atrophy A wasting away or arrested development.

Attacked To be fed upon by nematodes. To become infested with nematodes. See invasion, infective.

Attack Population The pathogenic nematodes per unit of substratum capable of invasion.

Attenuate Thin, Slender, Drawn-out, Extended

Attractant A chemical substance causing a positive migration response.

Attraction A positive response where the mean number of migrating nematodes reaching the plant end of a test chamber is significantly greater (at the 95 percent level of probability) than the mean number reaching the blank end and where this difference occurs within 3 hours in agar or 6 hours in sand.

Author The writer of a paper, article, treatise, etc. The authorâ€™s name usually accompanies the paper to gain credit for accomplishment and to affix responsibility.

Autoinfection The source of infection originates within body of the host. Infection from within by progeny of previous or current parasites. Occurs in both plant and animal hosts.

Autoinoculation The spread of an infestation from a focus of infection to other parts of an organism.

Autolysis Destruction of a cell by internal agents. Self digestion of a tissue.

Autopsy An examination and partial dissection after death. See postmortem, necropsy.

Autosome A chromosome which is not a sex chromosome.

Autotomy Self-amputation of the posterior part of the body by some nematode forms on gaining entrance to an animal host.

Autotype A specimen identified by the author as an illustration of his species and compared with the type or cotype. Heautotype. See type.

Awl-shaped See subulate, styliform.

Axial Belonging to, or around, in the direction of or along an axis.

Axial Stylet A buccal stylet aligned along the anteroposterior axis.

Axil The angle formed between a branch or petiole and the stem from which it arises.

Axis (pl. **Axes**) A straight line passing through a body about which the parts are symmetrically arranged.

Axon, Axone The long process of a nerve cell conducting impulses way from the cell body.

B

Bacillary Bands One or two laterally situated bands of cells of the lateral cords in the esophageal region. See hypodermal glands.

Bacillary Layer A non vibratile form of cilia which line the intestinal epithelium.

Bacteria Consuming See bacteriophagous.

Bacteriophagous Feeding on or consuming bacteria. See feeding.

Bacterivores Organisms that feed on bacteria.

Baculiform Having the shape of a rod or staff.

Baermann Apparatus The funnels and accessories used to implement the Baermann funnel technique.

Baermann Funnel Technique A method of isolating nematodes from soil, screening residue, plant tissue, or other matter where the material is placed in water and the nematodes by their own action move out into the water, settle and are drawn off from tubing attached to the funnel stem.

Bagging Matricidal vivipary (egg hatch within the body of the female), also known as Endotokia Matricida.

Bag-like See saccate.

Ballonets Cuticular inflations in the cephalic region assuming a swollen band shape immediately posterior to the lips. See head bulb.

Banks Groups of peripheral longitudinal muscle cells. Muscle fields.

Basal Related to, located at, or forming the base.

Basal Bulb An enlargement of the esophageal wall, muscular or glandular, at the posterior of the esophagus. See cardiac bulb, bulbous, isthmus, terminal bulb, true bulb, metacorpus, posterior bulb.

Basal Knobs The posterior knobs of the stylet. See stylet knobs.

Basal Lamella The inner-most cuticular strata which consists of a thin layer which in cross section appears to be striated. See cuticular layering.

Basal Plate The posterior radial apparatus of the cephalic framework.

Basal Ring The posterior circular base of the cephalic framework.

Base Pair

Batatiform Having the shape of a sweet potato.

Bathyal Fauna Organisms dwelling at oceanic depths of 600 to 6,000 feet, quiet water, little to no light. See benthonic fauna, pelagic fauna.

Beaded See moniliform.

Bean-shaped See reniform.

Bearding A condition of excessive root proliferation induced by nematode feeding.

Beet Sickness A disease of sugar beets incited by Heterodera schachtii.

Beet Weariness A disease of sugar beets incited by Heterodera schachtii.

Bell The bursa.

Bell-shaped See campanulate.

Benthonic Fauna Organisms which dwell on the bottom. They are further defined as littoral, neritic, bathyal or abyssal] by depth and nature of bottom. See pelagic fauna.

Benthos The sea bottom.

Besom A broom.

Besomiform Formed or-shaped like a broom.

Biconcave Possessing concave surfaces on opposite sides. See amphicoelous, amphicyrtic.

Biconvex Possessing convex surfaces on opposite sides, See amiphicyrtic, amphicoelous.

Bicorn With two horns. Crescentiform.

Bicornuate Bicorn.

Bicron One billionth of a meter (0.000,000,001).

Bifarious Orientated or pointed in two directions.

Bifenestrate In some species of the genus Heterodera the vulval bridge is stout and divides the fenestra so that the semifenestrae appear to be two nearly separate holes. See fenestrate, semifenestrate, ambifenestrate, circumfenestrate. .

Bifid Divided into two equal parts.

Bifilar Having two filaments or fibers.

Biflex Bending in two directions.

Biform Of two forms, Possessing characteristics of two forms.

Bifurcate Divided into two branches. Forked.

Bilaterally Symmetrical An organism in one plane with each side being approximately a mirror image of the other. This plane usually lies anteroposteriorly and dorsoventrally.

Binding Cells Three coelomocytic cells adjacent to the esophagus and opposite the fibril cells, two ventrolateral and one dorsal. See coelomocytes, fibril cells.

Bioassay Quantitative estimation of a biologically active substance by testing its effect under standardized conditions on living organisms or the effect of one organism upon another.

Biogenesis The doctrine that living organisms are produced only from pre-existing living organisms. See abiogenesis.

Biogeny The evolution of organic forms considered individually (ontogeny) or tribally (phylogeny).

Biological Control The limiting of pathogenic nematode population through depredation by natural enemies.

Biological Race A segment of total species population differing in some physiological characteristic. See biotype

Biomass The weight of living organisms in a given area; often applied as the total weight of organisms participating in a specific ecosystem function or service. For example, the biomass of bacteriphagous nematodes.

Biophagous Feeding upon living organisms or living tissue. See feeding.

Biosis Life. Vitality.

Biota The fauna and flora of a region.

Biotic Pertaining to life or the laws of animal and plant progress and evolution.

Biotope A region of uniform environment and biotic populations.

Biotrophic Organism An organism that requires a specialized feeding site to be kept alive so that it can complete ists life cycle. See biotrophy,

Biotrophy A relationship with the host whereby the feeding site must be kept alive to allow the parasite to complete its life cycle.

Biotype A subgroup of organisms which possess the same genetic characteristics. A subdivision of a race.

Biparasitic A parasite upon a parasite. See feeding.

Bipolarity Possessing two poles.

Bisexual The species having two sexes distinct and in separate individuals. See hermaphrodite, gonochoristid, reproduction.

Bisulcate Having two grooves.

Bivulvarity The occurrence of two vulvae on a single female.

Bivalents Paired homologous chromosomes associated with synapsis.

Blackhead Disease of Bananas A disease of banana plants incited by Radopholus similis.

Blackhead Toppling A disease of banana plants incited by Radopholus similis.

Black Root Rot A disease of strawberry incited by Pratylenchus penetrans, fungi and bacteria.

Blasting A failure to produce fruit or seeds.

Blastocoele The cavity of a blastula.

Blastoderm The cell layers formed by the repeated segmentation of the blastomeres.

Blastomere Cells produced by the first few cell divisions. A cell of the blastula.

Blastula The early embryo having the form of a fluid-filled sphere bounded by a single layer of cells. See gastrula.

Blind Bud An abortive flower bud especially one due to nematode infection.

Blind End The closed terminus of a tube or organ. See caecum.

Blinding Destruction of a plant growing point, stem, flower or root by nematode attack.

Blind Plant A plant with a bud or growing point killed by nematode infection.

Blister A vesicle of cuticle or epidermis.

Blister-like See pustule.

Bloat A disease of onions incited by Ditylenchus dipsaci.

Bloom The subcrystaline layer of a waxy nature found on the cysts of some Heterodera spp.

Blunt See obtuse.

Boat-shaped See scaphiform.

Body The nematode frame with its organs. The important and largest part of an organ. as the body of the spicule.

Body Cavity The hollow within the body which contains the internal organs. See pseudocoel.

Body Pores A series of minute depressions slightly submedian or lateral and apparently connected with the lateral cords via small canals. See caudal pores. ventromedian pores, ventrosubmedian pores, excretory pore.

Body Wall The structural frame consisting of cuticle, hypodermis and muscle layer.

Botryoid Resembling in shape a cluster of grapes.

Bottle-shaped See lageniform, ampullate, ampulliform.

Boundary Layer The cuticular stratum which is distinct only in certain instances and probably is a condensation layer between the matrix layer and the fiber layers. See cuticular layering.

Bounding Membrane A thin syncytial epithelium covering the esophagus.

Bow-like See arc, arcuate.

Brackish Water Saltish, slightly saline.

Brain The nerve ring.

Branched See ramified, ramiform.

Branched Gland A form with outgrowths far exceeding the diameter of the cell body in length.

Breaking Joint The site of autotomy.

Bridging Host The possible passing from one host to a previously unsuitable host by transition through the intermediation of another, the bridging host.

Brier Bud A disease of strawberry incited by *Aphelenchoides fragariae.* See dwarf.

Bristle Setae Ambulatory setae apparently non-functioning glands or sensory organs acting as organs of traction. See setae.

Bristle-shaped See arista, setiform.

Broadcast Application The treatment of soil with a nematicide in such a manner that the expanding vapors form a solid carpet throughout the field. Solid application. See row application, spot application, drench application, gas application.

Broken Striae Non-continuous configuration lines of the perineal pattern. See partial striae, lighter striae, intermediate striae.

Broom-shaped See besomiform.

Brown-Leaf Disease A disease of *Asplenium nidus* incited by *Aphelenchoides fragariae.*

Brown Ring Disease A disease of daffodils and hyacinth incited by *Ditylenchus dipsaci.*

Brown Ring Disease of Narcissus A disease of narcissus capable of being incited by *Ditylenchus dipsaci, Aphelenchoides fragariae, A. parientinus or Aphelenchus avenae.*

Buccal Aperture The oral aperture.

Buccal Capsule *Sensu lato,* the combined vestibulum and buccal capsule. The stoma. *Sensu stricto,* the anterior widening of the esophageal cuticular tube often distinguished from the lining of the latter by it more developed cuticularization.

Buccal Capsule Cephaloboid An elongated, chambered buccal capsule in which the cheilostom is broad and the protostom and telostom narrow while being enclosed by esophageal musculature. See buccal capsule panagrolaimoid.

Buccal Capsule Panagrolaimoid An elongated, chambered buccal capsule in which a short posterior part is narrowed and surrounded by the esophageal musculature. See buccal capsule cephaloboid.

Buccal Cavity A buccal capsule in the broad sense.

Buccal Ring Where present, a thickened ring delimited from the beginning of the buccal capsule by a groove.

Buccal Rod Where present, elongated cuticular thickenings of the buccal cavity.

Buccal Spear A stomatostyle.

Buccal Stylet A stomatostyle.

Buchner Funnel A funnel with an interior perforated plate and used for vacuum filtration.

Buffered Populations A situation existing whereby organisms affect one another in such a way that the population density of each remains within a certain range about a mean.

Bulb Flap One of three parts of the valvular apparatus. One in a dorsal and two in subventral positions. .

Bulb Flaps The valvular apparatus

Bulb Teeth The valvular apparatus.

Bulbar Valve The valvular apparatus.

Bulbiferous Having the form of a bulb.

Bulbiform Having the form of a bulb.

Bulboid Esophagus Having a cylindrical shape anteriorly and terminated by a basal bulb. Also termed oxyuroid esophagus.

Bulbous Bulb-like.

Bulbs The one or more muscular swellings of the esophagus. See true bulb, pseudobulb.

Bulbus The basal bulb of the esophagus.

Bulla (pl. **Bullae**) A blister-like prominence near the vulval fenestra of some Heterodera.

Bullate Having a blistered or puckered form or appearance. Having or bearing bullae. See bullae.

Burden The total number of parasites infecting an individual.

Bursa (pl. **Bursae**) Wing-like extensions of the lateral cuticle at the caudal end of the male. Caudal alae. See leptoderan, peloderan, rhabtitoid bursa, strongyloid bursa.

Bursa Copulatrix The bursa.

Bursa Enveloping Tail Peloderan bursa.

Bursal Musculature The arrangement of the muscles pertaining to the bursa. See musculus costae dorsalis, musculus costerum lateralium internis, musculus bursae basalis, musculus costae lateralis externus posterior, musculus costae lateralis externus anterior.

Bursal Nerve Innervated from bipolar sensory nerve cells forming a longitudinal strand on either side in or near the lateral cords.

Bursal Ribs See rays.

Bursate Having a bursa.

Byssus Elaborately-branched polar projections of the protein membrane on eggs of some nematodes.

C

Cabbagy Plant A disease of teasel incited by Ditylenchus dipsaci.

Caecum (pl. Caeca) A cavity open at one end, the blind end of a lumen or duct.

Caliciform Having the shape of a calyx or a cup.

Calvarium Subcuticular cephalic framework.

Camera Lucida An optical instrument for use with the compound microscope where by means of prisms and mirrors a virtual image appears as if projected upon a plane surface. Used for pencil sketching and measuring.

Campanulate Bell-shaped.

Campanuliform Having the shape of a bell.

Canal A tubular channel or passage. A duct.

Cane A thickening with a lack of ornamentation of the posterior cuticle. See helmet.

Cap Cell The terminal cell of the ovary. The primordial germ cell. .

Capitulum A sclerotized guiding piece for the gubernaculum on the ventral cloacal surface of some Hoplolaiminae. The manubrium of some authors. See telamon, Alternatively, ptoposed for dorylaimoid spicules, the head of the spicule from the proximal end to the start of the blade or lamina - the manubrium of some authors.

Capsula Var. of capsule.

Capsule A receptacle or bag. A membrane or sac-like form enclosing a structure or organ.

Cardia A muscular structure at the base of the esophagus opening into the intestine. See esophageal-intestinal valve, cardiac valve.

Cardiac Bulb The basal bulb of the esophagus. See cardia.

Cardiac Caecum The esophageal caecum.

Cardiac Esophageal Bulb The basal bulb of the esophagus.

Cardiac Glands Three glandular bodies located at the base of the esophagus.

Cardiac Region of the Intestine The anterior portion of the intestine as distinguished from the prerectum.

Cardiac Valve A thickening or complication of the esophageal lumen lining in the basal bulb. See cardia.

Cardiac Bulb The basal bulb.

Carina (pl. **Carinae**) An elevated ridge or keel-like structure not necessarily pronounced or acute.

Carinate Keeled. Having one or more longitudinal ridges.

Carnivore A flesh eater. A predaceous nematode. See feeding.

Carnivorous Preying or feeding on nematodes or other animals. See feeding.

Carrier An organism bearing internally an infectious agent of disease even though no marked symptoms are evident. See vector.

Castration Any process which prevents the development of mature ova or sperm in the gonads.

Catabolism The destructive processes in metabolism. The breaking down of more complex components in protoplasm. See metabolism, anabolism.

Cation An ion bearing a positive charge. See ion, anion.

Caudad Toward the tail. Opposed to cephalad.

Caudal Belonging to, or like, a tail. Situated on or near the tail.

Caudal Alae The bursa. Alae confined to the posterior part of the male nematode body. See alae.

Caudal Appendage In mermithids, the terminal portion of the larval tail. See terminal spine, caudal papillae.

Caudal Bursa A bursa which completely encloses the tail. Peloderan.

Caudal Flagellum A more or less filiform, long and thin elongation of the tail of uniform diameter, or becomes a caudal thread which is pointed at its tip.

Caudal Glands Usually three elongate unicellular cells in or near the tail, discharging by separate ducts into a common ampulla at the spinneret.

Caudal Papillae In mermithids, the terminal portion of the larval tail.

Caudal Papillae See submedian caudal papillae.

Caudal Pore The spinneret. The outlet of the caudal glands.

Caudal Wing The bursa.

Caudalid A single structure similar in size, morphology and body position as the hemizonion and located slightly anteriad of the anus. A parahemizonid.

Cauliflower Disease A disease of strawberry incited by a synergistic association of an *Aphelenchoides* nematode and bacterium, *Corynebacterium fascians*. Cauliflower dwarf.

Cauliflower Dwarf Cauliflower disease. See near cauliflower.

Causal Agent The parasitic organism responsible for a disease. Causal organism.

Causal Organism The organism that produces a given disease. Causal agent.

Causative Agent See causal organism.

Cavate Hollowed out.

Cavernicolous Cave-inhabiting. See arenicolous, hypogaeic, aquatic, edaphic, marine, rheophile, terrestrial, terricolous, xerophilus.

Cavernous Filled with caverns. Divided into small spaces.

Cecum (pl. **Ceca**) An outpocketing of the intestine at the anterior end, rarely composed of twists or loops. See caecum.

Ceiling The maximum population density for a given set of circumstances.

Cell The structural and functional unit of all plant and animal life. It consists of protoplasm, a nucleus and is surrounded by a membranous wall.

Cell Body The portion of a cell that contains the nucleus. In a nerve cell it is different from the region comprising the axon or dendrites. See soma.

Cell Constancy A situation in nematodes where the multiplication of cells ceases at hatching, except for the reproductive system, and growth is by enlargement of existing cells.

Cell Grazer A nematode that withdraws the cytoplasm from a cell and then moves to another cell; feeding from cels without requirineg development of a snurse cell system. See biotrophic organism.

Cell Membrane A differentially permeable membranous surface which surrounds the cell.

Cellules End organs or terminal bodies of some coelomocytes.

Central Area The nonstriated region anteriad of the vulva and posteriad of the cuticular striations in Meloidogyne perineal patterns.

Centromere That part of a chromosome to which the spindle fiber apparently attaches in mitosis.

Centrosome A minute protoplasmic body in the cytoplasm or nucleus that plays an important role in mitosis.

Cephalad Towards the head.

Cephalated Having the head set off in some manner.

Cephalic Belonging to, or situated in, on or near the head. Directed toward the head.

Cephalic Arches See ribs.

Cephalic Capsule A strongly cuticularized head set off by a groove.

Cephalis Cirrus (pl. **Cirri**) One of the elaborate cephalic appendages in front of the cephalic setae.

Cephalic Crown of Sensory Organs The circular arrangement of papillae, etc. about the oral opening See anterior cephalic crown, posterior crown, inner circle, outer circlet.

Cephalic Framework A rigid supporting structure which provides form to the head and axially acts as a stylet guide. See calvarium. .

Cephalic Glands Lateral glands opening near the cephalic papillae.

Cephalic Organ The head cleft.

Cephalic Papillae Papillae of the outer circlet about the mouth. .

Cephalic Setae Setae of the outer circlet about the mouth.

Cephalic Slits Paired pouch-shaped cephalic sense organs.

Cephalic Suture The posterior delimitation of the cephalic capsule.

Cephalic Tubers Prominences located in the head in labial and cephalic positions, often elaborately developed.

Cephalids Structures of a highly refractive nature, biconvex in longitudinal section and extending around the nematode body in the cephalic region. See anterior cephalid, posterior cephalid, caudalid, hemizonid, hemizonion, parahemizonid.

Cephalization: The tendency for sensory structures and the nervous system to be aggregated anteriorly.

Cervical Pertaining to, belonging to or situated in, on or near the neck.

Cervical Alae Alae confined to the â€œneck regional and are known only in the parasitic groups Strongylina, Ascaridina and Spirurina.

Cervical Duct The excretory duct.

Cervical Expansion Expansions of the cuticle in the cervical region. See cervical.

Cervical Gland The excretory gland.

Cervical Groove A pronounced transverse groove on the ventral surface of the cervical region, in some forms extending to completely encircle the cervical region.

Cervical Lacuna A space (lacuna) between the cuticle and epidermis probabl;y created by differences in contractility of the two surfaces See Lacuna.

Cervical Papillae Paired lateral tactile receptors situated near the nerve ring. Deirids.

Cervical Pore The excretory pore.

Cervical Vesicle Inflated cuticle anterior to the cervical groove.

Challange A test of immunity to a host by exposure to parasites after specific immunization.

Checkered See tessellate.

Cheek An ampulla of an amphid.

Cheeks In Meloidogyne kikuyuensis the heavily striated overlap of perineal cuticle on each side of the vulva.

Cheilorhabdions The walls of the cheilostom.

Cheilostom The lip cavity of the stoma, delimited anteriorly by the oral aperture posteriorly by the protostom. See protostom, telostom,

Cheilostome Alternative spelling of cheilostom.

Chelate A type of chemical compound in which a metalic atom is firmly combined with a molecule by means of multiple chemical bonds.

Chemokinesis Stimulation of random movement induced by a chemical agency.

Chemoreceptors Anteriorly the amphids, posteriorly the phasmids. See tactoreceptors.

Chemotropism A movement of turning or curvature induced by a chemical stimulus. See tropism.

Chemotaxis A change of position toward or away from a source of chemical stimulation. See taxis.

Chewing Plates The valvular apparatus. See grinder.

Chimney See soil chimney.

Chirotype The type specimen upon which a manuscript name is founded. See type.

Chisel A tillage machine with one or more soil-penetrating points that can be drawn through the soil to loosen the soil or subsoil.

Chitin The substance composing the exoskeleton of insects which is not equivalent to the cuticle of nematodes but is present in egg shells of some forms. See cuticle.

Chitin Plates The valvular apparatus - probably a misnomer in terms of the constituent material of the plates .

Chords Var. of cords,

Chorion The outer shell or covering of a nematode egg.

Chromatid During prophase and metaphase of mitosis or meiosis one of the two strands prior to separation which resulted from duplication of a chromosome. After separation each chromatid is known as a daughter chromosome.

Chromatin Minute granules of a nucleo-protein constituent of a cell nucleus.

Chromosome One of the fundamental bodies into which chromatin resolves itself prior to the mitotic or meiotic division of a cell.

Cicatrix (pl. Cicatrices) A scar or scar-like marking, especially in mermithids.

Cicatrization Tissue Tissue involved in the process of healing a wound.

Ciliate Provided with cilia.

Circa-Equatorial About the equator. About or near the middle.

Circinate Ring-shaped.

Circlet A small circular arrangement of structures or parts. A crown.

Circomyrian A type of muscle cell in which the muscle fibers completely surround the sarcoplasm. See platymyarian, coelomyarian.

Circumfenestrate In the genera *Heterodera* and *Globodera* where the vulva is lost when fenestration occurs by the breakdown of the thin cyst wall around the vulva leaving a circular hatching pore. See fenestrate, semifenestrate, bifenestrate, ambifenestrate. .

Circumenteric Ring The nerve ring.

Circum-Esophageal About or surrounding the esophagus.

Circum-Esophageal Commissure The nerve ring.

Circum-Oral About or surrounding the mouth.

Cirrus (pl. Cirri) One of the elaborate cephalic appendages in front of the cephalic setae.

Cirrus (pl. Cirri) A slender appendage formed by the eversion of a very long cloaca, frequently lined with spines.

Citron-shaped Having the form of a large lemon.

Classification The systematic arrangement of organisms into groups usually based on natural relationships. The common categories in diminishing rank are phylum, class, order, family, genus, species and variety.

Clavate Club-shaped.

Clavate Cells Modifications of the hypodermis which contribute to the formation of the lips, being two in each lip and they are long cells with a hollow interior, lamellate with expanded distal ends that fill the convexity of the lips.

Claviform Club-shaped. Clavate.

Clay As a soil separate, the mineral soil particles less than 0.002 mm. in diameter. As a soil textural class, soil material that contains 40 percent or more of clay, less than 45 percent of sand, and less than 40 percent of silt.

Clay Loam Soil material that contains 27 to 40 percent of clay and 20 to 45 percent of sand.

Clear To make the cuticle and other body parts transparent or translucent.

Clear Tail Length That portion of the tail peculiar to some nematode larvae in which no internal bodies, structures or arrangements can be seen. See tail terminal.

Cleavage Following fertilization, a repeated subdivision of the zygote cytoplasm with the corresponding nuclear mitosis. Segmentation.

Cleft The space caused by a slit, splitting, a crack, a fissure, etc. See suture.

Clintheriform Having a shape like a plate.

Cloaca In the male a common chamber lined with cuticle which receives the products of the intestinal and reproductive tracts and empties to the exterior via the cloacal orifice. The hind gut. See rectum.

Cloacal Cuticle The cuticular linings of the cloaca which differs in some respects from the external cuticle. See external esophageal cuticle, rectal cuticle, vaginal cuticle.

Cloacal Evaginations Spicule pouches which secrete and lodge the spicules.

Cloacal Orifice The opening of the cloaca

Cloacal Prominence A posterior projection of the region around the cloacal orifice in some nematode forms. The genital cone.

Clone A group of plants derived from a single plant by means of vegetative propagation. Each individual of a clone has the same heredity, so that in a uniform environment a clonal line of plants will respond rather uniformly. Soo otrain.

Clover Sickness A disease of clover incited by Ditylenchus dipsaci.

Club-shaped See clavate, waddiform.

Clustered See agminate.

Clypeiform Having the shape of a shield. See scutate, scutiform, escutcheon-shaped.

Coarse Root. A condition where most of the lateral roots are killed back by nematode attack resulting in an open system of mostly the main roots.

Cobb Formula A system of measurement utilizing decimal formula in which the measurements stated as percentages of the total length of the nematode are made at the specific body points as follows:

Stylet length	Nerve ring	Esophagus base	Vulva	Anus
as % of body length				
Body width at lip region	Body width at nerve ring	Body width at esophagus base	Body width at vulva	Body width at anus

For the male the letter â€œMâ€□ marks the middle of the body. See de Man formula.

Cockles Galls produced by some species of Anguina, especially A. tritici of wheat kernels. See purples, peppercorns.

Coelomate Having a body cavity. See pseudocoelom.

Coelom The main body cavity in which the intestine is suspended. Surrounded entirely by tissues of mesoderm origin.

Coelomocytes Cells lying in the pseudocoelom and appear not to be part of the general connective or isolation tissue. They apparently are of mesenchymatous origin. See athrocytes, fixed athrocytes, phagocytes, stellate cells, pseudocoelocytes.

Coelomyarian A type of muscle cell in which the muscle fibers are adjacent and perpendicular to the hypodermis and extend along the sides of the muscle cell for varying distances. See platymyarian, circomyarian.

Coexistence Living together in the same place.

Cohabitants Organisms dwelling together.

Collaret A band-like cuticular extension of the cephalic region opening anteriorly.

Collarette A posterior elongation of the annular ring.

Collum A neck or neck-like part or process.

Collum Testis A narrowing separating the testis from the deferent ducts.

Collum Uteri The constriction which separates the uterus from the vagina. In some forms it is elongated to form the ovijector.

Colonizer-persister (Bongers' c-p) series A series of categories of nematodes based on their life history chracteristics, response to enrichment and sensitivity to environmental perturbation.

Colony An assemblage of nematodes living in close association, especially of the same species and particular niche.

Column-like See rachis.

Combs Crests of cuticle which frequently bear spines.

Comb-shaped See pectinate.

Commensal An organism not truly parasitic capable of living harmoniously in, with or on another.

Commensalism The mode of different organisms living harmoniously together.

Comminute To breakdown into minute particles. Pulverize. See triturate, macerate.

Commissure A bundle of nerve fibers connecting ganglia.

Common Canal The excretory duct.

Compatible Host-Parasite Relationship Interaction between a nematode species and host plant that is favorable for nematode survival and reproduction. See incompatible.

Complete Metamorphosis The young hatch looking unlike the adult and through successive larval instars increase in size through several molts. The pupa or resting stage is entered where the body tissues break down and new structures for the adult arise. See metamorphosis, incomplete metamorphosis.

Concave Hollow. Incurved, as the inner surface of a curved surface or line.

Concavo-convex Having one surface concave the other convex with the convexity exceeding the concavity.

Cone The posterior segment of a female cyst of the genus Heterodera.

Cone A solid body having a circle for its base and terminating in an apex.

Congeneric Of the same genus.

Conical Cone-shaped.

Connective The cords of nerve fibers connecting one ganglion of a nervous system with another.

Conoid Having the shape of a cone.

Conspecific Of the same species. See infraspecific.

Constriction Said of the junction of the head and the cervical region when delimited by a sharply pronounced groove. See depression.

Constrictor Vulvae Muscles which function to close the vulva. See dilator vulvae.

Contaminate To infect or infest, with a pest an otherwise healthy organism. To pollute.

Contaminators Individuals of the inoculum. Organisms which bear the inoculum.

Contort To twist upon itself. To form into curves.

Contraction Swellings Numerous deeply staining enlargements formed at intervals along a muscle contractile fiber.

Control The reduction, by any means, of a plant pathogenic nematode population density which results in an economic increase in crop yield. See soil sterilant, spectrum.

Convergence A resemblance between individuals or structures derived from different ancestries or origins. Lines of development coming together from diverse origins to form similar characters. See divergence.

Convex A surface or line curved or rounded as the exterior of a sphere or circular form.

Coprophagous Feeding on dung. See feeding.

Copulation The act of sexual intercourse.

Copulatory Apparatus The copulatory armature.

Copulatory Armature The sex organs of the male employed in the act of copulation. See spicule, gubernaculum, supplementary papillae

Copulatory Muscles In most male nematodes specialized transverse muscles of varying numbers extending from the lateral cords to the subventral sides of the body, generally anterior to the cloacal orifice. The ventral curvature of the posterior part of the body in fixed material is caused by these muscles. See bursal muscles.

Copulatory Warts Greatly enlarged genital papillae of some male nematodes.

Cord-like See restiform.

Cordons Specialized, ribbon-shaped, paired bands of cephalic cuticle. See epaulets.

Cords Four longitudinal lines of hypodermal thickening lying on the inner side of the hypodermis and variously termed the dorsal, lateral and ventral cords. The cords contain the nuclei of the hypodermis. See intercordal areas.

Coriaceous Tough, leather-like.

Cornein The horny substance of corals and not equivalent to the cuticle of nematodes. See, cuticle.

Corniform An extended mucronate or pointed process like the horn of an ox.

Cornu Modified cephalic seta with leaf-shaped appearance and often with incisures of the edge, exytending forward and inward. See corona, corona radiata, perioral.

Corona A crown-like structure or process. A crown.

Corona Radiata Modified lips which form an upstanding collar with the inner surfaces having numerous lobes or cuticular teeth, sometimes compounded with an inner leaf crown. .

Corpus The most anterior segment of the esophagus of cylindrical form and of moderate width. Sometimes the base is swollen to form a bulb in such instances the corpus may be divided into procorpus, mesocorpus and metacorpus.

Corpus Esophagi The corpus.

Corridor A passage or tunnel made by an animal. See gallery.

Corrugated Wrinkled. Formed of alternate ridges and grooves.

Cortex Layer The outer layers of the cuticle. See matrix, fiber layers, cuticular layering.

Cortical Consisting of, or belonging to the cortex.

Coryogamy Var. of koriogamy.

Cosmopolitan Common everywhere. Widely distributed. Found in most areas of the world. See ecumenical.

Costa (pl. Costae) A rib or rib-like structure.

Costate Having the appearance, structure or form of a rib or ribs.

Cotype Any one of all the specimens available when a species is named, hence the type equals the sum of the cotypes. See type.

Coupled Annules Adjoining annules are inserted into one another providing added rigidly and allowing elasticity.

Crenate Having a margin formed of rounded scallops.

Crenation A rounded projection.

Crenature A rounded projection, also, the indentation between crenations.

Crenulate Minutely crenate.

Crescent Sickle-shaped, like the new moon with convex and concave edges.

Crescentiform Having the shape of a crescent.

Crest A ridge or linear prominence. See carinate.

Cristate Having a prominent crest. Crested.

Cristiform In the shape of a ridge or crest.

Cribriform Perforated like a sieve.

Crimp A pathological condition incited by nematodes where plant parts are crinkled, twisted and distorted.

Crimp A disease of strawberry incited by Aphelenchoides fragariae.

Crop Rotation The sequence of plants grown in a particular field.

Cross Breeding The mating of two varieties of the same species. To hybridize.

Cross Infection The infestation of another individual of the same species.

Cross Section A transverse cut. A cut of an organism or thing at right angles to the longitudinal axis.

Crowding Effect A situation where the average size of individuals is reduced in multiple infestations.

Crown A circular structure or process or arrangement of such structures or processes generally at or near the summit of an organ or part. See corona, circlet.

Cruciate Cross-shaped.

Cruciform Cross-shaped.

Crumb An aggregate of soil particles.

Cuff A widened glide path of the gubernaculum to help guide the spicule.

Cuirass A protective covering such as the cuticle.

Cuirasse A helmet-like appearance of the cuticle of the head.

Cultural Services Ecosystem services that provide non-material benefits, including recreational opportunities, aesthetic experiences and spiritual enrichment of humans.

Cuneiform Wedge-shaped. See sphenoid.

Cup-shaped See cupuliform, cyathiform.

Cupuliform Having the shape of a cup. Cyathiform.

Curds Dried masses of Ditylenchus myceliophagus and others similar to nematode wool.

Curly Leaf A disease of strawberry incited by Aphelenchoides fragariae. See dwarf, crimp.

Curly Tip A condition where nematode feeding near a root tip causes retardation of growth and the elongation on one side results in a curling effect.

Cusp A pointed process. See pointed.

Cuspate Pointed.

Cuspidate Terminating in an acute point.

Cuticula Cuticle.

Cuticle The noncellular external covering of the nematode and apparently connective tissue of the hypodermis. The cuticle lines the natural opening of the body i.e. the oral cavity, the anal aperture, the vulva, the vagina, the lumen of the esophagus, the rectum, the cervical pore, and the cuticular ampulla of the cervical pore. See hard cuticle, soft cuticle.

Cuticular Adornments Cuticular ornamentations.

Cuticular Framework See frame work.

Cuticular Layering Structural strata of which the cuticle is composed. See external cortical layer, internal cortical layer, fibrillar layer, matrix layer, boundary layer, external fiber layer, basal lamella.

Cuticular Ornamentation Markings of any type on the cuticle.

Cuticular Protuberance Any of a number of modifications of the cuticle, i.e. probollae.

Cuticular Rods In the lip region of some forms the stylet is surrounded by short longitudinal cuticularized rods forming a kind of guiding sheath. Modified cheilorhabdions.

Cuticular Wing A segment of a complicated Tylencholaiminae stylet.

Cuticularized Formed of cuticle.

Cutin A compound related to cellulose and not equivalent to the cuticle of nematodes. See cuticle.

Cutting Organ See onchia, odontia, jaws, denticle.

Cyathiform Concave. Cup-shaped. Cupuliform.

Cylindrical Esophagus Of a large diameter and uniform in width throughout. Characteristic of rapacious nematodes.

Cyst At maturity the body wall of the Heterodera female thickens, becomes resistant to decay, and turns brown as the worm dies, resulting in a protective shell for the eggs and is termed a cyst.

Cyst Cone The conical perineal region of cyst- forming nematodes.

Cyst-Forming Nematode Any species of the genera Globodera and Heterodera.

Cytogenetics The study of the structural basis of heredity and variation.

Cytokinesis The changes affecting cytoplasm of a cell undergoing mitosis, meiosis and fertilization. See karyokinesis

Cytology The study of the structure, organic processes, functions, etc., of cells.

Cytolysis The dissolution or destruction of cells.

Cytophagous Feeding on cells. See feeding.

Cytoplasm The protoplasm of a cell exclusive of the nucleus and cell wall.

D

Damage Year The growing season in which a crop sustained injury due to nematode parasitism.

Dauer Larvae A quiescent stage entered by larvae while enclosed in the cast cuticle of the previous stage.

Decant To gently pour off a liquid in such a manner as not to roil that remaining.

Decantation-Sieving Method. The gravity-screening method of nematode isolation.

Decapitate To remove the head. To behead, Decollate.

Decay A putrefactive change. The ultimate catabolic state. To rot.

Decline A condition of progressive deterioration of tree foliage and branches.

Decoction A condensing especially by steeping. An extract obtained by boiling.

Defecate To void excrement.

Definitive Complete. Fully developed.

Definitive Home The site of permanent attachment or lodging of parasitic nematodes after a period of larval wandering or migration.

Definitive Host The host in which the sexual life of a parasite takes place.

Degenerate Esophagus An esophageal region showing little or no evidence of its triradiate character or muscular tissue.

DEGO Dorsal esophageal gland opening, especially in the order Tylenchida. See dorsal esophageal gland orifice.

Degeneration Breakdown of the tissues of the internal organs. See devolution.

Dehydration The removal of water, See desiccation, exsiccate.

Deirids Exceptionally large cervical papillae. Lateral tactoreceptors situated near the nerve ring.

Delicate Of slight or thin characteristics.

Deltoid Triangular in shape.

De Man Formulae A system of measurement used in identification based on proportional morphometrics. Originally developed by De Man but other parameters have been added by subsequent users, some useful for specific taxa.

DeManian System A goneteric system of vessels connecting the intestine and uteri with one another and posteriorly with the exterior.

De Man Indices See De Man Formulae.

Demarcation Marked by bounds. Delimited. A separation.

Dendrites Branching cytoplasmic projections of a nerve having synapses with axons of other nerve cells.

Denematize To divest of nematodes.

Dentate Bearing teeth. See edentate.

Denticle A small tooth or projection.

Denticulate Having minute tooth-like projections.

Dentiform Having the shape of a tooth.

Dentigenous Ridges Elevations bearing small tooth like projections.

Dentoid Tooth-shaped.

Deoxyribonucleic Acid See DNA.

Dependent Hatching The need for an external stimulus for hatching to be affected.

Depresssion Said of the junction of the head and the cervical region when delimited by a more or less broad hollow. See constriction.

Depressor Ani The H-shaped muscle which serves to dilate the rectum and elevate the posterior lip of the anus thus permitting defecation.

Dermal LD50 The dosage of a chemical, based on the mass of the affected organism, at which 50% of the target organisms are killed when the material is in contact with the skin; for example, 10 mg of chemical per Kg of target organism tissue. See Lethal Dosage.

Desiccate To dry up. See dehydration. Exsiccate.

Desiccation The state or condition of being dried up.

Detritophagous Feeding on detritus. See feeding.

Detritus The disintegration products of decomposition.

Deutoplasm The food plasm of an ovum. The yolk.

Deutoplasmatic Inclusion A granule or crystal in the yolk.

Devoid Lacking something. Not bearing something.

Devolution Retrograde development. Degeneration.

Dew Point The temperature at which the liquefaction of a vapor begins. Usually the term is applied to condensation of moisture from water vapor in the atmosphere.

Diagnosis (pl. **Diagnoses**) A critical scrutiny or presentation of those characters which differentiate the species (or other group) from its closest relatives.

Didelphic Having two gonads. See amphidelphic, monodelphic, prodelphic, opisthodelphic, polydelphic, reproduction.

Didymous In pairs. Two fold.

Dieback A condition of progressive necrosis of branch terminals of trees and other plants.

Differentiation The diagnosis of a taxon.

Differentiation The act or process of characterizing or making different. Changing from general to special characters. Specialization.

Diffusate A compound, especially a stimulatory one, which spreads through the substratum from the point of origin. See root diffusate, leachate.

Digestion The process of preparing food by chemical breakdown for absorption and assimilation.

Digestive Diverticula Intestinal caecum.

Digger Hooks Modified lip structures of the genus Diploscapter.

Digitate Finger-like in shape or divided into finger-like processes.

Digitiform Having the shape of a finger.

Digonic Female A digonic hermaphrodite. See reproduction.

Digonic Hermaphrodite A situation where sperm and ova are produced in different gonads of the same individual. See syngonic female, reproduction.

Dilatator Ani The dilator ani.

Dilator Ani The muscle which expands the anus.

Dilator Vulvae Vulvar muscles inserted ventrolaterally in the hypodermis and functions to open the vulva. See vulvar muscles. .

Diluent A diluting agent.

Dimorphic Individuals of the same species occurring in two distinct forms i.e. male and female. See heteromorphic, monomorphic, polymorphic, reproduction.

Dioecious Existing as separate and distinct male and female. See dimorphic, reproduction.

Diorchic Having two testes. See monorchic, reproduction.

Diplogasteroid Esophagus Having a muscular cylinder anteriorly followed by a median bulb, terminated by a glandular base.

Diploid Having a double (2n) number of chromosomes. See haploid.

Direct Cycle Eggs of parasitic parents hatch into free-living larvae which develop directly into parasitic forms. Homogonic cycle. See heterogonic cycle.

Direct Straight growth to maturity through four successive molts.

Disc Plate-like with a flat circular surface.

Disciform Having the shape of a plate or disc.

Discoid Having a shape like a round plate. Disk- like.

Discoloration Changed to a different color. Stained. Disease A condition produced by some factor in which any part of a]iving organism is abnormal. See injury.

Dissemination The spread of infectious agents by any means.

Distad Toward the distal end.

Distal Remote from the point of origin or attachment. See proximal.

Distal Tubes The marginal tubes of esophageal radii. .

Distention The act or state of extending or stretching.

Diurnal Rhythm Having a 24 hour periodic cycle. Divergence The development of dissimilarity in forms from a common ancestor. See convergence.

Divergent Extended from a point in different directions or the trends of development.

Diversity The condition of being comprised of different types of entities, such as apecies. Diversity is greatest when all species present are equally abundant. See abundance

Diversity Indices Measures or indicators of the degree of diversity. See true diversity.

Diversity-weighted Abundance A product of diversity and abundance which recognizes that both may contribute to ecosystem services. See diversity, abundance, ecosystem services.

Diverticulum (pl. **Diverticula**) A blind tube or sac off from a cavity or canal. See caecum.

Divided Parted. Disunited. See partite, tripartite.

Dixenic Pertaining to the rearing of an organism with two known species of associated organisms. See gnotobiotic, agnotobiotic, xenic, axenic. synxenic, trixenic, polyxenic.

DNA Deoxyribonucleic acid. A nucleic acid containing the genetic information for development and functioning of living organisms. Consists of two strands of nucleotides which have sugar and phosphate backbones. One of four bases is attached to each sugar. The sequence of bases is copied into RNA through transcription and determines the sequence of amino acids in proteins. See RNA.

Dormancy The dormant state.

Dormant Inactivation as if in sleep. Quiescent See aestivation.

Dorsad In the direction of the dorsum. Toward the dorsal aspect. See ventrad.

Dorsal The back side aspect of a nematode. Belonging to the back side. See ventral, dorsum

Dorsal Cord The dorsal line of the four longitudinal lines of thickening on the inner side of the hypodermis. See cords, ventral cord, lateral cords. .

Dorsal Cone A prominent projection on the roof of the buccal capsule which contains the terminal orifice of the dorsal esophageal gland.

Dorsal Depressor Ani Muscle See depressor ani.

Dorsal Esophageal Gland The dorsal most of the three esophageal glands.

Dorsal Esophageal Gland Orifice The aperture of the duct from the dorsal esophageal gland into the lumen of the esophagus. See esophageal glands. .

Dorsal Gland Orifice The dorsal esophageal gland orifice.

Dorsal Gutter A longitudinal thickening in the in the roof of the buccal capsule which contains the terminal orifice of the dorsal esophageal gland.

Dorsal Lobe In males of some nematode forms the terminal section of the bursa lying between the dorsal rays.

Dorsal Nerve A nerve that originates from the mid-dorsal region of the nerve ring, extends posteriorly in the dorsal cord as a motor nerve devoid of, or poor in, ganglion cells.

Dorsal Rays The paired genital papillae of the dorsal lobe of the bursa.

Dorsal Somatic Nerve The dorsal nerve.

Dorsolateral The position on the nematode body situated 300 laterally from the dorsomedian line and perpendicular to the anteroposterior axis Subdorsal. See position.

Dorsolateral Muscle Fields A segment of the dorsosubmedian muscle field being delineated by a lateral cord and a subniedian thickening of the hypodermis. See dorsosubmedian muscle field, ventrosubmedian muscle fields.

Dorsolateral Nerves Nerves that arise between the dorsal and lateral nerves.

Dorsomedian The true middle line on the back side of an individual.

Dorsorectal Ganglion A ganglion which gives rise to a medium caudal nerve extending posteriad to the tail.

Dorsorectal Nerve A nerve that extends from the rectal commissure to the rectum.

Dorsosubmedian Muscle Fields The somatic muscle bounded by the dorsal cord and the lateral cords. See ventrosubmedian muscle fields, subdorsal muscle fields, dorsolateral muscle fields, ventrolateral muscle fields, subventral muscle fields,

Dorsoventral An imaginary line extending from the dorsal to ventral side. The dorsoventral axis. The median line. See position.

Dorsum (pl. **Dorsa**) The back. See venter.

Dorylaimoid Esophagus Having the shape of a long- necked bottle, being slender and thin at the anterior end then expanding towards the cardia.

Double Eggs Nematode ova of an abnormally large size and generally nonviable.

Double Guiding Ring A condition where the guiding sheath of the stylet has two thickened rings of cuticle.

Drench To wet thoroughly especially with a solution. To soak. To saturate.

Drench Application The treatment of soil or other material with a nematicide as a water emulsion or mixture. See broadcast application, gas application, row application.

Drop-like See guttate.

Duct A tube or channel, especially one for conveying the secretions of a gland. .

Ductule A small duct.

Ductus A duct.

Ductus Ejaculatoris A duct formed by the union of the vas deferens and the duct of the seminal vesicle for conveying the sperm to the spicula.

Ductus Masculinus The efferent duct.

Ductus Spermatic The vas deferens.

Duff The matted, partly decomposed organic surface layer of forested soils.

Dung Consuming See coprophagous.

Dwarf Nanism

Dwarf A disease condition of plants incited by Aphelenchoides spp. with varying degrees of deformation of stems, leaves, flowers and buds which may remain rudimentary. Often termed spring dwarf or summer dwarf.

Dwarf Plant Reduced in stature or form but retaining the same proportions as a normal plant. See stunted plant.

Dystrophy Nutrition faulty or imperfect.

E

Ecdysis (pl. Ecdyses) The act of shedding the outer cuticular layer. A molt. See instar.

Eclosion The act or process of hatching from the egg.

Ecology The relations between organisms and their environment. The reactions of organisms to the conditions of their existence.

Ecosystem An assemblage of living organisms interacting as a system in conjunction with each other and with their abiotic environment.

Ecosystem Function The biological, geochemical and physical processes that take place in an ecosystem. See ecosystem services.

Ecosystem Sevices The functions of the ecosystem, derived from the activities of living organisms in conjunction with their environment, that enable life on Planet Earth. They may be grouped into four broad categories: See Supporting Services, Provisioning Services, Regulating Services and Cultural Services.

Ecotone Transition areas between different communities occupied by mixed and fluctuating populations.

Ectoferment Saliva of extraoral digestion.

Ectoparasite An organism living upon and/or feeding on the host from the exterior. See endoparasite.

Ectopore The excretory pore.

Ecumenical Worldwide in extent. General. Cosmopolitan.

Edaphic Relating to, or belonging to the soil. See arenicolous, aquatic, hypogaeic, marine, rheophile, terrestrial, terricolous, xerophilous.

Edentate Toothless. See dentate.

Eelworm A vernacular term for any individual of the phylum Nematoda, especially for free-living and plant-parasitic forms. See nema, nematode, roundworm, threadworm.

Eelworm Disease of Hyacinth A disease condition of hyacinth incited by Ditylenchus dipsaci.

Effective Number of Species A measure of species diversity. See true diversity.

Effector A molecule that selectively binds to a protein and regulates its biological activity. It may increase or decrease enxyme activity, gene expression or cell signalling. Molecules secreted by the esophageal glands of plant-feediong nematodes that affect the development or function of plant cells are examples of effectors.

Efferent Bearing or conducting away from an organ or position. See afferent.

Efferent Duct The conveying duct of the male connecting the testis with the rectum. It is composed of the vas deferens, ductus ejaculatorius, vesicula seminalis and the distal sphincter.

Effete No longer capable of producing ova. Egg The reproductive body produced by the female. An ovum.

Egg Chamber A reservoir in the ovijector.

Egg Mass In the broad sense, the gravid female, Meloidogyne spp. with her external matrix containing extruded eggs. In the narrow sense the extruded eggs contained within the matrix. See external egg mass, internal egg mass.

Egg Sac An egg mass.

Eight-shaped (**8-Shaped**) Structure A panduriform structure surrounding the vestibule extension at the approximate latitude of the cephalic constriction. See lemniscate.

Ejaculatory Duct The ductus ejaculatorius.

Ellipsoidal Elliptical. Oval.

Elliptical-shaped like an ellipse. Oblong-oval.

Elongate To lengthen. To stretch out.

Elutriator An apparatus for washing soil for the extraction of nematodes. See Seinhorst apparatus.

Elutriator Apparatus An apparatus for quantitative extraction of nematodes from soil.

Embedded Buccal Capsule A condition when the buccal capsule is totally surrounded by the esophageal musculature.

Embryo An organism in the early stages of develop went. The organism before hatching from the egg.

Embryonated Said of an egg containing an embryo or larva. See ovic embryo, juvenile.

Embryonic Found in, or belonging to the embryo. Incipient and rudimentary.

Embryonic Development Growth and maturation of the egg.

Emend To free from a defect or a deficit.

Encyst To become enclosed in a cyst, capsule, molted cuticle or by structures which function as such. See ensheath.

End Bulb The basal bulb.

Endemic Native to a certain region. Indigenous. See exotic.

Endocuticle A section of homogenous material internal to the exocuticle. A 5-banded layer in some Heterodera.

Endogenous Nemin Nemin obtained from the whole of the nematode. See nemin.

Endoparasite An organism living within and feeding on the host. See ectoparasite.

End Organs The nerve endings of the amphidial nerve coiled within the amphidial pouch. Fibrillar terminals.

End Organs Phagocytic structures of some coelomocytes. See terminal bodies, cellules.

Endotokia Matricida The failure of normally oviparous nematodes to deposit their eggs which may then accumulate, continue to develop, and hatch within the female body. (But see Taylor, Netscher, Chen and Caswell-Chen, 2003)

Enemy Crops A control technique where a nonhost crop is planted on Heterodera infested land. These crops produce active root diffusates induce hatching but do not support nematode development. See trap crops.

Ensate Sword-shaped. Having two edges, broad based and tapering to a point.

Ensheathed A situation where a molted cuticle remains as a protective covering investing the larva thus affording protection against adverse conditions. Encysted.

Ensiform Sword-shaped. Having two edges, broad based and tapering to a point.

Entad An inward extension from without.

Enteric Belonging to, or relating to the digestive tract or enteron.

Enteron The digestive tract as a whole.

Entoparasite An internal parasite. Endoparasite.

Entrails Viscera, Internal organs.

Environment All external conditions that may act upon an organism or thing to influence its development or survival.

Epaulets Specialized, ribbon-shaped, paired bands of cephalic cuticle. See cordons.

Epiblast The outer layer of the blastoderm. See hypoblast.

Epidermal Fields The cords.

Epidermis The outer cylinder of cells consisting of a single layer of epithelium which secretes the cuticle. The hypodermis.

Epiptygma A vulval flap.

Epizoite A sedentary nonparasitic animal living attached to another animal.

Epizootic A disease affecting many animals of the same kind simultaneously. Epidemic.

Equatorial Situated at the middle or central region of a body or part. See midbody.

Eradicate Complete destruction especially of a population.

Eructation The ejection of contents from the intestine via the mouth.

Escutcheon-shaped Having the shape of a shield. See scutate, scutiform, clypeiform.

Esophageal Bulb Any of the several enlargements of the esophageal wall, muscular or glandular. See cardiac bulb, bulbus, terminal bulb, basal bulb, true bulb, metacorpus, posterior bulb.

Esophageal Caucus A blind diverticulum at the posterior part of the esophagus.

Esophageal Collar Muscle tissue which surrounds the stoma.

Esophageal Cuticle The cuticular lining of the esophagus lumen which differs in some respects from the external cuticle. See external cuticle, rectal cuticle, cloacal cuticle, vaginal cuticle.

Esophageal Cylinder The esophagus proper.

Esophageal Funnel A situation anterior in which the tips of the esophageal sectors are parted infundibuliformly. See infundibuliform.

Esophogeal Gland Nuclei and Orifices The position of the nuclei and orifices of the five esophageal glands of the Dorylaimina may be described in relation to the length of the esophagus:

Esophageal Glands Elongated glands of simple or branched tubules located in the esophageal sectors, the secretions of which are apparently of an enzymatic nature, one gland being situated dorsal and two submedian or modified. See salivary glands, pharyngeal glands.

Esophageal Glands Enclosed A situation where the salivary glands are contained within the esophagus body proper.

Esophageal-Intestinal Canal That segment of the esophageal lumen leading from the haustrulum through the esophageal-intestinal valve to the intestine. .

Esophageal-Intestinal Valve A muscular structure at the base of the esophagus opening into the intestine. See cardia. .

Esophageal Lumen The cavity bounded by the esophageal musculature serving to convey food from the stoma to the intestine.

Esophageal Nerve Ring The nerve ring.

Esophageal Radii Triradiate expansion folds of the esophageal lumen which may converge terminally or become rounded to form marginal tubes. .

Esophageal Sectors The three sections of the esophageal musculature as divided by the esophageal radii, one dorsal and two ventrolateral

Esophago-Intestinal Cells The esophageal-intestinal valve.

Esophago-Intestinal Junction The esophageal-intestinal value.

Esophago-Intestinal Valve Var. of esophageal-intestinal valve.

Esophagus The muscular tube that leads from the stoma or stylet base to the intestine. The pharynx, in a strict sense since a true esophagus is lacking in Nematoda.

Esophagus Degenerate A situation where the esophageal region shows little or no evidence of its triradiate character or musculature.

Etiological Agent The parasite inciting a disease in a plant or animal.

Etiology Inquiry into the causes of a disease.

Etymology The origin and derivation of the genus and species name of an organism, or of names applied at other taxonomic levels.

Euploid Having an exact multiple of the haploid chromosome number. See aneuploid, haploid, monoploid, heteroploid, hyperploid, polyploid.

Euryhaline Capable of tolerating a wide variation of osmotic pressure in an environment. See Stenohaline.

Eurystomatous Descriptive of diplogasterid nematodes with a toothed stoma.

Eutrophic Being in a well nourished state.

Evaginate To turn inside out or to cause an internal structure to protrude or project externally.

Eversible Capable of being turned outward or inside out. Capable of being everted.

Evert To turn inside out.

Eviscerate To disembowel. To gut.

Excrescence An appendage or outgrowth. A protuberance. See process.

Excrete To void waste products from the body.

Excretory Canal The excretory duct.

Excretory Duct A tube or canal by which the excretory products of the excretory system are conveyed to the exterior via the excretory pore.

Excretory Pore A ventral opening in the cuticle by which the waste products of the excretory system are emitted to the exterior. Cervical pore.

Excretory Sinus A common cavity of the excretory ducts communicating to the exterior by the excretory pore via a cuticle-lined reservoir or an elongated terminal duct.

Excretory System All the structures concerned in ridding the body of waste products other than the intestine.

Excurrent Attenuate. Narrowly elongated.

Exocuticle The cortical layer, fiber layer, fibrillar layer, inner matrix and boundary layer of the cuticle.

Exotic Introduced into the locale. Not native, See endemic.

Exsiccate To dry up. See dehydration. desiccate.

External Amphid The cuticular manifestation of the amphid.

External Circlet The combined cephalic crown and the outer labial crown by forward migration of the former. The outer circlet.

External Cortical Layer The outermost stratum of the cuticle which is subdivisible into parts, the outer denser part and the inner less dense and thinner part. See cuticular layering.

External Cuticle The cuticle covering the exterior of the nematode body which differs in some respects from that which lines parts of the digestive tract and vagina. See esophageal cuticle, rectal cuticle, cloacal cuticle, vaginal cuticle.

External Egg Mass Eggs of Meloidogyne deposited in matrix outside the gall tissue. See egg mass, internal egg mass.

External Genital Armature The bursa.

External Parasite An ectoparasite.

External Protractor Spiculi Muscles extending posteriorly from the external side of the spicule head.

Extogenous Capable of growth outside of the body.

Extorted Turned or twisted away from the median axis. See intorted.

Extracellular Digestion Digestion of food within the intestinal cavity allowing the simpler molecules to be absorbed by the cells. See predigestion, extraoral digestion, intracellular digestion.

Extrados The exterior curve of an arch. See intrados.

Extraintestinal Digestion Extraoral digestion.

Extraoral Beyond or outside of the mouth.

Extraoral Digestion The digestion or partial digestion of food outside of the body (for instance, the salivary injection into cell protoplasm prior to ingestion). See predigestion, extracellular digestion, intracellular digestion.

Extrorse Turned or facing outwards. Toward the outside. See introrse.

Eyes Structures present in some freshwater and marine nematodes which consist of a lenticular cuticular body situated on a pigment cup.

F

F1 The first-generation offspring resulting from a given mating. F2 the second generation, etc.

Facultative Aerobic Capable of living without the presence of atmospheric oxygen, though capable of utilizing it when available.

Facultative Anaerobic Normally living in the presence of atmospheric oxygen, but capable of becoming anaerobic. See aerobic, anaerobic, facultative aerobic, obligatory aerobic, obligatory anaerobic.

Facultative Parasite Organisms ordinarily saprophytic or saprozoic which may become parasitic under certain conditions. See obligate parasite, facultative saprophyte.

Facultative Saprophyte Organisms ordinarily parasitic which may subsist as saprophytes during part of their life cycle but do not complete their cycle away from a host. See facultative parasite, obligate parasite, obligate saprophyte.

Falcate Curved like a sickle.

Falciform Having the shape of a scythe or sickle.

Fallow Cropland left idle in order to restore productivity, mainly through accumulation of water, nutrients, attrition of pathogens; probably a combination of these.

Family In the systematic arrangement of organisms into groups or categories denoting natural relationships a category ranking above a genus and below an order. See classification.

Fascicle A bundle. A cluster.

Fat Body A trophosome.

Faulted The interruption then continuance of a line parallel to the original but on a different plane.

Fauna Animals or animal life. See flora, biota.

Faunal Analysis Assessment of the relative proportions of nematodes of different feeding habits and life history characteristics to provide inference of the state of the soil ecosystem and its potential for providing Ecosystem Services.

Fecundation Fertilization, as of eggs by spermatozoa.

Feeding The intake of substances essential to growth, sustenance and maintenance of an organism. See bacteriophagous, bacterivore, biophagous, carnivorous, coprophagous, cycophagous, detritophagous, fungivore, fungivorous, herbivorous, lancinate, microbivorous, microphagus, monophagous, mycohelminths, mycophagous.nematophagous, oligophagous, ominivorous, parenteral feeding, phyllophagous, phytophagous, polyphagia, polyphagous, polyphagy, predaceous, predator, rapacious, rypophagous, saprobiont, saprophage, saprophagous, saprophyte, saprozoic, saprozoite, sarcophagous, scavenger, voracious.

Female Gangs The female genitalia.

Female Genital Aperture The vulva.

Female Intersex An individual which begins development as a female but finishes development as a male. See intersex, male intersex.

Fenestra (pl. Fenestrae) A window. A transparent spot. In the genus Heterodera a thin walled transparent region of the vulval cone which may breakdown to form the hatching pore. The thin- walled region about the anus in some Heterodera. See fenestrate, semifenestrate, bifenestrate, ambifenestrate, circumfenestrate, vulval fenestra, anal fenestra.

Fenestration The process of forming the hatching pore by the breakdown of the thin areas of cuticle of the anal or vulval cone.

Ferment Saliva.

Fertilization Membrane The lipoid membrane.

Fiber Cells Modifications of the hypodermis which contributes to the formation of the lips, two in the dorsal lips and one in the lateral and ventral lips.

Fiber Layers Three cuticular strata of dense connective tissue which is oblique, ribbon-ike, possibly spiral and delimited by the boundary layer and the basal lamella. See cuticular layering.

Fibril (pl. Fibrillae) The fibers which constitute the actual contractile substance of the muscle cell.

Fibril Cells Three coelomocytic cells which lie adjacent to the esophagus and opposite the three esophageal radii. See coelomocytes, binding cells.

Fibrillae Thread or fiber-like supporting structures of the hypodermis.

Fibrillar Layer The cuticular stratum consisting of a condensation of spongy matrix forming a closely woven network between the internal cortical layer and the matrix layer. See cuticular layering.

Fibrillar Terminals The nerve endings of the amphidial nerve coiled within the amphidial pouch. End organs.

Fibroplasm That portion of the muscle cell differentiated into muscle fibers. See sarcoplasm.

Fibrous Root System Adventitious roots which arise from the base of the stem, bulb or corm. See Lap root system.

Field Capacity The amount of moisture remaining in a soil after the free water has been allowed to drain away into drier soil material beneath; usually expressed as a percentage of the oven dried weight of soil. It is the highest amount of moisture that the soil will hold under conditions of free drainage after excess water has drained away following rain or irrigation.

Filament A slender, thread-like object, process or appendage.

Filiation Descent from a parent. Lineage.

Filiform Having the shape of a thread or filament.

Filling Cells The cells of the lip pulp other than the clavate cells, fiber cells or arcade cells.

Fimbriate Fringed.

Final Host The definitive host.

Finger-like See digitate.

First Bulb The metacorpus.

Fissiparous The production of individuals by fission. See reproduction.

Fissura Cephalica Head cleft.

Fissure A cleft or groove separating adjacent tissues, parts or organs.

Fix To harden and preserve relaxed specimens especially for permanent mounting and microscopic study.

Fixative Generally a mixture of chemicals which harden and preserve the bodies of relaxed specimen s.

Fixed Athrocytes Stationary coelomocytes capable of absorbing foreign material from the body fluids and storing it in crystalline form. See coelomocyte.

Flaccid Limp. Feeble. Limber.

Flabellum. Fan- or feather-shaped dorsal or ventral extensions of the cuticle.

Flagellum In differential staining of mermithid larvae bodies with carmalum, the posterior region stains very light and is termed â€œflagellum.â€

Flange A rib, rim or expansion developed for strength, guiding or attachment to another part or organ.

Flap Apparatus The valvular apparatus.

Flask-shaped See ampullate, ampulliform, lageniform.

Flavate Bodies Where present, four elongate, yellowish, gland-like structures situated submedianly near the base of the stylet.

Flesh Consuming See biophagous, carnivorous, feeding.

Flooding A control measure against plant parasitic nematodes where soil is kept saturated or submerged.

Flora Plants or plant life. See fauna, biota.

Flush The development of a generation of nematodes.

Flushing Glands Secretion supplying glands of the amphids and plasmids

Focus (pl. **Foci**) A localized region of disease or the principal site of a general disease.

Foregut The esophageal region of the alimentary tract. See midgut, hindgut.

Forked See furcate, trichotomous, tricuspid, tricuspidate, tridentate, trifid, trifurcate.

Fornent In an opposite position. Facing. See apposition, juxtaposition.

Fossa A ditch-like depression.

Fossorial Equipped for, or capable of digging.

Framework A cephalic structure which provides rigidity to the head. It is subcuticular and axially forms a stylet guide in stylet-bearing forms. See calvarium.

Fovea A pit or depression, a small fossa.

Frass Solid excrement of larval insects.

Free-living Living independently of any host organism. A nonparasite. See parasite.

French Bud A disease of strawberry incited by Aphelenchoides fragariae. See dwarf.

Frenching A disease of strawberry incited by Aphelenchoides fragariae. See dwarf.

Fringe An edging of processes extending beyond the margin and are usually of equal length. See fimbriate.

Frontal Disc The labial disc.

Fumigant Nematicide A nematicidal chemical with high volatility that moves through the soil primarily in the gaseous phase.

Functional Guild A group of species that contribute to the same ecosystem function.

Fungivorous Feeding on fungus mycelium. See feeding.

Fungivores Organisms that feed on fungi.

Fungus (pl. **Fungi**) A low form of plant life which lacking chlorophyll and being incapable of manufacturing its own food, lives off dead or living plant or animal matter. The body of a fungus consists of delicate threads (hyphae) which in mass form branched systems called mycelia.

Funnel-shaped See infundibuliform.

Funnel Technique See Baermann funnel technique.

Furcate Forked. Having two divergent branches with a common base.

Fusiform Spindle-shaped. Tapering toward each end.

G

Gall A pronounced localized swelling. An outgrowth of unorganized cells generally more or less spherical or fusiform.

Gallery A passage or tunnel made by an animal.

Galvanotaxis A tactic response in which an electric potential is the stimulus. See Taxis.

Gamete A mature reproductive cell. See spermatozoon, ovum.

Gametogenesis The development or formation of reproductive cells.

Gamogonic Pertaining to sexual reproduction.

Ganglion (pl. **Ganglia**) A well defined concentration of nerve cell bodies forming a nerve center.

Gas Application The treatment of soil or other material with a nematicide in a vapor phase. See broadcast application, drench application, row application.

Gaseous Aeriform.

Gastrula The early embryo, consisting of an open mouthed sac-like body with a wall of two layers of cells, an outer epiblast and an inner hypoblast. The opening is called the blastopore and the central cavity the archenteron. See blastula.

Gavage Introducing contents into the stomach by means of a tube.

Gene The basic unit of heredity in a living organism. A segment of genomic information encoded by DNA that specifies a trait. See DNA..

Gene-for-Gene A hypothesis proposed by Flor that for each gene conditioning a reaction in a host, there is a corresponding gene in the pathogen that codes for pathogenicity. See gene.

Generalist Predators Nematodes that feed on other living organisms but not necessarily restricted to a single type iof prey. See specialist predators.

Generation A single set of progeny in the succession of natural descent. The period of time in which one set of progeny follows another. See life cycle..

Geniculate Bent to an obtuse angle.

Genital Aperture The vulva in females, The anus leading to a cloaca in males.

Genital Armature All the apparatus involved in copulation.

Genital Cone A posterior projection of the region around the cloacal orifice. The cloacal prominence.

Genital Organs The reproductive organs. See ovary, testis.

Genital Papillae Tactile papillae or setae in the anal region of the male and may be preanal, postanal or caudal in position. See pedunculated papillae, rosettes, plectanes, copulatory warts, preanal sucker.

Genital Primordium The initial cells leading to the development of the reproductive system.

Genital Setae See genital papillae.

Genital Tract The gonads.

Genital Tubes The gonads.

Genitalia The organs of generation. In the male these consist of the testis, seminal glands with their ducts, gubernaculum and spicule. In the female the vulva, vagina, ovary, oviduct, seminal receptacle and uterus.

Genoholotype A genotype specifically named as generic type by the author. See type.

Genolectotype The type of a genus selected from a series of species placed in the genus after the description. See type.

Genosyntype One of a series of species upon which a genus is based with no one species having been established as type. See type.

Genotype The species which is designated as the type species of a genus. See type.

Genotype The total genetic makeup of an individual or group. See phenotype.

Genus (pl. **Genera**) In the systematic arrangement of organisms into groups or categories denoting natural relationships a category ranking above a species and below a family. In binominal nomenclature the first word of the scientific name of a species is the generic name and is capitalized. The second term is the trivial name and is not capitalized. See classification.

Geographical Distribution The inhabited range of a species.

Geotaxis A tactic response in which gravitational attraction is the stimulating factor. See taxis, tropism.

Geotropism A tropic response in which gravitational attraction is the orienting factor. See tropism, taxis.

Germarium The germinal zone.

Germ Layers The layers of cells that are distinguishable in the embryo immediately after gastrulation. In most animals, the layers give rise to similar tissues and organs. Diploblastic animals, such as Coelenterata, have two layers, ectoderm and endoderm. The ectoderm gives rise to external structures and the endoderm to internal structures such as intestine and reproductive organs. Triploblastic animals have a third layer, mesoderm, which gives rise to muscle and skeletal structures and which surrounds the body cavity.

Germinal Zone That part the ovary where the oogonia divide. See growth zone.

Giant Cells A structure formed most commonly in root tissue by the coalescing of several cells through dissolution of common walls under stimulation of salivary secretion by gall-forming nematodes and some other kinds. Lysigenoma. See hypertrophy, hyperplasia, nurse cell.

Gland Nuclei and Orifices See Esophogeal Gland Nuclei and Orifices.

Gland-like Adeniform.

Glandular Esophagus A ventricular esophagus.

Gliding Ring The guiding ring of the stylet.

Globate Spherical.

Globular Globe-shaped. Spherical.

Glottoid Apparatus The metastom of Rhabditids and Diplogasterids formed of one dorsal and two subventral metarhabdions. See isoglottid, isomorphic.

Glottoid Teeth Metarhabdion projections of the glottoid apparatus.

Glucoprotein A segregate of the cuticle (a mucoid).

Glucosamine The chief constituent of chitin.

Glucose A simple 6-carbon sugar.

Glycogen Animal starch. A soluble polysaccharide made up of numerous glucose molecules. The form in which carbohydrates are stored in the animal. Helminth glycogen is a mixture of compounds varying considerably in molecular weight.

Gnathi Jaws.

Gnotobiotic Pertaining to the rearing of an organism with no other or only known species of associate organisms. See Agnotobiotic, axenic, xenic, synxenic, monoxenic, dixenic, trixenic, polyxenic.

Gonad A primary sex gland. An ovary or testis.

Gonochoristic Sexually reproducing species in which there are at least two distinct sexes. The sex of an individual is genetically determined and does not change throughout its lifetime. Gonochorism.

Gonoduct A reproductive tube. A gonad.

Gonopore The opening to the exterior of the gonads; the vulva in the female and the anus in the male.

Gradual Metamorphosis Incomplete metamorphosis.

Grape Cluster-like See aciniform, botryoid.

Gravid Bearing an egg or eggs.

Gravity Flow System A method of applying liquid nematicides to soil utilizing liquid head, liquid viscosity, tubing diameter and tractor speed for regulating the rate of application.

Grinder Cuticular Plates in the posterior bulb of the esophagus of many bacterial-feeding nematodes. Have the function of crushing bacteria prior to digestion in the intestine. See chewing plates, valve plates.

Grouped See agminate, aggregated.

Growth Zone That part of the ovary where the oogonia develop. See germinal zone.

Gubernacular Muscles Specialized muscles for the functioning of the gubernaculum. See retractor gubernaculi, protractor gubernaculi, seductor gubernaculi.

Gubernaculum In male nematodes a grooved cuticularized structure, sometimes paired, which guides the spicule and is formed by sclerotization of the dorsal wall of the spicular pouch. See cuff.

Guide Collar The guiding ring of the stylet.

Guide Ring The guiding ring of the stylet.

Guiding Ring of Stylet The anterior end of a guiding sheath of the stylet cuticularized or muscular thus appearing to be more dense than surrounding tissue. See guiding tube of stylet. .

Guiding Tube of Stylet A sheath about the stylet extending from the â€œguiding ringâ€ to the stylet base.

Guild A group of species with similar behavior, life history characteristics or ecosystem function. See functional guild, structural guild.

Gut The intestine.

Guttate Drop-like.

Guttiform Having the shape of a drop. Drop-like in form.

Gynandromorph An individual in which the secondary male and female characters are present in the same individual but different parts of the body, sometimes an hermaphrodite but not necessarily so, See intersex, reproduction.

Gynodioecious Descriptive of a population expressing the characteristics of gynodioecy.

Gynodioecy The condition in which a population comprises both females and hermaphrodites.

Gynotype A female type. See type.

H

Habitat The natural environment of an organism, specifically, the locale in which it grows and lives.

Haemolymph The watery lymph-like nutritive fluid of the nematode pseudocoel.

Hairy See piliferous, trichoid.

Hairy Root A condition of excessive root branching induced by nematode feeding.

Half Turned See hemitrope.

Halophilous, Halophilic Living in a salty environment.

Hamabiosis A form of symbiosis in which two species live in association without obvious motive or advantage to either symbiont.

Hamate Having a hook or hook shaped. See adunc.

Hand-shaped See palmate.

Haploid Single in appearance or arrangement. Having the basic (n) number of chromosomes for the species. See diploid.

Hard Cuticle The material that builds up the jaws teeth, cuticular lining of the oral cavity and the male genital armature. It has consistency of chitin. See soft cuticle.

Hatching The breaking of the egg shell by a larva during the process of emergence. Eclosion. See spontaneous hatching.

Hatching Factor An external hatching stimulus which acts on eggs. See root diffusate, leachate.

Hatching Pore The fenestra.

Hatch Plant A host plant capable of inducing larval emergence from eggs.

Haustrulum The spheroidal cavity of the valvular apparatus which functions as a pump. .

Head Bristles The cephalic crown of papillae or setae.

Head Bulb Specialized cuticular inflations which form a band-likc structure immediately posterior to the lips. See ballonets.

Head Cap A distinctively set off head region.

Head Cleft A pouch-like cephalic sensory organ, lateral in position, which possesses, in comparison with the amphids, a simpler innervation. Also termed cephalic organ, organe cephalique, fissura cephalica.

Head Offset Head region delimited by a depression or a constriction.

Head Shields Cuticular collar extensions that arch over the lips, considered interlabia by some authors.

Heautotype A specimen indentified by the author as an illustration of his species and compared with the type or cotype. Autotype. See type.

Hectare A. unit of land area equivalent to 2.471 acres.

Helix Having a spiral form.

Helmet A thickening in the labial region set off by a groove and denoted by a lack of ornamentation of the anterior cuticle. See cane.

Helminth A worm.

Helminthiasis A disease incited by worms in or on a host.

Helminthic Relating to, or belonging to, worms.

Helminthology The branch of zoology dealing with worms, especially parasitic worms.

Hemispheroid Having the shape of half a globe or sphere. Semiglobose.

Hemitrope Half turned around. Half inverted.

Hemizonid A nerve commissure of a highly refractive nature, generally biconvex in longitudinal section and which extends ventrally from lateral cord to lateral cord. See cephalids, hemizonion, caudalid,parahemiizonid.

Hemizonion A small nerve commissure with many characteristics of the hemizonid and apparently always shortly posterior to the hemizonid. See cephalids, hemizonid, caudalid, hemi. .

Herbivores Organisms that derive their sustenance entirely from vegetable matter. See feeding.

Herbivorous Feeding on vegetable matter. See feeding.

Heredity Transmission of genetic characters of parents to their progeny

Hermaphrodite An individual with both functional male and female reproductive organs. See protandrous hermaphroditism, digonic hermaphrodite, syngonic hermaphrodite, intersex. gyandromorph, androgynous, reproduction.

Heterogeneous Differing in kind, qualities or nature. See homogeneous.

Heterogenetic Development where the larvae do not resemble the parents. See monogenetic reproduction.

Heterogonic Cycle A situation where eggs of parasitic parents develop into free-living males and females, their progeny then proceed to the parasitic phase. Indirect cycle. See homogonic cycle.

Heterogony Alternation of two sexual generations or of a sexual and a hermaphroditic generation. See reproduction.

Heterology The lack of similarity between structures from being composed of different components or of a different derivation, See anology, homology.

Heterolysis Destruction of a cell by an external agent, especially, by lysins or enzymes from an outside source. See autolysis.

Heteromorphic Having different forms at different life stages. See dimorphic, monomorphic. polymorphic.

Heteroploid Having a greater or smaller chromosome number than the diploid number but not a simple multiple of the haploid chromosome number. See monoploid, aneupolid, euploid. hyperploid, polyploid.

Hexamerous Having six radially-arranged parts or structures.

Hexebesen A witches broom condition on a plant.

Hiatus A natural opening in the egg shell. See plug. May also refer to a space or gap in a sequential arrangement of structures such as supplements.

Hibernation The anabiotic state. See anabiosis.

Hill Application Spot application of a nematicide.

Hindgut The rectum. See foregut, midgut.

Hispid Rough with minute spines.

Hispidulous Minutely hispid.

Holarctic Realm A faunal area comprising Europe North Africa, to the Sahara desert, Asia to the Himalaya Mountains and the North American continent down to Mexico. See Nearctic region.

Holidic Pertaining to a medium whose intended constituents, other than purified inert materials, have exactly known chemical structures before compounding. See meridic, oligidic.

Hologonia A gonad where proliferation of primordial germ cells extend along its entire length.

Hologonic Gonads See hologonia.

Holomyrian A situation where the longitudinal muscle cells are indistinguishable as individual cells especially as viewed in transverse section. See meromyarian, polymarian.

Holotype The individual specimen selected as type of a species by the author. See type.

Homogeneous Similar in kind, qualities or nature. See heterogeneous.

Homogenous Layer The matrix layer of the cuticle.

Homogonic Cycle A situation where eggs of parasitic parents hatch into free-living larvae which develop directly into parasitic forms. Direct cycle. See heterogonic cycle.

Homologue A part or organ of the same relative structure, position or origin as another. The same in different individuals varying in form and function. See analogue.

Homologous Similarity of structure, but independent of function, See analogous.

Homology Structure of fundamental similarity, but independent in function, derived from descendence through some common ancestral form. See heterology.

Homonym A generic or specific name already occupied and therefore rejected due to the law of priority which requires the use of the earliest published name. Two or more entities having the same name.

Homotype A specimen compared with the type and determined to be conspecific with the type. See type.

Honeycombed See alveolate.

Hook-like See adunc, hamate, unciform, uncinate. Horn-shaped See bicorn, corniform.

Horizontal Gene Transfer (HGT) The transfer of genetic material between organisms in a manner other than sexual or asexual reproduction, e.g. by transformation or viral infection. See vetrical gene transfer.

Horizontal Resistance Resistance to races or biotypes of the nematode species is uniform, not race specific - reduces selection pressure. Usually controlled by several genes. See resistance.

Host The organism which is invaded or parasitized by a disease-producing agent and from which the parasite obtains its sustenance. See host range, suscept.

Host List The range of organisms infected by a specific parasite. See host range.

Host Plant The plant which affords sustenance to a nematode parasite and allows reproduction.

Host Preference The selection of a host most suitable for the survival of the parasite.

Host Race Nematodes of the same species differing only in their preference of host plants. See race, physiologic race.

Host Range The number of organisms parasitized by a specific nematode. See host list, stenoxenous.

Host Selection The food preferences of parasitic nematodes.

Hot water Treatment The steeping of bulbs, seeds and other plant parts in a water bath at a temperature lethal to the infecting nematodes and leaving the plant material undamaged.

Humidity The weight of water vapor in a given quantity of air, compared with the total weight of water vapor which the air is capable of holding at a given temperature.

Humus The well decomposed organic matter in mineral soils.

Hyaline Clear. Transparent. Without color. Glassy.

Hydrogen-ion Concentration A measure of the acidity of a chemical in solution. The greater the concentration of hydrogen ions, the more acid the substrate. The hydrogen-ion concentration is expressed in terms of the pH of the substrate. See pH.

Hydrolysis A chemical decomposition process of a compound which involves the addition of the elements of water.

Hydrotaxis A taxis in which a moisture gradient is the response initiating stimulus.

Hygroscopic Readily absorbing and retaining moisture.

Hyperparasitism A situation where parasites in turn are parasitized by other organisms.

Hyperplasia The abnormal increase in the number of cells of a tissue. See hypertrophy.

Hyperploid Having a chromosome number slightly greater than an exact multiple of the haploid number. See monoploid, euploid, aneuploid, heteroploid, hyperploid, polyploid.

Hypersensitive A violent reaction to parasitic attack resulting in sudden death of invaded tissues providing a barrier against further invasion.

Hypertonic A solution having an osmotic pressure higher than an isotonic fluid such that it gains water by osmosis across a membrane. See isotonic, hypotonic.

Hypertrophy The abnormal enlargement of cells generally by dissolution of common cell walls. See hyperplasia.

Hypha (pl. Hyphae) One of the thread-like strands, or filaments, that constitute the body or mycelium of a fungus. It may be divided into cells by cross walls, a septate condition, or be one elongated cell with several nuclei, a nonseptate or coenocytic condition. The hypha may be coarse or fine, aerial or submerged, stiff or flexible and exhibit different types of branclung. Some hyphae are specialized for producing spores or for penetrating host tissue.

Hypoblast The inner layer of the blastoderm. See epiblast.

Hypodermis A thin cell layer beneath the cuticle with longitudinal thickenings protruding between the longitudinal muscles to form the cords which contain the nuclei of the hypodermal cells, See lateral cord, ventral cord, dorsal cord. .

Hypogaeic Subterranean. Living underground. See arenicolous, aquatic, edaphic, marine, rheophile, terrestrial, terricolous, xerophilous.

Hypothesis (pl. Hypotheses) A tentative explanation of a phenomenon. See postulate, theory.

Hypotonic A solution having an osmotic pressure lower than an isotonic fluid such that it loses water by osmosis across a membrane. See isotonic, hypertonic.

Hypotygma Papilla-like processes on cloacal aperture of male.

Hypotype A specimen, other than the type, upon which a subsequent or emended description or figure is based. Apotype. Plesiotype. See type.

I

Ideotype A specimen named after comparison with the type, to be distinguished from a topotype. See type.

Imbricate Arranged or appearing lapped over one another like shingles on a roof.

Imbrication An overlapping of the edges as of tiles or shingles.

Immature Not yet at the state of full sexual development.

Immobile Larva The development of the male of cyst-forming nematodes within the cuticle of the third stage larva.

Immunity The ability of an organism to remain free from a disease or of parasitism by virtue of inherent properties of that organism. An immune organism is exempt from the-particular disease. See tolerant, resistant.

Impermeable Not permitting passage. Impervious.

Impregnation The act of rendering pregnant. Fecundation.

Incisures See involutions.

Incompatible Host-Parasite Relationship Interaction between a nematode species and host plant that is unfavorable for nematode survival and reproduction. See compatible.

Incomplete Metamorphosis A growth phenomenon where young through successive stages, or instars acquire adult characteristics which become functional at maturity. See complete metamorphosis.

Increase Factors The several formulae used to calculate the gain in population of cyst-forming nematodes as follows:

Increase factor X1 = (total cysts obtained at seasonâ€™s en)/(original number of cysts)

Increase factor X2 = (number new cysts produce at seasonâ€™s end)/(original number of cysts)

Increase factor X3 = (total eggs obtained at seasonâ€™s end)/(original number of eggs)

Increase factor X4 = (number new eggs produced at seasonâ€™s end)/(original number of eggs)

Incubation The period of time and conditions of environment between inoculation of an organism by a disease-producing agent and the appearance of symptoms.

Indentate Having an irregular margin. See serrate.

Indicator Plant A suitable suscept plant used to signify the survival or reproduction of plant-parasitic nematodes after some treatment or condition of population stress.

Indigenous Living in its natural or original locality. Native. See endemic.

Indirect Cycle Eggs of parasitic parents develop into free-living males and females, the offspring then proceed to the parasitic phase. Heterogonic cycle. See homogonic cycle.

Indirect Development Complete metamorphosis.

Infect To invade and establish a parasitic relationship within the host proper. See infest.

Infection Incidence The population density of attacking parasitic nematodes.

Infectious Larvae Infective larvae.

Infective Having the qualities necessary to enter a host and produce a disease condition.

Infective Larvae Nematode larva at a stage of development capable of penetrating and infecting a host.

Infective Juveniles Infective larvae.

Infective Stage The period of development in the life cycle of a parasitic nematode in which it possesses the qualities enabling infection of a host.

Infective State The infective stage.

Infest To attack externally. To contain the parasites, said of nonliving material. See infect.

Inflated Annules Enlarged annules often with a lumen or other modification of an intra-annule cavity.

Inflation of the Cuticle A blister-like swelling of the cuticle.

Inflect To curve or bend inwards or towards the axis. See infiexed.

Inflexed Curved, bent or directed abruptly inward, downward or toward the body axis. See inflect.

Infraspecific Included within a species. See con- specific.

Infundibuliform Funnel-shaped.

Ingest To take substances, especially food, into the body.

Inhalation LC50 The concentration of a chemical at which 50% of the target organisms are killed when the chemical is inhaled; for example, 10 mg of chemical per liter of air. See lethal concentration.

Injury Damage of relatively short duration to an organism. See disease.

Inlet Valve The esophageal lumen just anterior to the valvular apparatus and opening into the haustrulum

Inner Body The inner mass

Inner Circlet The crown of inner labial papillae. The internal circlet. See outer circlet: Inner Cuticular Layer This layer corresponds to the fiber layers of the cuticle. .

Inner Labial Situated on that part of the lips located closest to the longitudinal axis. See inter labial, outer labial.

Inner Labial Bristles Inner labial papillae which have been altered to bristle-like structures.

Inner Labial Papillae Sensory papillae situated toward the apex of the lip segment forming a circlet of six in primitive forms. See outer labial papillae. .

Inner Leaf Crown An amplification of the corona radiata by the formation of a second more posterior crown of lobes. .

Inner Mass A darkly staining structure slightly post-equatorial in microfilariae.

Inner Membrane The lipoid membrane.

Innervation The nerve distribution to or in a part or organ.

Inoculate To cause an infectious agent (the maculum) to come in contact with a host.

Inoculant Inoculum.

Inoculation The communication of an infective agent to the host to cause a disease.

Inoculum The infectious agent used in inciting a disease.

Inorganic Substances occurring as minerals in nature or obtainable from them by chemical means. All matter except the compounds of carbon, but including carbonates, Inanimate. See organic,

Inseminate To introduce spermatozoa into the vagina of female nematodes.

Insemination The introduction of semen into the female reproductive tract. See copulation.

Instar The nematode larval condition of growth assumed between successive molts. See molt, ecdysis, larval stages, stadium.

Integument Cuticle. Covering. Investment. Skin.

Intercallary Additional. Interpolated. Inserted between.

Intercordal Areas The nonthickened regions of hypodermis devoid of nuclei lying between the cords. See cords. .

Interface The common surfaces of two bodies.

Interior Mass Inner mass.

Interkinesis The interval between mitoses of a nucleus.

Interlabia Small lobes situated between the lips.

Interlabial Situated between the lips. See inner labial.

Interlabial Clefts The space or fissure separating the lips.

Intermediate Host The host in which the asexual stages of a parasitic life cycle take place.

Intermediate Striae The lesser lines situated between the major configuration lines of a perineal pattern. See partial striae, lighter striae, broken striae.

Internal Circlet The crown of inner labial papillae. The inner circlet.

Internal Cortical Layer The cuticular stratum bounded by the external cortical layer on the outer most side and the fibrillar layer on the inner side. See cuticular layering.

Internal Egg Mass Eggs of Meloidogyne spp. deposited in matrix within the gall tissue. See egg mass, external egg mass.

Internal Parasite Endoparasite.

Interphasmidal Line An imaginary line connecting the phasmids in a Meloidogyne perineal pattern.

Intersex An individual which exhibits a blending of male and female characters and which function as one or the other or neither sex, but never as both. See gyandromorph, male intersex, female intersex.

Interstrial Region The area between two striae.

Intestinal Caecum A blind diverticulum arising at the anterior of the intestine and extending anteriorly next to the esophagus.

Intestinal Tract The alimentary canal. See Gut. .

Intestine A simple tube, composed of a single layer of epithelial cells in which digestion of food takes place. Gut. .

Intestino-Rectal Sphincter The valve of intestinal tissue separating the intestine from the rectum or cloaca.

Intestino-Rectal Valve The intestino-rectal sphincter.

Intima An internal lining.

Intima of the Esophagus A cuticularized tube of triradiate form circumscribing the lumen of the esophagus.

Intorted Turned or twisted inward. See extorted.,

Intracellular Occurring or situated within a cell.

Intracellular Digestion The digestion of food particles within the cell. See extra-oral digestion, predigestion, extracellular digestion.

Intrados The interior curve of an arch. See extrados.

Intraoral Within the mouth.

Intrauterine Within the uterus.

Intrauterine Generations The development of mature males and females within the uterus of the mother; then yet another generation may develop within the progeny.

Intravial Stain See intra vitam.

Introduced Species See adventive.

Introrse Facing or directed inward, See extrorse.

Invalivate To mix with saliva.

Invasion Growth or movement of an infectious agent into a host with its subsequent establishment as a disease. See attack, infective.

Invasive Larvae Infective larvae.

Invertebrate Without a backbone or vertebral column. A â€œlowerâ€ animal.

Involutions The longitudinal cuticular clefts which divide the lateral fields. Incisures.

Ion An atom or molecule bearing an electrical charge. See anion, cation.

Isoglottid Having the metarhabdions of the glottoid apparatus at the same level, See anisoglottid.

Isolation Tissue The pseudocoelomic membranes and mesenteries.

Isomorphic Of identical or like form especially pertaining to the metarhabdions of the glottoid apparatus See anisomorphic.

Isotonic Having equal osmotic pressures. See hypotonic, hypertonic.

Isthmus The segment of musculature between the medium bulb and basal bulb of the esophagus. See corpus, basal bulb.

J

Jaws A cuticular framework around the mouth which functions in grasping and holding; fundamentally 3 partite. Teeth in the broad sense.

Juvenile A nematode in a developmental stage which does not yet have functional gonads. Any immature nematode.

Juvenile Female A fourth stage female in which the vagina and uterus are functional but the ovaries have yet to mature. See koriogamy.

Juxtaposition Situated adjacent to another. The act of being placed side by side. Apposition. See fornent.

K

Karyokinesis Mitosis.

Karyoplasm The protoplasm of the nucleus. Nucleoplasm.

Keel-like See carina.

Keratin A segregate form of the cuticle corresponding to the external cortical layer. See cuticular layering.

Kidney-shaped See reniform.

Kinesis Movement induced by a stimulus and is not necessarily orienting. See taxis.

Knobs See stylet knobs.

Koriogamy The impregnation of a female nematode which possesses a fully developed vagina and uterus but an immature ovary.

L

Labial Annule The labial disc.

Labial Crown of Papillae The anterior most circlet of papillae located on the lips. In forms with six lips two are lateral, two subdorsal and two subventral in position. In forms with other lip conditions they are accorded their respective positions. See anterior cephalic crown.

Labial Disc (or Labial Disk) The more or less circular font of cuticle about the oral opening and delimited posteriorly by the first transverse striation.

Labial Papillae Papillae located on the lips. See labial crown of papillae.

Labial Pulp Cavity See lip pulp.

Labial Sclerolization See framework.

Labial Muscles Specialized lip muscles apparently of the same origin as somatic muscles.

Labial Setae The setae of the inner circlet and located on the lips or close to the mouth.

Labium (pl. **Labia**) Lip.

Lacuna A space; generally applied to a space between the cuticle and the epidermis and underlying muscle created as a consequence of the cuticle being less contractile and the muscle more contractile See cervical lacuna(e).

Lageniform Bottle-shaped. A more or less cylindrical body with a narrow elongate neck.

Laggard An individual which matures very slowly or remains under the mature size.

Lamella (pl. **Lamellae**) A thin sheet, plate or layer. Sheet-like process, part or organ.

Lamellate Composed of thin sheets, plates or layers.

Lamelliform Having the form or composed of thin sheets, plates or layers.

Lanceolate Oblong and tapering to a point.

Lance-shaped. Spear-shaped. See oblanceolate.

Lancets Shall sharp teeth in the buccal cavity of some nematodes.

Lancet-shaped Formed like a lance commonly sharply pointed and two edged. See lanceolate.

Lancinate To pierce a cell or organ, etc. with a tooth or stylet.

Larva (pl. **Larvae**) In the nematological sense, the immature form of a nematode. See nymph, juvenile, neotony.

Larval Stages The periods of growth between molts. See Instar.

Larvated Eggs Embryonated eggs.

Larviparous Reproducing by bringing forth living larvae. Ovoviviparous. See viviparous, oviparious, fissiparous, reproduction.

Laterad Toward the side and directed away from the median line.

Lateral Relating to, belonging to, attached to, or situated at the side. The position on the nematode body situated 90Â° from the median line and perpendicular to the anteroposterior axis. See position.

Lateral Alae Lateral or sublateral extensions of the cuticle which extend along the body in both males and females. See caudal alae, bursa.

Lateral Cephalic Inflations Cuticular inflations confined to a limited lateral region just posterior to the lips. See inflation of the cuticle..

Lateral Cords Longitudinal hypodermal thickenings lying in the lateral position. See cords, dorsal cord, subdorsal cord, ventral cord, subventral cord. .

Lateral Fields A form of cuticular configuration above the lateral cords.

Lateral gene transfer See horizontal gene transfer

Lateral Glands Lateral hypodermal glands.

Lateral Grooves Involutions.

Lateral Guiding Pieces Cuticularized structures which guide the spicule. .

Lateral Hypodermal Glands Two sublateral rows of unicellular glands lying in the lateral cords and communicating to the exterior by short ducts through pores in the cuticle. See hypodermal glands.

Lateral Lines Involutions.

Lateral Membrane A cuticular flap situated on both sides of the vulval slit in some nematodes. See epiptygma, vulvar flap.

Lateral Nerves Nerves originating mostly at the lateral ganglia. Of a sensory nature with ganglionic swellings along the lateral cords, supplying a sensory branch to the cervical papillae when present and enter the lumbar ganglia posteriorly.

Lateral Organs The amphids. See organs of the lateral cords.

Lateral Papillae The amphids.

Lateral Pore An opening to the lateral hypodermal glands. .

Lateral Ridges A broad longitudinal ridge which covers the lateral cord and in some forms may be subdivided superficially by involutions. See longitudinal ridges.

Lateral Wings Longitudinal lateral elevations of cuticle which extend along all or part of the body, especially from the base of the esophagus to the anal region. See lateral field.

Lateroanal In the lateral position at the level of the anus. See position.

Laterodorsal The position on the nematode body situated 45Â° from the dorsomedian line and perpendicular to the anteroposterior axis. Submedian. See position, lateroventral.

Laterodorsal Retractor Spiculi A muscle extending anteriorly from the manubrium, to the body wall between the laterodorsal somatic muscle field and the lateral cord.

Laterosubdorsal The position on the nematode body situated 60Â° from the dorsomedian line or 30Â° dorsad from the lateral line and perpendicular to the anteroposterior axis. See position.

Lateroventral The position on the nematode body situated 45Â° from the ventromedian line and perpendicular to the anteroposterior axes. Submedian. See position, laterodorsal.

Lateroventral Retractor Spiculi A muscle extending anteriorly from the manubrium to the body wall between the lateroventral muscle field and the lateral cord.

Latitude A site in a nematode measured on the meridian of a cross section.

L.A.Value The logarithm of the concentration in arbitrary units of the active root diffusate factor and is proportional to its activity. See log activity values.

Layered See lamelliform.

LC50 The concentration of a chemical at which 50% of the target organisms are killed; for example, 10 mg of chemical per liter of air or water. See lethal concentration.

LD50 The dosage of a chemical, based on the mass of the affected organism, at which 50% of the target organisms are killed; for example, 10 mg of chemical per Kg of target organism tissue. See lethal dosage.

Leachate The dissolved soluble constituents of a root diffusate and its solvent. See root diffusate.

Leaching The result of percolating a solvent through a substrate to dissolve its soluble constituent.

Leaf Blotch Disease A pathological condition of Chrysanthemum incited by Aphelenchoides ritzemabosi.

Leaf Crown The corona radiata.

Leaf Drop A disease condition of fig trees incited by Paratylenchus hamatus and Xiphinema index.

Leaf Galls Leaf swelling which serve as structures for the maturation and reproduction of the nematode parasites and are comparable to seed galls.

Leather-like See coriaceous.

Lectins Sugar-binding proteins which are highly specific for the sugar component to which they bind. Often involved in biological recognition between cells and proteins.

Lectin Binding The use of sugar-binding proteins to adhere to specific regions of cell walls. For example, some bacteria use lectins to attach themselves to the cells of the host organism during infection.

Lectotype The specimen selected as holotypc from the syntypes upon which a species revision is based. See type.

Legs Hollow, glandular ambulatory setae.

Lcmniscate Ribbon-like, having the shape of an â€œ8â€□. See panduriform, eight-shaped structure.

Lemon-shaped See citron-shaped. Prolate.

Lenticular Lens-shaped. Having the form of a biconvex lens.

Leptoderan Caudal alae which do not meet posterior to the tail. See peloderan.

Lesion An injury, wound or morbid structural change. A localized spot of diseased tissue.

Lethal Concentration The concentration of a chemical at which the target organisms are killed; for example, 10 mg of chemical per liter of air or water. See LC50.

Lethal Dosage The dosage of a chemical, based on the mass of the affected organism, at which the target organisms are killed; for example, 10 mg of chemical per Kg of target organism tissue. May be expressed based on ingestion (Oral Lethal Dosage) or skin contact (Dermal Lethal Dosage). See LD50.

Lethargy A state of inaction.

Life Cycle The successive series of changes through which an organism passes in the course of its development. See generation.

Life History The record of events in the development of an individual.

Liganientum Cephalo-esophageal arcade cells.

Lighter Striae Intermediate striae.

Ligneous Wood-like. Woody.

Linear Resembling a line. Having a form long and uniform in width.

Lineate Having lines or stripes. Having a form long and uniform in width.

Linguiform Tongue-shaped.

Lingulate Tongue-shaped.

Ligulate Strap-shaped.

Lip Cap The anterior most cuticular annulation, disc-like in shape, circumoral and usually thicker than adjacent head annules.

Liplets Diminutive lip-like structures frequently projecting.

Lipoid Membrane A thin, delicate covering membrane immediately beneath the egg shell of fertilized eggs. The inner membrane. See the fertilization membrane.

Lip Pulp The interior tissue of the lips composed of several large, elongate cells. See clavate cells, fiber cells, arcade cells, filling cells.

Lip Region The cuticular area from the basal ring forward.

Lip Sclerotizations The cephalic framework.

Lip Sector The several regions of the anterior cephalic region as delimited by the radial blades.

Littoral Fauna Organisms dwelling between high and low tide lines. See benthic fauna, pelagic fauna.

Lively See alacrious.

Loam The textural class name for soil having a moderate amount of sand, silt and clay. Loam soils contain 7 to 27 percent of clay, 28 to 50 percent of silt and less than 52 percent of sand.

Lobe A generally rounded part or projection of a part or organ separated from neighboring parts by fissures or constrictions.

Lobiform Shaped like a lobe or rounded process.

Lobulate Consisting of or with many small lobes or lobules.

Log Adivity Values A L. A. of zero is that concentration of hatching factor of root diffusate which just fails to give a hatch significantly higher than water. A L. A. of I represents a value of 10 times, a L. A. of 2 represents a value of 100 times, etc. The logarithm of the concentration of the active factor of the root leachate.

Longitudinal Extended in length. Lengthwise.

Longitudinal Alae Longitudinal thickenings of the cuticle, generally lateral or sublateral, extending the length of the body. See alae.

Longitudinal Axis The long axis from head to tail. Anteroposterior axis.

Longitudinal Cords The longitudinal thickenings of the hypodermis. See cords.

Longitudinal Fibrils Noncontractile or static fibers of the muscle sarcoplasm. See transverse fibrils.

Longitudinal Fields The longitudinal ridges.

Longitudinal Lines The longitudinal ridges of some authors and the longitudinal cords of others.

Longitudinal Lip Groove A cephalic cuticular depression above each of the radial blades of the cephalic framework.

Longitudinal Lip Striae Generally faint, irregularly spaced markings of the cephalic cuticle which terminate at the cephalic constriction, See constriction.

Longitudinal Markings This form of cuticular ornamentation may he ridges, alae or the result of interruptions in the transverse markings.

Longitudinal Plane The lengthwise plane or dimension ala body. See sagittal plane, transverse plane.

Longitudinal Ribbon The lateral field.

Longitudinal Ribs The longitudinal ridges of the cuticle.

Longitudinal Ridges Local longitudinal thickenings of the two external cuticular layers which extend the whole body length. See lateral ridges.

Longitudinal Striation In the cuticle a groove parallel to the longitudinal axis. See transverse striation.

Longitudinal Thickenings The lateral fields. The lateral cords.

Lumbar Ganglia Large ganglia in the anal region receiving the lateral nerves and ventrolateral connectives.

Lumen The cavity delimited by the walls of a tubular vessel.

Lumen Rays The esophageal radii.

Lysigenoma (pl. Lysigenomata) A structure formed most commonly in root tissue by the coalescing of several cells through dissolution of common walls under stimulation of salivary secretions by gall-forming nematodes and some other kinds. Giant cell. See hypertrophy, hyperplasia, nurse cell.

Lysis Cell destruction. A tissue dissolution process.

M

Macerate To waste away. To soften, separate and wear away. See comminute, triturate.

Macrophagous Feeding on objects of a relatively large size. See feeding, microphagous.

Macroscopic Visible to the unaided eye. See microscopic.

Macula A spot, especially one not elevated above the surrounding level. See ocellus.

Male Intersex An individual which begins development as a male but, completes development as a female See intersex, female intersex.

Male Plant A disease of strawberry incited by Aphelenchoides fragariae. See dwarf.

Malformation An abnormal development or formation of a part or structure.

Mammalian Toxicity The effect of a pesticide on non-target human and warm-blooded animals. See Toxicity.

Mammiform Nipple-shaped. Breast-shaped.

Mammillate Having nipple-like protuberances or processes. Digitate.

de Man Formula See De Man Formula.

de Man Indices The De Man formula.

Mandibles Hard, strong cuticularized structures about the mouth which function in biting or grasping.

Manubrium The enlarged, cephalated proximal portion of the spicule.

Marginal Fibers Fibers, probably of an elastic nature, which extend from the ends of the lumen of the esophageal radii to the bounding membrane of the esophagus.

Marginal Tubes The terminal cylindrical endings of some esophageal radii; the radii of other forms may have convergent terminals.

Marine Living in the sea. See aquatic, arenicolous, edaphic, hypogaeic, rheophile, terrestrial, terricolous, xerophilous, cavenicolous, fossorial.

Mass Invasion A situation where large numbers of nematode parasites attack a host simultaneously.

Mass Hatching The phenomenon of large number of nematode eggs hatching within a short period of time in response to some stimulus.

Massive Gland A simple roundish pear-shaped cell or an elongate tubular cell of uniform diameter and devoid of internal cavities. Salivary gland.

Masticatory Apparatus Organs or structures adapted to the chewing of food.

Matricin A fibroid segregate of the cuticle corresponding to the matrix layer. See cuticular layering.

Matrix The enveloping substance within which something originates or develops.

Matrix Glands In Meloidogyne spp. six glands, two subdorsal, two lateral and two subventral which secrete the gelatinous matrix through the anus and in which the eggs eventually are imbedded.

Matrix Layer The cuticular stratum consisting of a spongy mass between the fibrillar layer and the boundary layer. See cuticular layering.

Maturation The entire process by which diploid primordial germ cells are transformed into haploid gametes.

Maturation Division Mitosis.

Mediad Toward the median plane or line.

Medial Pertaining to, directed toward, situated at, or occurring in the middle.

Median A midway point in position. Situated in the middle position. A plane of division dividing a bilateral organism into right and left halves. See position.

Median Bulb The metacorpus.

Median Esophageal Bulb The metacorpus.

Median Pseudobulb The metacorpus.

Medioventral Ventromedian. See position.

Medium (pl. Media) Substratum.

Meiosis The process by which a nucleus halves its chromosome number from the original diploid state to the reduced haploid state. Typically meiosis consists of two nuclear divisions: in one, without prior longitudinal splitting the chromosomes separate resulting in half of the original diploid number being in each resulting daughter nucleus; in the other the process is essentially mitotic in character. Four haploid cells are thus produced by the two divisions.

Meiotic Parthenogenesis Reproduction without fertilization and with a meiotic reduction division of the unfertilized egg cell. Diploid number may be re-established by fusion of nucleus with polar body. SeeParthenogenesis, Automictic Parthenogenesis.

Mensural Of or pertaining to measure. Measurable.

Menstruum A solvent. Any substance capable of dissolving a solid body.

Mentek A disease of rice incited by Hirschmanniella oryzae.

Meridic Pertaining to a medium in which the chemical identity of certain, but not all, of the absolutely essential molecules has been established. Media in which most of such requirements are known are highly meridic; those in which but one or a few are known are slightly meridic. See holidic, oligidic.

Mermithaners Male ants infected with a species of mermithids. See mermithogynes, mermithostratiotes, mermithergates.

Mermithergates Worker ants infected with a species of mermithids. See mermithogynes, mermithergates, mermithostratiotes, mermithaners.

Mermithogynes Female ants infected with a species of mermithids. See mermithergates, mermithostratiotes, mermithaners.

Mermithostratiotes Soldier ants infected with a species of mermithids. See mermithogynes, mermithergates, mermithaners.

Mermithoid Esophagus Having a degenerate musculature and possessing a long cuticularized tube with the cells of the esophagus and glands distributed in an irregular manner.

Merogony A mode of reproduction where the spermatozoon enters the egg cell and provides stimulus to further embryonic development but does not fuse with the nucleus to form a zygote.

Meromyarian Having eight longitudinal muscle cells. See holomyarian, polymyarian.

Mesenteries A delicate membrane which invests the intestine and the gonads and extends from the esophageal membrane to pseudocoelomic membranes covering the musculature. See pseudocoelomic membranes, isolation tissue.

Mesenteron The intestine.

Mesorhabdions The walls of the mesostom.

Mesostom A subdivision of the protostom as distinguished by clefts in the rhabdions and is preceded anteriorly by the prostom and posteriorly by the metastom. See prostom. metastom.

Messenger Ribonucleic Acid See mRNA.

Metabolic Footprint A measure of the contribution of an individual, species or guild of organisms to ecosystem function or service based of their utilization of carbon and energy. See metabolism, abundance, ecosystem services.

Metabolism The chemical changes occurring in living organisms by which energy is provided for vital processes and activities and new material is assimilated to repair the waste. The sum of the processes of anabolism and catabolism which take place in living matter.

Metacentric Having two equidistant arms due to the median position of the centrome. See telocentric.

Metacorpus The posterior subdivision of the corpus taking an ovate form and being preceded by a cylindrical anterior segment. The median bulb. See procorpus.

Metagenesis Reproduction where the immediate descendants are dissimilar from the parents with the alternate type produced only with every other generation or after a period of several generations. The alternation of sexual and asexual reproduction in the life cycle of certain organisms. Alternation of generations.

Metameric Segmentation The repetition of elements of the main organ systems of the body along the length of the body. For example, in the earthworms (Annelida), each of the externally visible rings marks a segment (or metamere) of the body that contains a similar pattern of blood vessels, nerves, excretory organs, external features, etc.; similar patterns are evident in Arthropoda. In the Vertebrata, segmentation is most obvious during embryonic development; in adults it is confined to mesoderm derivatives such as muscle and skeleton, e.g., repetition of vertebrae and ribs.

Metamorphosis A pronounced reorganization of body tissues and form in changing from one stage of development to another, as from a larva to an adult. See complete metamorphosis, incomplete metamorphosis.

Metaneme Filamentous organs located in or near lateral hypodermal chords. Interpreted as stretch receptors, known in some nematode species, but not C. elegans. Occur in nematodes of the order Enoplida. Have a central scapulus and anterior and posterior filaments

Metaphase A stage in mitosis or meiosis when chromatids are arranged on the equatorial plane of the spindle. See prophase, anaphase, telophase.

Metarhabdions The walls of the metastom. .

Metastom The posterior subdivision of the protostom as distinguished by clefts in the rhabdions. See mesostom, prostom.

Metatype In the broad sense, a specimen compared by the author of a species with the type and determined to be conspecific. In the narrow sense, a topotypic specimen compared by the author of a species with the type and determined by him to be conspecific. See type.

Myceile, Micella, Micell In colloids a unit of structure derived from complex molecules.

Microclimate The local climatic condition of a particular niche resulting from the modification of the general climate condition by local differences in relief, exposure and cover.

Microfilariae Young or embryonic filarioids of the Filarioidea.

Micron (pl Âµm, Micra) A unit of measurement equal to 1/1,000 of a millimeter, designated by the Greek letter mu

Microphagous Feeding, on objects of a minute size. See feeding, bacteriophagous, macrophagous.

Microplot A small duplicate of a field plot designed to be as similar as possible.

Micropyle The minute opening in the investing membrane of an egg permitting the entry of a spermatozoon.

Microscopic Small in size. Not visible to the unaided eye. See macroscopic.

Midbody The equatorial region of the nematode body. Median. See equatorial.

Middle Cuticular Layer This layer corresponds to the matrix layer of the cuticle, See cuticular layering.

Middorsal The true dorsal line. Dorsomedian, See position.

Middorsal Nerve The dorsal nerve.

Midgut The intestine proper, from the posterior of the esophagus to the anterior of the rectum See foregut, hindgut.

Midintestine The intestine proper.

Midventral The true ventral line. Ventromedian, See position.

Midventral Glands Supplements.

Migrate To move from one feeding site or location of development to another.

Migratory Nematodes Endoparasitic and ectoparasitic nematodes which are capable of feeding and moving about in roots or soil.

Mitosis The process by which a nucleus divides into two nuclei. Typically mitosis consists of four successive stages

Mitotic Parthenogensis Reproduction without fertilization and by mitotic division of egg cells. See Parthenogenesis, Apomictic Parthenogenesis.

Modified Changed somewhat in form or other characteristics.

Molt, Moult To cast off the cuticle. See ecdysis, instar.

Moniliform Constricted at intervals so as to resemble a string of beads. Beaded.

Moniliform Glands In the demanian system posterior glands which accompany the ducts which open to the exterior shortly anterior to the anus.

Monodelpbic Having one ovary. See amphidelphic didelphic, prodelphic, opisthodelphic, polydelphic.

Monogenesis Development from a single hermaphroditic parent. Asexual reproduction. See reproduction.

Monogenetic Development where the larvae resemble the parents. See heterogenetic, reproduction.

Monogony The development of eggs without benefit of fertilization by spermatozoa. See parthenogenesis, reproduction.

Monograph A treatise in detail on a particular subject.

Monomorphic Exhibiting only one form, especially applied to species which contain only one sex (female). See dimorphic, heteromorphic, polymorphic,.

Monophagous Feeding on only one kind of food, especially nematodes feeding on plants. See feeding.

Monophyletic Derived, developed or of a single stock or common ancestral form. See polyphyletic.

Monoploid Having the gametic number of chromosomes. Haploid. See aneuploid, euploid, heteroploid. hyperploid, polyploid.

Monorchic Having one testis. See diorchic, reproduction.

Monotype A holotype of a species based on a single specimen. See type.

Monoxenic Pertaining to the rearing of an organism with only one known species of associated organism. See gnotobiotic agnotobiotic, xenic, axenic, synxenic, dixenic. trixenic, polyxenic.

Monoxenic Culture A nematode population containing only one other species of organism. See pure culture, aseptic culture, agnotobiotic culture, axenic culture.

Moon-shaped See bicorn, crescentiform, seleniform, semilunar.

Mordant In staining, any substance which when added to the dyestuff combines to form an insoluble compound to produce a fixed color in the material stained.

Moribund In a dying condition. Near death.

Morphogenesis Development or evolution of morphological characters.

Morphology The study of form and structure of organisms.

Morphometry Measurement of external form. See De Man Formulae.

Morphometric Parameters Measurement of external form. See De Man Formulae.

Morphotype The type of the dimorphic form of a species. See type.

Morula The globular mass of cells formed by holoblastic cleavage of the egg in its early development.

Morular Organ A region at the ovarial end of the uterus where the egg shell is formed.

Mosaic-like See tessellate.

Moult Var. of molt.

Mouth The oral aperture.

Mouth Capsule The buccal cavity.

Mouth Spear See stomatostyle, ondontostyle. onchiostyle.

mRNA Messenger Ribonucleic Acid. The molecule of ribonucleic acid (See RNA) that is transcribed from a DNA (See DNA) template and carries the coding information for the structure of a protein to the ribosomes (See Ribosome), the sites of protein synthesis.

Mucivorous Feeding on the protoplasm of plants. See feeding.

Muck Soil Highly decomposed organic soil material developed from peat. In gencral, muck has a higher mineral or ash content than peat and is decomposed to the extent that the original plant parts cannot be identified.

Mucoid A segregate of the cuticle (glucoprotein).

Mucro An abrupt point, tip or process which terminates the nematode tail.

Mucron A small knob-like ending on a terminus.

Mucronate Ending abruptly in a point.

Mucus The gelatinous matrix.

Multispire Two or more turns of a spriral. See spiral.

Mural Stylet A stylet derived from the wall of the stoma.

Mural Tooth A cutting or piercing structure situated on the pharyngeal wall but formed further back in the esophagus.

Muscle Fibers See fibroplasm.

Muscle Fields The somatic muscles divided into groups by the cords and named according to position. See dorsosubmedian muscle fields, ventrosubmedian muscle fields. .

Muscles of Special Function Muscles of the same origin as somatic muscles but limited as to region and function. See somato-esophageal muscles, somato-intestinal muscles, intestinal muscles, anal muscles, vulvarmusculature, bursal musculature, spicular muscles.

Musculature The muscles of a body with special reference to their arrangement and function.

Musculus Bursae Basalis A bursal muscle arising from the ventral side of the bursa extending dorsally to the root of the dorsal ray. See bursal musculature.

Musculus Costae Dorsalis A much branched bursal muscle arising mediodorsally in the dorsal ray, extending anteriorly to become trifurcate then the median arm splits into four parts. See bursal musculature.

Musculus Costae Lateralis Externus Anterior Bursal muscles arising anterior to the musculus costae lateralis externus posterior, extending posteriorly to the base of the ventral rays. The muscles probably function to extend the bursa. See bursal musculature.

Musculus Costae Lateralis Externus Posterior A bursal muscle arising anteriorly dorsad of the lateral cords, extending posteriorly and becoming trifurcate at the base of the lateral rays. The muscle probably functions to extend the bursa. See bursal musculature.

Musculus Cestrum Literalism Internist A bursal muscle arising as paired sub median muscles at the body wall anterior to the intestino-rectal valve; each laterally extended branch enters the root of the lateral ray. The muscle probably functions to bend the bursa inwards. See bursal musculature.

Mushroom-shaped Stalked with an umbrella top.

Mutant An organism that undergoes mutation.

Mutation A genetic variation with the progeny differing from their parents in one or more characters.

Mycophagous Feeding on fungi. See feeding.

N

Nanism Dwarfishness. Being small in stature.

Natatory Parts Structures adapted or used for swimming.

Natural Enemies Organisms detrimental to the survival of others.

Natural Selection The process of population or specific stress by the environment tending to eliminate those forms least adjusted to survive under the conditions in which they live.

Neallotvpe An allotype described after previous publication of the original description. See type.

Near-Cauliflower Intermediate forms of disease expression between the healthy plant and cauliflower dwarf.

Nearctic Pertaining to or belonging to the nearctic region.

Nearctic Region That part of the Holarctic Realm which covers most of the North American continent and Greenland. See Holarctic Realm.

Neck That portion of the nematode body from the base of the lips to the base of the esophagus. See cervical.

Necrobiosis The death of an individual cell. See necrosis, nematosis.

Necrobiotic Food Tendencies The preference of plant cells for food at an early stage of necrosis by pathogenic nematodes.

Necromeny Feeding on microbes that are exploiting the cadavers of dead organisms.but not being actively involved in killing the organism.

Necropsy A post-mortem examination. See autopsy.

Necrosis The death of cells surrounded by living tissue, specifically death to cells in mass in contrast to necrobiosis. See necrobiosis, nematosis.

Necrotic In a dead and decaying condition. Affected with or characterized by necrosis.

Nectaries Plant cells near the heads of sedentary nematodes which have been stimulated to enlarge by saliva excretions and from â€˜which the nematodes derive their sustenance. Nurse cells, giant cells.

Needle-shaped See acerate, acerose, acicular, aciform, spiculiform, spicule.

Nekton Fauna Animals that swim freely by their own efforts. See palagic, plankton, benthos.

Nema A term for any individual of the phylum Nematoda.

Nemacide See nematicide.

Nemata An alternative name for the phylum Nematoda.

Nematicide Any agent lethal to nematodes.

Nematization The state of being populated, infested or infected with nematodes.

Nematize To populate, infest or infect with nematodes.

Nematocide Var. of nematicide.

Nematoda The name of the phylum constituted by nematodes.

Nematode A term for any individual of the phylum Nematoda. See eelworm, nema, roundworm, threadworm.

Nematode Index The numbers of nematodes per unit of substratum:

0 = none = 0

1 = rare = 1

2 = infrequent = 10

3 = frequent = 100

4 = abundant = 1,000

5 = very abundant = 10,000

See root-knot index.

Nematode wool A mass of resistant, desiccated Ditylenchus larvae frequently found on bulb crops and others. Wool.

Nematodologist Var. of nematologist.

Nematological Pertaining to, or having the character of, nematology.

Nematologist One versed in nematology.

Nematology The bodey of science centered on nematodes.

Nematophagous Feeding on, or deriving sustenance from, nematode bodies. See feeding.

Nematosis A morbid condition due to parasitism by nematodes. See necrobiosis.

Nematostat A chemical, situation or phenomenon which holds a nematode population in equilibrium.

Nematotoxic Pertaining to a nematotoxin.

Nematotoxin Any substance lethal to nematodes.

Nemic Belonging to, or pertaining to, nematodes.

Nemic Index See nematode index.

Nemicide See nematicide.

Nemin The substance or substances that cause trap formation by predaceous fungi. See Endogenous nemin.

Neontology The study of recent organisms as distinguished from paleontology.

Neoteny Larval characteristics persisting in the adult. See reproduction.

Neotony Sexual maturity in the larval stage. See paedogenesis, reproduction.

Neotype A specimen from the original type locality designated as a new holotype. Plesiotype. See type.

Nepo Virus A plant virus with polyhedral particles transmitted by nematodes in the Longidoridae. See Tobra Virus.

Neritic Fauna Organisms dwelling below the low tide line to depths of about 600 feet and subject to some wave action. See benthonic fauna, pelagic fauna.

Nerve Ring A belt, broad and flat in slender nematodes but narrow in forms with a large esophagus, containing cell bodies of neurons distributed around it. The nerve cells are generally too diffuse to be termed true ganglia. The nerve ring represents the dorsal and ventral connections between the lateral ganglia. See circumenteric ring, circum-esophageal commissure. .

Nests Clusters of nematodes within plant tissue.

Nests Centers of nuclear division of a syncytial hypodermal cell.

Net-like See reticulate.

Neuron (pl. **Neurons**) A nerve cell and its series of processes.

Niche A habitat or environment best suited for the survival of the organism.

Nipple-like See mammillate, digitate.

Noncoelomate Lacking a true body cavity. See pseudocoelom.

Non-host Usually used when no varieties of a plant species support reproduction of a nematode species. For example, Citrus spp. are non-hosts to Globodera rostochiensis. See host.

Northern See septentrional.

Noxious Injurious. Destructive. Baneful.

Nuclear Nests See nests.

Nucleoplasm The protoplasm of the nucleus as distinguished from cytoplasm.

Nucleus (pl. **Nuclei**) A well defined and specialized portion of cellular protoplasm held as being of crucial importance to cell physiology and heredity. It usually consists of nucleoplasm bounded by a nuclear membrane and composed of 1mm, chromatin and differentiated structures such as nucleoli.

Nurse Cells Modified plant cells about the heads of sessile nematodes which feed on the cells. See giant cells, nectaries

Nursery A center of nematode infestation in plant tissue well isolated from many population inhibiting pressures.

Nymph An immature individual having the form characteristic of the adult. See juvenile.

O

Oar-shaped See remiform, spatulate.

Oblanceolate Inversely lanceolate.

Oblate Flattened, as a spheriod depressed at two opposite poles. See prolate.

Obligate Parasite An organism only capable of deriving its food from living organisms. See facultative parasite, facultative saprophyte, obligate saprophyte.

Obligate Saprophyte Organisms which subsist on dead organic matter or from available inorganic material and have no relationship with living cells. See obligate parasite, facultative parasite, facultative saprophyte.

Obligatory Aerobic Requiring the presence of atmospheric oxygen in order to live. See anaerobic, facultative aerobic, facultative anaerobic, obligatory anaerobic.

Obligatory Anaerobic Growing only in the absence of atmospheric oxygen. See aerobic, anaerobic facultative aerobic, facultative anaerobic, obligatory aerobic.

Oblique Slanting. Inclined.

Oblique Cuticular Markings Conspicuous oblique beneath the cortical layer of the cuticle.

Oblong Elongated.

Obovate Inversely ovate. See ovoid.

Obtuse Not pointed. Blunt. See pointed.

Ocellus (pl. **Ocelli**) An â€œeyeâ€□ spot. A colored spot. See macula.

Ochlesis Any detrimental condition due to overcrowding.

Ocular Micrometer Disc A ruled glass plate used in the eyepiece of a microscope for measuring minute objects.

Odontium (pl. **Odontia**) A labial tooth (teeth) situated on the pharyngeal wall but formed in the esophagus and moved forward. See onchium.

Odontostyle A stylet terminating with a dorsally oblique aperture. The stylet was derived through evolutionary development from an odontium and originates in the esophagus wall. See onchiostyle.

Oecology Var. of ecology.

Oesophagus Var. of esophagus.

Oligidic Pertaining to a medium, consisting wholly or largely of crude materials in which no molecule (other than water) has been established as an absolute nutritional requirement. See holidic, meridic.

Oligophagous Feeding on a few kinds of food, especially nematodes with a limited host range. See feeding.

Oligozoic A habitat having few kinds or numbers of animals.

Omnivorous Subsisting on all types of food, especially feeding on both animal and vegetable material. See feeding.

Omnivore-Predators A grouping of nematodes with feeding habits that include feeding on other living organisms. See Generalist Predators, Specialist Predators.

Onchiophore The curved posterior extension of the onchiostyle feeding structure in Trichodoridae.

Onchiostyle A feeding structure with an onchium or tooth anteriorly and an extension of onchiophore posteriorly.

Onchium (pl. **Onchia**) A solid tooth (teeth) which was formed in place as compared to the development of an odontium in the esophagus then moved forward. .

Onchus An onchium.

Onion Bloat A disease of onions incited by Ditylenchus dipsaci.

Ontogenetic Pertaining to the life history or development of the individual organisms. See phylogenetic.

Ontogeny The life history or development of an individual as distinguished from that of the species and higher groups. See phylogeny.

Oocyte A female gamete prior to maturation.

Oogenesis The formation of the egg, its preparation for fertilization and development.

Oogonium (pl. **Oogonia**) The first stage in the differentiation of an egg cell from a primordial germ cell.

Ooplasm The cytoplasm of an egg.

Oosperm A fertilized ovum. A zygote. Oosphere An unfertilized ovum. A female reproductive cell.

Opercular Plug A removable covering from the hiatus of some nematode egg shells allowing the escape of the larva.

Opisthodelphic Having a single ovary posterior to the vulva. See amphidelphic, monodephic, prodelphic, didelphic.

Opposed Pertaining to structures that are opposite each other, either located on each side of the body or directed anteriorly and posteriorly.

Orad Toward the oral aperture.

Oral Aperture The anterior entrance into the stoma.

Oral Capsule The stoma.

Oral Cavity The stoma.

Oral LD50 The dosage of a chemical, based on the mass of the affected organism, at which 50% of the target organisms are killed when the material is ingested; for example, 10 mg of chemical per Kg of target organism tissue. See Lethal Dosage.

Oral Opening The oral aperture.

Oral Orifice The oral aperture.

Order In the systematic arrangement of organisms into groups or categories denoting natural relationships a category ranking above a family and below a class. See classification.

Ordinal Belonging to, or characteristic of an Order.

Ordinate The vertical line perpendicular to the transverse line (abscissa) to show by graphic design the relations of two series of facts. See abscissa.

Ordinate Arranged in rows, such as cuticular ornamentations, etc.

Organe Cephalique The head cleft.

Organic All carbon compounds derived from living organisms.

Organism An organic entity so constituted as to carry on the various functions of life. A living creature.

Organs of Sense See tactile organs, amphids, phasmids, ocellus.

Organs of the Lateral Cords A conspicuous series of gland-like structures giving the esophagus an appearance of alternate tissues.

Organ Z A muscular organ of unknown function, bearing three or four inner apophyses, lying between the spermatheca and the uterus. .

Orifice An aperture. Opening of a tube, duct, etc. Ornamentation Sculpturing or markings on a body or part.

Orthokinesis Variation in generalized, undirected, random locomotory activity caused by variations in the intensity of stimulation.

Oscillation One complete body swing from one extreme limit to the other of a nematode serpentine movement.

Osmosis The tendency of two solutions to equalize by diffusion through a selectively permeable (semi-permeable) membrane. See euryhaline, stenohaline.

Outer Circlet The combined cephalic crown of papillae and the outer labial crown by forward migration of the former. The external circlet. See inner circlet. .

Outer Cuticular Layer This layer corresponds to the external and internal cortical layers of the cuticle. See cuticular layering. .

Outer Labial Sited on the distal lip segments.

Outer Labial Bristles Outer labial papillae which have altered to bristle-like structures. The circlet is considered cephalic by some authors. .

Outer Labial Papillae Sensory papillae situated distally on each lip segment forming a circlet of six in primitive forms. The circlet is considered cephalic by some authors. See inner labial papillae.

Outer Leaf Crown The corona radiata.

Outlet Valve The esophageal lumen at the junction of the haustrulum with the esophageal- intestinal canal. See inlet valve.

Outstretched Ovary Where all segments of the reproductive tube are in a straight line. See straight ovary, recurved ovary, reflexed ovary.

Oval Ellipsoidal. Having the outline of an egg.

Ovarium The ovary.

Ovary The reproductive gland of the female which produces the ova. See testis.

Ovate Oval in shape with the broader end basal. See ovoid, obovate.

Ovejector Var. of ovijector.

Overall Application Broadcast application.

Ovic Juvenile Unhatched first stage larva contained within a deposited egg.

Ovic Embryo Term for eggs containing an embryo or larva. An embryonated egg.

Oviduct A tube which serves for the passage of ova from the ovary to the uterus. .

Oviform Having the shape of an egg. Egg-like. See ovoidovoidal.

Ovijector That portion of the oviduct modified to aid in the expulsion of the egg, generally heavily muscularized.

Oviparous Producing eggs that hatch after expulsion from the body. See ovovipiparous, viviparous, larviparous, fissiporous, reproduction.

Oviposit To lay eggs.

Ovocyte An oocyte.

Ovoid Having a shape like an egg. See ovate.

Ovoidovoidal Having the shape of an egg. Egg-like. See ovate, oviform.

Ovoviviparous Producing thinly shelled eggs which hatch within the uterus. Larviparous. See viviparous. oviparous, fissiparous, reproduction.

Ovum (pl. **Ova**) The reproductive cell of the female nematode. An egg.

Oxyuroid Esophagus Having a cylindrical shape anteriorly and terminated by a basal bulb. Also termed bulboid.

P

Paddle-shaped See spatulate, remiform. spathulate.

Paedogenesis Reproduction in a larval stage. See neotony, reproduction.

Paedogenetic Reproducing in the larval stage.

Paired See didymous.

Palmate Hand-shaped. Resembling a hand with the fingers spread.

Palps Specialized labial appendages, apparently for touching, grasping or guiding material to the mouth.

Panduriform Obovate with a concavity on either side much like a coke bottle.

Papilla (pl. **Papillae**) Minute elevations of the cuticle. Any small nipple-like projection or part. In general sensory organs.

Papillary Nerves Six nerves, two dorsolateral, two lateral, two ventrolateral from the nerve ring to the sense organs of the anterior end consisting of three circlets of sixteen members.

Paplliform Having the shape of a papilla.

Parahemizonid The hemizonion and similar structures.

Paralectotype All types remaining after selection of a lectotype. See type.

Paraphasmids Thick walled ampullae at the sites of the phasmids.

Parasite An organism that obtains its sustenance wholly or in part from another living organism. See free-living feeding.

Parasitic Castration A situation where parasitic nematodes destroy the gonads or arrest their development, especially in insects.

Parasiticide Any agent lethal to parasites.

Parasitism An association where one individual lives at the expense of another, makes no return and is destructive to its host. The state of being parasitic.

Paratype All specimens remaining after the selection of the holotype and allotype. See type.

Paravulvae Specialized ventral body pores in females, located anterior and posterior to vulva.

Parenteral Feeding The absorption of food other than by way of the intestine. See feeding.

Pars Distalis Vaginae The distal part of the vagina with cuticular lining contiuous with that of the exterior of the body. .

Pars Proximalis Vaginae The proximal section of the vaginal canal with hyaline walls.

Pars Refringens Vaginae The intermediate section of the vaginal canal with sclerotized walls.

Parthenogenesis The development of eggs without benefit of fertilization by spermatozoa. See monogony, reproduction.

Partial Sterilization An incorrect term for selective killing of organisms in a substrate. See Pasteurization.

Partial Striae Intermediate striae.

Partite Parted. Separated.

Passive Migration The transport of nematodes by water, wind, on hoofs, feet and bodies of animals, the activities of man, etc.

Pasteurization The process of selectively killing a particular organism or group of organisms in a substrate leaving other organisms alive. See sterilization.

Patent Open. Spreading. Affording passage. Expanded. Distended. See patulous, prepatent.

Pathogen A disease causing agent.

Pathogenic Disease causing.

Pathogenicity

Pathological, Pathologic A diseased condition. A morbid state due to a disease. Unhealthy.

Pathology The study of disease, their nature, causes, etc.

Patterns Cuticular ornamentations.

Patulous Open. Spreading. Expanded. Distended. See patent.

Pearls White Heterodera cysts.

Pear-shaped See piriform, pyriform.

Pea-shaped See pisiform.

Pectinate-shaped like a comb. Having narrow parallel projections.

Peduncle A stem or stalk. A stem-like structure supporting an organ or other structure.

Pedunculate Set on or attached by a slender stem or neck. Petiolate.

Pedunculate Papillae Genital papillae of male nematodes elevated on little stalks.

Pelagic Fauna The plankton and nekton animals of the open sea.

Pelagic Zone The open sea, especially beyond the littoral zone and above the abyssal zone.

Pellicle A thin membrane or film.

Peloderan Caudal alae which meet posterior to the tail. See leptoderan.

Pellucid Transparent.

Penetrate To pass beyond the surface. To pass into or through a thing. To pierce. See lancinate.

Peppercorns Galls produced by some species of Anguina, especially by A. tritici in wheat kernels, See cockles, purples.

Percutaneous Penetration through the skin.

Perforate To pierce. See lancinate.

Perianal Situated or occurring around the anus.

Periaxial Surrounding an axis.

Peribuccal Around, enclosing or surrounding the buccal cavity.

Perienteric Fluid Fluid of the pseudocoel.

Perineal Belonging to the perineum.

Perineal Pattern Configurations, especially of Meloidogyne spp., on the cuticle surface of the perineum; specific designs are common to each species. See posterior cuticular pattern.

Perineum The superficial region about the anus.

Perioral Around, enclosing or surrounding the oral aperture.

Perioral Cuticle Cuticle of the lip region surrounding the oral aperture, often demarcated by ridges, incisures, flabella and cornua. See flabellum, cornu.

Petaloid Having the form or appearance of a flower petal.

Petiolate Stalked. Situated upon a stalk.

pH A symbol of a scale used to designate the relative acidity of a solution. The scale ranges from 1 to 14. pH 7, the midpoint, represents a neutral solution. Numbers less than 7 indicate increasing acidity; those more than 7 increasing alkalinity.

Phagocyte A coelomocytic cell which ingests or absorbs noxious organisms, tissue or detritus.

Phagocytosis The engulfing and destruction of microorganisms, tissues or detritus by phagocytic action.

Pharyngeal Bulb A muscular swelling of the esophageal wall around the buccal capsule.

Pharyngeal Caecum The esophageal caecum.

Pharyngeal Glands The esophageal glands.

Pharyngeal Gland Nuclei and Orifices See Esophogeal Gland Nuclei and Orifices.

Pharyngo-Intestinal Valve The esophageal-intestinal valve.

Pharynx The stoma. The mouth cavity and its walls. The muscular esophagus of some authors.

Pharynx Lumen The esophageal lumen.

Phasmidial Glands The precaudal glands.

Phasmids The lateral caudal papillae connected with the lateral precaudal glands. Paired postanal lateral chemoreceptor sensory organs. See precaudal glands, scutellum.

Phenotype The visible characters of an organism resulting from the interaction of genotypic characters and environment. See genotype.

Phoresy A symbiotic relationship in which one organism transports another organism of a different species. Often exhibited in arthropods and also in some fishes.

Phoretic Relationship A symbiotic relationship in which one organism is transported by another. See phoresy.

Phospholipid Fatty Acid Fatty acids derived from phospholipids which are the primary lipids of cell membranes. Their analysis may be used to determine the composition of a microbial assemblage.

Phyllophagous Feeding on leaves, See feeding.

Phylogenic, Phylogenetic Relating to the lineage history of the development of a natural group.

Phylogeny The lineage history of the development of a genus, family, class or other natural group. See ontogeny.

Phylum (pl. **Phyla**) In the systematic arrangement of organisms into groups or categories denoting natural relationships a category ranking above a class and below a kingdom. See classification.

Physiological Race, Physiologic Race See race, hostic race.

Phytocidal Lethal to plants. See phytotoxic.

Phytocide Any agent lethal to plants.

Phytonematode A plant parasitic neniatode.

Phytoparasite Nematodes capable of obtaining sustenance from plants.

Phytophagous Feeding on the tissues or juices of plants. Herbivorous. See feeding.

Phytotoxic Injurious or lethal to plants. See phytocide.

Phytoxic Phytotoxic.

Piercing Organ See stylet, odontia, jaws, spines, denticle, lancinate.

Pilliferous Hairy.

Pioneer A biotype. A mutation or an adaptation through selection of a nematode population feeding on a nonhost.

Piriform Pear-shaped. See pyriform.

Pisiform Having the size and shape of a pea

Plankton Organisms that float and move passively with winds and currents, generally of microscopic size. See pelagic, nekton, bethos.

Plant-Inhabiting Nematodes Two different biological groups which can be distinguished as parasites and saprozoites. See feeding.

Plaques An inflated condition of the cuticle with the general appearance of looking like deep annulations divided longitudinally. See inflation of the cuticle, warts.

Plate-like See disc, clintheriform.

Platymyarian A type of muscle cell in which the muscle fibers are all adjacent to the hypodermis and are perpendicular to it. See coelomyarian, circomyarian. .

Plectanes Cross striated cuticular plates which function to support genital papillae of some males.

Plectoid Having the characteristics of the genus Plectus.

Plesiotype A specimen upon which subsequent or additional description or figure is based. See type.

PLFA See phospholipid fatty acid.

Plicate Bearing folds or parallel ridges. See ridge.

Plicatulate Minutely ridged.

Ploidy The condition of being a multiple of a basic haploid number.

Phiricellular Many celled.

Point The apex of anthing having a tapering end. See apex, mucro.

Pointed See acunate, acute, acuminate, acurninose, cusp, aciforin, obtuse.

Polar Body One of the daughter nuclei derived from the first or second division of meiosis containing practically no cytoplasm and is expelled from the oocyte.

Polar Filament A projection of the protein membrane on some nematode eggs, probably serving in attachment.

Polydelphic Having three or more gonads. See amphidelphic. monodelphic, prodelphic, opisthodelphic, didelphic.

Polygamous Mating with two or more females. See reproduction.

Polymorphic, Polymorphous Occurring in two or more forms. See dimorphic, heteromorphic, monomorphic.

Polymyarian The presence of many longitudinal muscle cells; best viewed in transverse body section. See holomyarian, meromyarian.

Polyphagia The ability to subsist on various kinds of food. See feeding.

Polyphagous Feeding on many kinds of food, especially nematodes feeding on decaying organic matter. Plant pathogenic nematodes feeding on two or more hosts.

Polyphagy The ability of plant-pathogenic nematodes to subsist on two or more host plants. See feeding.

Polyphyletic Derived from two or more ancestral types, races, families, etc. See monophyletic.

Polyploid Having a chromosome number in multiples of the basic haploid number. See monoploid, aneuploid, euploid, heteroploid, hyperploid, diploid.

Polyploidy A polyploid condition..

Polyxenic Pertaining to the rearing of an organism with many other known species of associated organisms. See gnotobiotic, agnotobiotic, xenic, axenic, synxenic, monoxenic, dixenic. trixenic.

Ponding A problem of puddling in applying large quantities of nematicide carrying liquids due to the compacting of the soil surface layers.

Population The individuals inhabiting a specific unit of substratum.

Population Density The number of individuals per unit of substratum.

Pore A minute opening or orifice; generally the outlet of a gland.

Pore Space The space within soils not occupied by solid particles.

Port Of Entry Point of penetration of a parasitic nematode through either natural or unnatural openings.

Position The site of attachment of an organ or part. See lateral, laterodorsal, dorsoventral, submedian, anteroposterior, dorsolateral, subdorsal, laterosubdorsal, subventral, ventrolateral, lateroventral, sublateral, dorsomedian, median, ventromedian, postanal, posterior, anterior, anteroventral, medioventral, preanal, postanal, lateroanal. anterodorsal.

Possum Ears A disease of strawberry incited by Aphelenchoides fragariae. See dwarf.

Postanal Posterior to the anus. See position.

Postanal Pulvillus A satiation where the ventral cord passes o both sides of the rectum and is then continuous with swollen cells on the dorsal surface of the rectum.

Postalar Tail That segment of the tail posterior to the bursa.

Postembryonic Pertaining to the life stage after hatching from the egg.

Postembryonic Development Growth after hatching from the egg.

Posteriad Directed backward. Opposed to anteriad.

Posterior On or toward the hind end of a body or part. See anterior.

Posterior Bulb The basal bulb.

Posterior Cephalic Crown A circlet of four sub- median setae or papillae located immediately posterior to the anterior cephalic crown or fused with it in some forms. See anterior cephalic crown, labial crown of papillac.

Posterior Cephalid The second cephalid at which the two median cords arise. See cephalids.

Posterior Cuticular Pattern The perineal pattern.

Posterior Uterine Branch The postvulvar uterine branch.

Post Mortem After death, especially an examination. See necropsy, antemortem.

Postrorse In a backward or downward direction. See antrorse. retrorse,

Postuterine A position posterior to the uterus.

Postuterine Rudiment The postvulvar uterine branch.

Postuterine Sac See postvulvar uterine sac.

Post-uterine Sac See postvulvar uterine sac.

Postvulvar Posterior to the vulva.

Postvulvar Ganglion A ganglion which receives the paired ventral nerve cords at a point posterior of the vulva where fusion into a single nerve cord occurs.

Postvulvar Rudiment The postvulvar uterine branch. Postvulvar Sac The postvulvar uterine branch.

Postvulvar Uterine Branch A rudimentary extension of the posterior uterus, .

Postvulvar Uterine Sac A rudimentary extension of the posterior uterus which sometimes functions as a spermatheca.

Potato Sickness A disease of potato incited by Globodera rostochiensis.

Praeapical Glands The phasmids.

Praecaudal Glands Var. of precaudal glands.

Praecorpus Var. of procorpus.

Praerectum Var. of prerectum.

Preadult The last larval stage before becoming adult.

Preanal Anterior to the anus. See position.

Preanal Ganglion A ganglion which receives the Ventral nerve cord and from which the anolumbar connectives extend to the lumbar ganglia.

Preanal Sucker A greatly enlarged genital papilla having a sucker-like form.

Preanal Tubes The cuticularized ducts of the preanal ventral glands.

Precaudal Glands Two organs situated in the lateral sectors of the tail. The openings to the exterior are termed the phasmids. Phasmidial glands.

Precociousness The presence of reproductive structures in preadult forms. See neotony, paedogenesis, reproduction.

Predaceous Living by preying on other organisms. See feeding.

Predacious Fungi Fungi with the capacity to capture nematodes and derive sustenance from them.

Predator A nematode which lives by preying on other nematodes or organisms. See generalist predator, specialist predator.

Predigestion The digestion or partial digestion of food outside the intestine. See extraoral digestion, extracellular digestion, intracellular digestion.

Preferred Host A plant affording a more favorable environment to a nematode parasite than other suscepts.

Prepatent The condition or situation preceding the patent condition.

Preputal Fold A campanulate fold of cuticle directed posteriorly over the tail of some male forms.

Prerectum The segment of the alimentary tract between the intestine and the rectum, separated from the intestine by a stricture in the lumen and from the rectum by a sphincter muscle. .

Pressure The stresses endured by a nematode population in its struggle for survival. See stress.

Pressure Orifice System A method of applying liquid nematicides to soil utilizing pressure, orifice diameter and tractor speed for various rates of application.

Primary Muscle Fields The somatic muscles as divided into four main groups by the cords. The dorsosubmedian muscle fields and the ventrosubmedian muscle fields. .

Primitive Persisting in structure little advanced from ancestral form. See primordial.

Primordial Growth of an individual or organ in its earliest or primitive form.

Primordium (pl. **Primordia**) The initial development of an organ, tissue or individual.

Probolae Prominent and elaborate specialized appendages encircling the oral aperture; situated in some nematode forms in labial and cephalic positions. See pseudolabia, cephalic tubers. .

Process A prominence or prolongation of a surface. An outgrowth or extension. See apophysis, corniform, excrescence.

Procorpus Tue anterior subdivision of the corpus taking a cylindrical form and generally being terminated by an oval posterior segment. See metacorpus. .

Proctodaeum The invagination of the epiblast that produces the anus and intestine.

Prodelphic Having a single ovary anterior to the vulva. See amphidelphic, monodelphic, opisthodelphic.

Progeny Offspring. Young. See reproduction. Prolate Elongated, s a spheroid stretched out in the direction of the two opposite poles. Lemon-shaped.

Proliferative Cell The terminal cell.

Prolongation Stylet extensions.

Pronucleus (pl. **Pronuclei**) A gamete nucleus after maturation is completed. Two haploid gamete pronuclei fuse to form a diploid zygote nucleus in fertilization.

Propagate To continue or multiply by bearing young, seeds or cuttings and the like. See reproduction.

Prophase The initial stage of mitosis or meiosis in which chromosomes appear. See metaphase, anaphase, telophase.

Prorhabdions Walls of the prostom.

Prostatic Glands Glands in some male nematodes which produce an adhesive secretion at the distal end of the ejaculatory duct.

Prostom The anterior subdivision of the protostom. See mesostom, metastom. .

Protandrous Hermaphroditism Spermatozoa and later ova are produced by the same gonad. A syngonic female. See reproduction.

Protandry The production of spermatozoa and later ova by the same gonad. See reproduction.

Protein A very complex organic compound built of numerous amino acids and which form an essential part of the structure and diet of all animals. Protein contains nitrogen, carbon, hydrogen, oxygen and usually sulfur.

Protein Membrane The outer covering of the egg surface sculpturing of some and it is deposited by the uterine wall.

Proterotype The primary type which includes all the material upon which the original description is based. See type.

Protoplasm The basic substance of which all living matter is made, being grayish, semitransparent, viscid and a complex colloidal physiochemical system that constitutes tile living matter of plant and animal cells.

Protorhabdions The walls of the protostom.

Protostom The middle region of the stoma delimited anteriorly by the cheilostom and posteriorly by the telostom. The protostom is subdivisible into three parts: the prostom, mesostom and metastom.

Protostoma Var. of protostom.

Protract To extend or protrude.

Protractor Gubernaculi Paired gubernacular muscles which extend from the subventral body wall anterior to the proximal part of the gubernaculum. See gubernacular muscles.

Protractor Muscle A muscle which functions to extend an organ. .

Protractor Onchii The protractor muscle of the stylet.

Protractor Spiculi Paired spicular muscles extending obliquely from the spicules to the ventral side of the body. See spicular muscles.

Protractores Spiculorum See protractor spiculi.

Protuberance An elevation above the surface, A protrusion or bulge.

Prouterus A situation where the proximal end of the uterus is separate and may function as a receptacle or shell gland.

Provisioning Services Ecosystem services that provide products whicht support or maintain life. For example, food, fiber, fuel, energy, pharmaceuticals.

Proximad Toward the proximal end.

Proximal Designating that end of a part or organ nearest to the point of attachment or origin.

Pseudobulb A swelling of the esophageal musculature in which the lumen does not widen to form a cavity. A nonvalvate esophageal bulb. See true bulb.

Pseudocoel The pseudocoelom.

Pseudocoel Cells Coelomocytes.

Pseuocoelocytes Coelomocytes.

Pseudocoelom The body cavity of the nematode. See pseudocoelomic membranes, mesenteries, coelomocytes, body cavity.

Pseudocoelomate Possessing a body cavity that is not entirely surrounded by mesoderm.

Pseudocoelomic Membrane A delicate sheath which invests and supports the internal organs, specifically it surrounds the esophagus, lines the internal surface of muscle cells and extends between each pair of muscle cells to the hypodermis. See mesenteries. isolation tissue.

Pseudoganty Activation of an egg by a sperm without nuclear fusion.

Pseudolabia Prominences of the head often elaborately developed. They are situated in labial and cephalic positions in some nematode forms. See probolae, cephalic tubers.

Pseudonchs Structures in the pharynx which resemble onchia in certain views.

Pulvillus A cushion-like group of cells. See postanal pulvillus.

Punctations Small pits or deep depressions on the surface of the cuticle; the shape is usually rounded but may wary.

Purples Galls produced by some species of Anguina, especially by A. tritici of wheat kernels. See cockles, peppercorns.

Pure Culture A nematode population containing a single species only; it differs from aseptic culture by not being free of other organisms. See pure population, agnotobiotic culture, monoxenic culture. axenic culture, pure line population.

Pure Line Population A nematode population derived from a single parent or set of parents.

Pure Population A nematode population containing a single species only; it differs from aseptic population by not being free of other organisms.

Pustule A small papilliform or blister-like elevation.

Putrifaction The process of organic decomposition especially the anerohic breakdown of proteins.

Pyriform Pear-shaped. See piriform.

Q

Quadricolumella In the female gonad probably the gland region that secretes the egg shell.

Quarantine A restraint upon goods, animals, plants or other materials which may bear pathogenic organisms.

R

Race A hiotype or groups of biotypes which differ from one another in certain physiological characters which function to segment the species into subspecific categories. A population which reacts differently than other populations of apparently the same species. See hostic race.

Rachis An axial structure or column.

Radial Developed or situated around a central axis. See actiniform.

Radial Bars In the cephalic framework six rigid supports connecting the vestibule of the stylet guide and the basal ring and ribs.

Radial Blades The radial bars.

Radial In Form See actiniform, stellate, triradiate.

Radial Muscles Muscles of the esophagus which dilate the lumen of the esophagus.

Ramified Branched..

Ramiform Resembling or-shaped like branches, Ramified.

Random Movement Having a haphazard course or direction in contrast to a tactic response. See taxis.

Rapacious Feeding on prey. Predaceous. Voracious. See feeding.

Rays Genital papillac and their accompanying muscles embedded in the bursa of some nematode forms.

Rays The esophageal radii.

rDNA Ribosomal DNA are sequences encoding for ribosomal RNA and which regulate transcription of DNA to RNA. They contain transcribed and nontranscribed spacer segments. See DNA, RNA. 18S rDNA is the ribosomal DNA sequence containing genes that encode for ribosomal RNA (See rRNA). Sequence data from these genes is widely used in molecular analysis to reconstruct the evolutionary history of organisms as it is conserved and considered suitable for constructing evolutionary divergences.

Receptaculum Seminis The spermatheca.

Reciprocally Infective Capable of transferring from one crop to another then return to infect the original crop.

Rectal Commissure A commissure arising from the preanal ganglion and extending to the dorsorectal nerve.

Rectal Cuticle The cuticular lining of the rectum which differs in some respects from the external cuticle. See external cuticle, esophageal cuticle, cloacal cuticle, vaginal cuticle.

Rectal Glands Large unicellular glands, usually three in females and six in males.

Rectal Matrix Glands See matrix glands.

Rectal Muscles Specialized muscles which function to widen the anal opening and are attached to the dorsal wall of the rectum.

Rectal Sinus A cavity immediately anterior to the anus and to which the rectum is connected.

Rectilinear Formed of straight lines or characterized by straight lines.

Rectum A narrow tube, flattened in dorsoventral direction and separated from the intestine by a sphincter muscle. It is derived as an invagination of the epidermis and is lined by a cuticularized intima. The hindgut. See cloaca.

Recurved Reflexed.

Recurved Ovary A situation where the distal portion of the gonoduct is turned back on itself. Recurved ovaries are considered to belong to the more primitive type of ovary. See outstretched ovary. .

Red Plant A disease of strawberry incited by Aphelenchoides fragariae.

Red Ring A disease of coconut palm incited by Bursaphelenchus (Rhadinaphelenchus) cocophilus.

Reduction Division See meiosis.

Reflexed Turned or folded back upon itself, especially applied to configuration of ovaries and testes of nematodes. .

Regenerative Power The ability to heal wounds.

Regulating Services Ecosystem services that regulate other ecosystem processes. For example, activities and abundance of pest and disease organisms, decomposition of organic wastes, purification of water and air.

Relax To inactivate nematodes, generally by heat prior to fixing. To kill.

Remiform Having a shape like an oar. See spatulate, spathulate.

Renette Generally one but, in some forms, two ventral excretory cells. The excretory gland.

Reniform Having the shape of a kidney.

Replacement Spear Present in juvenileDorylaimoidea. Located in the anterior, slender portion of the esophagus.

Reproduction The process of perpetuating the species by giving rise to offspring. See neotony, paedogenesis, monogony, parthenogenesis, digonic hermaphrodite, syngonic female, syngony, protandrous hermaphroditism, syngenesis, polygamous, progeny, protandry, sex determination, sexual dimorphism, syngenesis. androgynous, dioecious, diorchic, monorchic, dimorphic, hermaphrodite, recocious, monogenetic, hetero genetic fissiparous, bisexual, gyandromorph,, heteronomy, larviparous, ovoviviparious viviparous, oviparious, monogenesis.

Reproductive System All the organs and structures concerned in the production and delivery of ova and sperm.

Reservoir A plant parasitic nematode population surviving on a â€œweedâ€ host. A cavity or ampulla for the storage of a fluid or secretion.

Resistance The ability of an organism to remain relatively unaffected by a disease due to possession of certain inherent properties. A host may be slightly, moderately or highly resistant. The reduced rate of parasite reproduction on a host.

Resistant Stage That period of a nematode life cycle in which environmental extremes have little or no affect.

Resister The species biotype capable of withstanding the adverse affects of parasites, climate, resistant hosts, etc.

Respiration The absorption of oxygen from the surrounding medium and giving off carbon dioxide and other products formed by oxidation in the tissues.

Restiform Corded or cord-like.

Reticulate Having cross markings like a net.

Rectractor Gubernaculi Muscles extending from the distal part of the gubernaculum to the dorsal or lateral walls of the body.

Retractor Muscle Muscles which serve to return an organ to its original position. .

Retractor Spiculi Paired muscles extending from the manubrium of the spicule anteriorly to the hypodermis near the lateral cords. See spicular muscles. .

Retroarcuate Curved backwards.

Retrocession A recession or movement backward.

Retrogressive Development A simplifying evolutionary trend of an organism usually expressed as the complete or partial loss or amalgamation of structures.

Retrorse In a backward or downward direction.

Retrovesicular Ganglion The union immediately posterior to the excretory pore where the dorsolateral nerves fuse.

Reviviscent Capable of reviving, especially some nematode forms after a period of desiccation.

Rhabdions The walls of the stoma. See cheilorhabdions, protorhabdions, telorhabdions.

Rhabditiform Having the shape of a rod.

Rhabditoid Having characteristics of the genus Rhabditis.

Rhabditoid Bursa A condition where the caudal alae meet posteriorly and anteriorly forming a complete oval. A wide caudal bursa.

Rhabditoid Esophagus Antcriorly of a cylindrical shape, often with a pseudobulb, followed by the isthmus and terminated with a basal bulb.

Rheophile Inhabiting rivers and streams. See hypogaeic, aquatic. edaphic, marine, terrestrial, terricolous, xerophilous, cavernicolous, fossorial, arenicolous, xiphilous.

Rheotaxis A situation where the tactic response is dependent upon the mechanical stimulation of a moving fluid, especially water, for the directive factor. See taxis.

Rheotropism A situation where the tropism is dependent upon the mechanical stimulation of a moving fluid, especially water, for the orientating factor. See tropism.

Rhizosphere The area immediately surrounding the plant roots.

Ribonucleic Acid See RNA

Ribosomal DNA See rDNA.

Ribosomal RNA See rRNA.

Ribosome The cell structure that is the site of protein synthesis. Ribosomal RNA (See rRNA) decodes messenger RNA (See mRNA) into amino acids

Ribs In the cephalic framework, six rigid supporting structures radiating transversely, then posteriorly to fuse with the anterior edge of the basal ring and the six radial bars. Four ribs are sublateral, one dorsal and one ventral. See basal ring, radial bars, cephahc arches. .

Ridge A raised line or strip on a surface.

Ringed See annulate. annulose. .

Ring-shaped See circinate.

Ripe Pertaining to mature ova or spermatozoa.

River Inhabiting See rheophile.

RNA Ribonucleic acid. A nucleic acid consisting of a chain of nucleotide units, each with a nitrogenous base, a ribose sugar and a phosphate.

RNAi RNA interference. A mechanism inhibiting gene expression or by hindering gene transcription. See RNA.

rRNA Ribosomal ribonucleic acid is the RNA component of the ribosome (See ribosome). It provides a mechanism for decoding messenger RNA (See mRNA) into amino acids by interacting with transfer RNA (SeetRNA) which transfers the amino acids that correspond to the mRNA codon. 18S rRNA is the structural RNA which codes for the small subunit of ribosomes in eukaryotes. The genes coding for 18S rRNA are referred to as 18S rDNA (See rDNA).

Rod Border Structures on intestinal epithelium which are probably modified immobilized cilia.

Rod-like See baculiform, rhabditiform.

Rogue To remove undesired individual plants from a crop.

Root Ball The interwoven network of plant roots of a potted plant.

Root Diffusate A root excretion that affects the behavior of plant parasitic nematodes, especially as a hatching or tactic stimulus, See leachate.

Root Galls A pronounced localized swelling of root tissue due to hypertrophy and hyperplasia incited by nematode feeding.

Root-Knot Index An index of the amount of root galling as a measure of the Meloidogyne population:

1 = zero percent roots galled
2 = 1 to 30 percent galled
3 = 30-100 percent galled
4 = 100 percent and above

See nematode index.

Rootlet A small root.

Root Rot Complex A morbid condition of roots and root tissues incited by multiple attacks of nematodes, fungi, bacteria, etc, in stages or nearly simultaneously.

Root Zone See rhizosphere.

Rosette A pathological condition where the internodes of plants are shortened and the normally separated leaves are closely clustered.

Rosette The uvette of the deManian system.

Rosettes Punctation patterns of the cuticle surrounding genital papillae.

Rostrum (pl. **Rostra**) A beak-like projection ventrad near the proximal end of the aphelenchoid spicule, joining the dorsal and ventral spicular shafts.

Rot Destruction of tissues by decay organisms; frequently after nematode injury.

Roundworm A vernacular term for any individual of the phylum Nematoda, especially as applied to parasites of vertebrates. See eelworm, nema, nematode. threadworm.

Row Application The placing of a nematicide in such a manner that the lethal vapor is effective in a broad band where the future row crop is to be planted. See broadcast application, spot application, drench application, gas application.

Rudimentary A condition of imperfect development. Development on a primitive level. Development arrested at an early stage. Vestigial.

Rudimentary Posterior Uterine Branch A vestigial extension of the posterior uterus. A postvulvar uterine branch.

Rudimentary Posterior Uterus A vestigial extension of the posterior uterus. A postvulvar uterine branch.

Rugose Wrinkled.

Rugosissimus Extremely rugose or wrinkled.

Rugula (pl. **Rugulae**) A small wrinkle.

Rugulose Having fine wrinkles.

Rypophagous Filth-eating. See feeding.

S

Sac A bag or sack-like structure. A pouch.

Saccate Sac-shaped. Contained in a sac.

Sacciform Having the shape of a sac.

Saccular Sac-like.

Saccule A small sac or pouch.

Sack-like See sac.

Sagittate Arrowhead-shaped. Triangular-elongate.

Sagittal Plane The median-vertical-longitudinal plane in a bilaterally symmetrical animal or a section parallel to that plane. See longitudinal plane, transverse plane.

Sagittiform Arrowhead-shaped.

Saint Johnâ€™s Disease A disease of peas incited by Heterodera goettingiana.

Saliva A secretion o the esophageal glands presumably containing a digestive enzyme.

Salivary Gland The esophageal glands.

Sand Individual rock or mineral fragments in soils having diameters ranging from 0.5 mm. to 2.0 mm. Usually sand grains consist chiefly of quartz but they may be of any mineral composition. The textural class name of any soil that contains 85 percent or more of sand and not more than 10 percent of clay.

Sand Inhabiting See arenicolous.

Sandy Clay Soils of this textural class contain 35 percent or more of clay and 45 percent or more of sand.

Sandy Clay Loam Soils of this textural class contain 20 to 35 percent clay, less than 28 percent silt and 45 percent or more of sand.

Sandy Loam Soils of this textural class have 50 percent sand and less that 20 percent clay.

Sap The juices of a plant cell. The semiliquid constituents which compose the protoplasm of the plant cell. The watery solution which circulates through plant vascular tissue.

Saprobiont An organism which subsists as a saprophyte or saprozoite. See saprophage, feeding.

Saprophage An organism which derives its sustenance from decaying organic matter. See feeding.

Saprophagous Feeding on dead or decaying organic matter. See feeding.

Saprophyte Any organism deriving sustenance from dead or decaying organic matter. See feeding.

Saprozoic Pertaining to an animal which derives sustenance from dead or decaying organic matter. See feeding.

Saprozoite An animal which derives its sustenance from dead or decaying organic matter. See feeding.

Sarcophagous Flesh eating. See Feeding.

Sarcoplasm The undifferentiated protoplasm of the muscle cell. See fibroplasm.

Scaffold An abnormally heavy, refractive guiding ring.

Scalariform Bearing markings like the rounds of a ladder.

Scale-like See imbricate.

Scalloped See crenate.

Scaphiform Having the shape of a boat.

Scavenger An organism which derives sustenance from the refuse, offal and decaying matter of other life, See feeding.

Sclerite A hard plate or piece of cuticle.

Sclerotized A noncommittal term for hardened. dense, refractive parts of the nematode body.

Sclerotized Vagina See Pars refringens vagianae.

Sculate Having the shape of a shield. See scutiform, escutcheon-shaped. clypeiform.

Scutellate Having a surface divided into small plate-like areas.

Scutellum (pl. **Scutella**) Greatly enlarged phasmids common to some species of the Hoplolaiminae. .

Scutiform Having the shape of a shield. See scutate, escutcheon-shaped, clypeiform.

Second Bulb The terminal bulb.

Secondary Invaders Any and all organisms found in a host after the initial attack and injury by the primary parasite.

Secretion A useful product produced by a cell or gland. The act of producing or passage of this product. See excretion.

Sector The interval between the cords.

Sedentary Endoparasite See sedentary nematodes.

Sedentary Nematodes Nematodes which become established within or on the roots of host plants i.e. root-knot, cyst-forming, Cobbâ€™s root- knot, reniform, etc., where the females remain fixed in position throughout their life cycle.

Seductor Gubernaculi Paired muscles extending from the proximal end of the gubernaculum to the dorsolateral walls of the body.

Seed Galls Enlarged ovules converted to a gall and containing nematode larvae, especially on grasses and cereals.

Segging In Oats A disease of oats incited by Ditylenchus dipsaci. See tulip root.

Segmentation The repetition of elements of the main organ systems of the body along the length of the body.

In embryological development the subdivision of the single-celled egg into blastomeres. Cleavage.

Seinhorst Extraction Technique A method for the isolation of nematodes from infected plant material utilizing funnels, a constant fresh water spray and suitable settling basins.

Seinhorst Elutriator Apparatus An apparatus employed for the isolation of nematodes from soil resulting in a sample relatively free of debris.

Seinhorst Technique An elutriation method for quantitative extraction of nematodes from soil.

Seleniform Having the shape of a full moon.

Semifenestrate A situation in some species of the genus Heterodera where the vulval bridge divides the fenestra into two parts. See fenestrate, bifenestrate, ambifenestrate, circumfenenstrate.

Semilunar Having the shape of a halfmoon.

Seminal Receptacle The spermatheca.

Seminal vesicle In male nematodes enlarged tube- or pouch-like structures which function to store the semen.

Semivagrant Nematode forms with mobile larval stages and sedentary parasitic adults.

Senescence Old. Aged.

Senility Old age. Passed the reproductive stage.

Sensilla Pouch An expansion of the amphidial tube containing the sensory elements of the sensilla and is situated posterior to the amphidial pouch. .

Sensillum (pl. Sensilla) A simple sense organ, or one of the structural components of a compound sense organ. The nerve ending connected with the amphidial nerve. .

Sensory Elements Stimuli preceptors collectively making up the sensilla.

Sensory Organ A body structure which functions as a preceptor of stimuli.

Sepsis The state or condition of being septic. See asepsis.

Septentrional Northern. Boreal.

Septic Contaminated with microorganisms. - See aseptic.

Septicemia A morbid condition caused by pathogenic bacteria and their toxins within the body.

Seriate Occurring in a series or periodically.

Serrate Having marginal teeth or notches. See dentate.

Sessile Obligate feeding and reproducing at one fixed site. Not free to move from one site to another.

Seta (pl. Setae) Tactile sensory organs of elongate cuticular structure and articulate with the cuticle proper. A bristle-like protuberance. An aciculum. See tubiform setae, ambulatory setae, bristle setae, adhesion tubes.

Setaceous Bearing setae. See setose.

Setiform Having the form of a seta. Bristle-shaped.

Setose Bearing setae.

Sex-Determination The factor or factors which cause the development of an egg into a male or female nematode. See reproduction.

Sex Ratio The proportion of males to females in a nematode population.

Sexual Dimorphism A difference in form or stricture as males and females of the same species.

Sexual Reproduction Reproduction requiring tile union of male and female gametes. See asexual reproduction, syngenesis.

Shaft The main body of the spicule, usually curved and often bearing a thin flange of cuticle termed the velum.

Sheath A covering enclosing a body, part or organ and generally elongated in form.

Shields Interlabia.

Shield-shaped See scutate, scutiform, escutcheon-shaped, clypeiform.

Sickle-shaped See falcate, crescentiform, falciform.

Sieve-like See cribriform.

Sigmoid A two directional curve-shaped like the letter S.

Silt Individual mineral particles of soil that range in diameter between the upper size of clay, 0.002 mm., and the lower size of very fine sand, 0.05 mm. Soil of the textural class silt contains 80 percent or more of silt and less than 12 percent of clay.

Silt Loam Soil material having 50 percent or more of silt and 12 to 27 percent of clay or 50 to 80 percent of silt and less than 12 percent of clay.

Silty Clay Soil of this textural class has 40 percent or more of clay and 40 percent or more or silt.

Silty Clay Loam Soil of this textural class has 27 to 40 percent of clay and less than 20 percent F of sand.

Single Egg Mass Line A nematode population derived from eggs produced by a single female. See pure line population, single larva line, pure population, pure culture.

Single Larva Line A nematode population derived from an individual juvenile nematode.

Sinuate Wavy. Sinuous.

Sinuous Having a serpentine or undulant form.

Sinus A curvilinear indentation. An opening or depression. A cavity, See antrum.

Sinus Canals Greatly enlarged ducts.

Sinusoidal Wavy, serpentine, winding in form.

Skeletal Fibrils The longitudinal and transverse fibrils of muscle cells.

Skeleton The cuticle to which the musculature is attached.

Skin The cuticle.

Skin Glands See lateral glands, cephalic glands, cervical glands, preanal tubes, cement glands, hypodermal glands.

Slimeflux Wound sap of trees.

Slow Decline A disease of citrus incited by Tylenchulus semipenetrans.

Soft Cuticle A gelatinous, elastic, extensible material that makes up most of the nematode body covering. See hard cuticle.

Soil Borne Nematodes living within the confines of the soil during at least part of their infective stage.

Soil Chimney Any passage leading to the soil surface allowing nematicidal vapors to become diluted and escape.

Soil Fumigation The use of a nematicide to disinfest soil of plant parasitic nematodes.

Soil Inhabiting See edaphic, hypogeal, terrestrial, terricolous.

Soil Mass Application The fumigation of soil in depth as contrasted â€˜with a surface application.

Soil Porosity The degree to which the soil mass is permeated with pores or cavities. Soil Sterilant A nematicide with a broad lethal I spectrum i.e. will control nematodes, fungi, weeds, etc. See control.

Soil Structure The arrangement of primary soil particles into compound particles or clusters that are separate from adjoining aggregates and have properties unlike those of an equal mass of unaggregated primary soil particles.

Soil Texture The relative proportions of the various size groups of individual soil grains in a mass of soil. Specifically, it refers to the proportions of sand, silt and clay.

Soil Type A category under soil series based on the texture of the surface soil. See textural class.

Sole The site of attachment of the special ambulatory setae or adhesion tubes.

Solid Application The treatment of soil with a nematicide in such a manner that the expanding vapor forms a solid carpet throughout the field. Broadcast application See row application, drench application, gas application.

Solubilization A form of colloidal solution where a relatively insoluble chemical is dispersed in the form of micelles, consisting of alternating molecules of detergent and water with the inner core being the chemical.

Soma (pl, **Somata**) The body proper of an animal as distinguished from the germ cells. Alternatively, the portion of a cell that contains the nucleus.

Somatic Muscle Muscles of the body proper as distinguished from muscles of special function. Unspecialized musculature. .

Somatic Cells Body cells of an individual as distinguished from germ cells.

Somatic Setae The setae in general on the body cuticle as distinguished from cephalic setae, caudal setae, etc.

Somato-Esophageal Muscles Specialize muscles common to the esophageal region though of diverse form and origin.

Somato-Intestinal Muscles Specialize muscles common to the region of the intestine which extend from the body wall to the intestine.

Sorb To take up and retain by adsorption or absorption.

Sorption The act or process of sorbing. A general term covering adsorption and absorption.

Spathulate Paddle-shaped. Narrow and flat at the base and broad at the top. See reniform, spatulate.

Spatulate Spoon or paddle-shaped, Broad and rounded at the top, attenuate at the base. See reniform, spathulate.

Spear Frequently used to describe an odontostyle, distinguishing it from a stomatostyle. See stylet.

Spear-shaped See lanceolate, oblanceolate.

Specialist Predator A nematode with feeding structures adapted to feeding on a specific group of organisms, usually on other nematodes.

Species (pl. Species) A group of organisms below generic rank and above sub specific rank which possess characteristics in common distinguishing them from similar or other groups and which are transmissible to their progeny.

Species Concept That a group of organisms of similar kind form a primary biological unit.

Species Name Under the system of bionomial nomenclature the genus and trivial name used together. See trivial name.

Species Diversity The condition of being comprised of different types of apecies. Diversity is greatest when all species present are equally abundant. See true diversity.

Species Richness The number of species present or the number of species in a particular guild. See guild, functional guild.

Specimen A unit of sample. A single example of nematode. The individual unit of a collection.

Spectrum (pl. Spectra) The range of parasitism, activity, host range, control, etc.

Spermagonium (pl. **Spermagonia**) The sperm- producing structure of a digonic female.

Sperm Spermatozoa.

Spermatheca The enlarged portion of the female reproductive system which functions as a reservoir in receiving and holding spermatozoa from the male. Receptaculum seminis.

Spermatid One of the cells which arise by division of the secondary spermatocytes and develops into a spermatozoon.

Spermatocytes The cells which arise from the spermatogonia; two generations evolve from the last generation cell of the spermatogonia form the spermatids being the primary spermatocytes and secondary spermatocytes, respectively.

Spermatogenesis The development or formation of spermatozoa.

Spermatogonium (pl. **Spermatogonia**) The first generation in the differentiation of spermatozoa from primitive male germ cells.

Spermatozoon (pl. **Spermatozoa**) The matured and functional sex cell or gamete of the male which surface unites with the ovum. Sperm.

Sperm Cell Spermatozoon.

Sperm Duct The vas deferens.

Sphenoid Wedge-shaped. See cuneiform.

Spherical See globate.

Sphincter A circular muscle which functions by closing a duct Or orifice.

Sphincter Z The circular muscles of the Z organ.

Spickels A disease symptom of Ditylenchus dipsaci infection of daffodils, hyacinths and others; typified by small, pale green swellings containing the stem nematodes.

Spicate Bearing a spike or spikes. Having the form of a spike.

Spicula See Spicule.

Spicular Chord The direct line connecting the proximal and distal ends of the spicule.

Spicular Muscles Specialized muscles for the protraction and retraction of the spicules. See laterodorsal retractor spiculi, lateroventral retractor spiculi, protractor spiculi.

Spicular Pouch A muscular pouch surrounding the spicules that functions as the spicule protractor musculature in triplonchid nematodes.

Spicule Pouch See Spicular Pouch.

Spicular Sheath A conical or tubular extension of the cuticle distally about the spicules. The spicular pouch of some authors.

Spicular Cytoplasmic Core The inner contents of the spicules.

Spiculate Bearing spicules.

Spicule (pl. **Spicules**) Male intromittent organs functioning during copulation for the transfer of sperm, often paired; each an elongate cuticularized piece, extrusible through the cloacal opening.

Spicule Needle-like in form. A slender, pointed object.

Spicule Cephalated A situation where the proximal end of the spicule is expanded more or less like a hollow bulb. See manubrium.

Spicule Pouches Sac-like gland which secrete and hold the spicules.

Spicule Rostrum The distal projection elements of the spicule.

Spiculiform Having the form of a spicule. Needle-shaped.

Spiculum The term applied to fused spicules.

Spike-shaped Spicate. See needle-shaped.

Spiked-Tail Stage Where Meloidogyne larvae retain, until the final molt, the pointed tail of the second stage larva.

Spikkels A disease condition of narcissus incited by Ditylenchus dipsaci. The spikkels appear as small yellowish, slightly raised pimple-like spots on the leaves.

Spindle-shaped A cylindrical body tapering towards both ends. See fusiform.

Spination The development of spines on the body.

Spine A firm sharp pointed process generally of the cortical and matrix layers of the cuticle.

Spiniferous Bearing spines.

Spinigerous Bearing spines.

Spinneret The terminal pore of the caudal glands, See caudal pore.

Spinose Bearing spines.

Spinous Spine-like,

Spinule A minute spine.

Spinulose Bearing minute spines.

Spiral Winding or circling about a central axis in a closing or broadening manner. See helix, unispire, multispire.

Spoil, Spoile The cast cuticle at molting.

Spontaneous Generation The theoretic arising of living from nonliving matter. Abiogenesis.

Spontaneous Hatching The hatching of nematode eggs in the absence of an active root diffusate. See hatching.

Spoon-shaped See spatulate, spathulate.

Spot See macula.

Spot Application The treatment of soil with a nematicide only at the future planting site i.e. a melon hill, a tree site. See row application, broadcast application, drench application, gas application.

Spraing Corky ringspot disease of potatoes; a virus disease transmitted by species of Trichodoridae. See Tobra Virus.

Spreading Decline A disease of citrus incited by Radopholus similis,

Spring Dwarf See dwarf.

Squared See truncate.

Stadium (pl. **Stadia**) The interval between molts of larvae. instar. Any period in the development or process of a thing.

Staff-like See baculiform.

Stalked See pedunculate, petiolate.

Star-shaped See stellate, radial, asteriform.

Steep To obtain an extract by soaking. See decoction.

Stellate Resembling a star. Star-shaped. Radiated.

Stellate Cells Stelliform coelomocytic cells in the anterior third of some nematodes. See coelomocytes.

Stelliform Having the shape of a star.

Stem Sickness A diseased condition of plant stems incited by Ditylenchus spp.

Stenohaline Incapable of tolerating a wide variation of osmotic pressure in an environment. See euryhaline.

Stenostomatous Descriptive of diplogasterid nematodes with a stoma lacking teeth.

Stenoxenous Having a narrow host range.

Sterotaxis A taxis where contact with a firm substrate is the response initiating stimulus. See thigmotaxis, taxis, tropism.

Stereotropism A tropism where contact with a firm substrate is the response initiating stimulus. See thigmotropism, taxis, tropism.

Sterilization The process of rendering a substrate devoid of all living organisms. See pasteurization.

Stichocyte An individual cell of a stichosonie.

Stichosome A serial duplication of esophageal glands exterior of the esophagus musculature.

Stilt Bristles Ambulatory setae,

Stimulus (pl. Stimuli) Any agent that incites a nerve or body activity. See taxis, tropism.

Stirrupform Circular with a horizontal base.

Stoma The mouth cavity. That segment of the digestive tract between the oral opening and the triradiate beginning of the esophagus including the oral aperture and the stoma walls. See buccal capsule.

Stomatal Pertaining to or belonging to the stoma.

Stomatostyle A stylet which was derived through evolutionary development from a stoma; the cheilorhabdions lie about the oral opening, the cephalic framework is derived from the prorhabdions, the subulate portion of the stylet is derived mesorhabdions (the prorhabdions of some authors), the shaft is derived from the metarhabdions (the meso-metarhabdion of some), and the sylet knobs are derived from the telorhabdions. See onchiostyle, odontostyle.

Stomodeum The anterior or oral portion of the alimentary tract.

Stone Cell A total encasement of a female root- knot nematode and egg mass within host tissue as the result of the growth of wound periderm.

Straight Female Genital Tube See outstretched ovary.

Straight Ovary See outstretched ovary.

Strain A subgroup within a species differing in virulence, symptom production, or to sonic extent in host range from the rest of the species; probably more properly termed a race.

Strap-shaped See ligulate.

Strawberry Bunch A disease of strawberry incited by Aphelenchoides fragariae.

Stream Inhabiting See rheophile.

Stress The oppressions endured by a nematode population in its struggle for survival. See pressure.

Stria (pl. Striae) An elongate depression or groove. A minute banding. .

Striation A stria. The arrangement of striae.

Stricture A contraction of the lumen of a passage. .

Strongylatid Bursa A modification of the oval rhabditoid bursa into three lobes; one dorsal and two lateral, with muscular tissue accompanying the caudal papillae forming rays.

Structure Any organ, appendage or part.

Structural Guild A group of species with similar life history characteristics, See colonizer-persister series.

Stubby Root A disease condition of arrested root growth in celery, sweet corn and other crops incited by Trichodorus christiei in which the root system is composed of numerous short stubby branches often arranged in a cluster.

Stunted Plant A plant reduced and deformed in size and shape. See dwarf plant.

Stylet A relatively long, rather slender, hollow feeding structure; usually axial in the adult. A spear. See mural stylet, odontostylet, onchiostyle stomatostyle, buccal stylet, spear.

Stylet Aperture The anterior opening into the stylet. In Tylenchida the stylet aperture is obliquely ventral and short, in Dorylaimida the stylet aperture is obliquely dorsal and up to approximately half the stylet length.

Stylet Extension A rigid section of the alimentary tract front the base of the stylet to the beginning of the esophagus musculature; from this region the knobs, flanges and various other basal developments of stylets have been derived in Dorylaimoidea. .

Stylet Guide Where the vestibule and vestibule extension form a sheath to hold the stylet. .

Stylet Knobs Three swellings, derived from the telorhabdions, in triradiate positions at the base of the stylet.

Stylet Protractor The muscles which function to extend the stylet.

Stylet Shaft The subulate portion of a stylet, derived from the metarhabdions. .

Styliform Having the shape of a .stylet. Having a subulate terminus.

Subcephalic Setae Setae located on the head but posterior to the cephalic circlet.

Subclass A grouping less inclusive than a class but more than an order.

Subcrystalline Layer A waxy, translucent substance on the cysts of Heterodera spp. persisting only a short time in the soil.

Subcuticular Layer The hypodermis.

Subcuticle The hypodermis.

Subcuticular Striation Longitudinal markings due to musculature underlaying the cuticle.

Subdorsal The position on the nematode body situated 30Â° laterally from the dorsomedian line and perpendicular to the anteroposterior axis. Dorsolateral. See position. .

Subdorsal Muscle Fields A segment of the dorsosubmedian muscle field being delimited by the dorsal cord and the laterodorsal cord. See dorsosubmedian muscle fields, ventrosubmedian muscle fields. .

Subdorsal Nerves The dorsolateral nerves.

Subfamily A grouping less inclusive than a family but more than a genus.

Subgenus A grouping less inclusive than a genus but more than a species.

Subinterlabial Projection A process or processes which extend into the oral aperture.

Sublateral The position on the nematode body 30o ventral of the lateral position and perpendicular to the anteroposterior axis. See position. .

Sublateral Wings The bursa.

Submedian The position on the nematode body situated 45Â° from the dorsomedian or ventromedian line and perpendicular to the anteroposterior axis. Laterodorsal. Lateroventral. See position. .

Submedian Caudal Papillae A posterior continuation of the submedian papillae. .

Submedian Papillae Two series of papillae sub- median in position extending from the anal region to a point anterior to the supplements; these may be continued posteriorly and are known as the submedian caudal papillae. .

Suborder A grouping less inclusive than an order hut more than a family.

Subordinal Belonging to. or characteristic of, a suborder.

Subphylum A grouping less inclusive than a phylum but more than a class.

Subspecies A grouping less inclusive than a species but more than a variety or strain. See topomorph.

Substrate A substratum.

Substratum (pl. **Substrata**) The medium in or on which an organism lives or grows.

Subterminal Position less than terminal. Almost terminal.

Subulate linear at the base and attenuate at the terminus. Awl-shaped. See styliform.

Subventral The position on the nematode body situated 30Â° laterally from the ventromedian line and perpendicular to the anteroposterior axis. Ventrolateral. See position. .

Subventral Cuticular wings The bursa.

Subventral Esophageal Glands Two of three esophageal salivary glands whic are located in subventral positions. See dorsal esophageal gland.

Subventral Esophageal Gland Orifice The aperture of the subventral salivary gland duct opening into the lumen of the esophagus.

Subventral Muscle Fields A segment of the ventrosubmedian muscle fields being delimited by the ventral cord and the lateroventral cord. See dorsosubmedian muscle fields, ventrosubmedian muscle fields.

Subventral Nerves The ventrolateral nerves.

Sucker A special oval or rounded modification of the cuticle only on some males and functions in copulation.

Summer Dwarf See dwarf.

Superclass A grouping more inclusive than a class but less than a phylum.

Superfamily A grouping more inclusive than a family but less than an order.

Superficial Organs Appendages involving the cuticle and hypodermis i.e. setae, supplements.

Supergenus A grouping more inclusive than a genus but less than a family.

Superinfection A severe infection.

Supernatant The separate floating on the surface of a fluid.

Superorder A grouping more inclusive than an order but less than a class.

Supplementary Glands The supplements.

Supplements In males, organs of secretion and attachment ventromedian in position, derived of cuticle and functioning during copulation.

Supplements Contiguous Where the supplements are situated in a continuous row in contact or near contact with one another.

Supplements in Fascicles Where the supplements are situated in a clusters or groups, with space between the clusters.

Supporting Services Ecosystem services that are necessary for the functioning of otjher services. For examople, the Supporting Service of nutrient cycling is necessary for the service of primary production which, in turn is necessary for the Provisioning Service of food production.

Supraanal Above or anterior to the anus.

Suranal Supraanal.

Survival The continuation of living and reproducing.

Survival Value The total utility of the characters of an organism in its environment.

Suscept A plant subject to a given disease. See host.

Susceptible Lacking an inherent ability to resist disease.

Suture A seam or impressed line formed by the union of two adjacent inarticulate margins. See cleft

Swarm An aggregation of free moving nematodes. To migrate, move or join in a crowd.

Sweetpotato-shaped See batatiform.

Sword-shaped See xiphiiform, xiphoid, anceps, ensate, ensiform.

Symbiosis An association mutually beneficial to two different kinds of organisms.

Symbiont An organism living in a state of symbiosis.

Symmetry The mode of arrangement of parts, organs, etc. about the body axis.

Symptom Any clue which leads to the detection of a diseased condition.

Synaeresis, Syneresis The separation of a liquid from a gel; caused by the contraction of the gel.

Syncytium A multinucleate cell,

Syndrome The aggregate expression of signs and symptoms which characterize a disease.

Syngenesis Reproduction requiring sexually dimorphic parents. Sexual reproduction. See reproduction.

Syngonic Pertaining to, or having the characteristics of syngony.

Syngonic Female Where sperm and later ova are produced by the same gonad. Protandrous hermaphroditism. See digonic female, reproduction.

Syngony Protandrous hermaphroditism.

Synkaryon A nucleus formed by the fusion of the two pronuclei.

Synonym A systematic name replaced due to priority of another name, reclassification, etc. See tautonym.

Synonymous Having the character of a synonym. A like in fact. A continuum of sameness,

Synoptical Key An arrangement of the most essential nematode characters so that a particular identity may be determined by selecting only those that apply.

Syntype Any one individual of all the specimens available to the describer of a new species when no holotype has been selected. Cotype. See type.

Synxenic Pertaining to the rearing of an organism with one or more known species of associated organisms. See gnotobiotic, agnotobiotic, xenic, axenic, monoxenic, dixenic, trixenic, polyxenic.

Systematics The arrangement of species according to a distinct plan or system. The science of classification. Taxonomy.

T

Tactic Pertaining to, belonging to, or characteristic of taxis.

Tactile Organs Sensory organs of touch.

Tactoreceptors Sensory organs of touch.

Tail The posterior portion of the nematode body delimited proximally by the anus and distally by the body terminus.

Tail Spot An internal structure indicating the future site of the anus in microfilariae.

Tail Terminal The hyaline portion of the tail peculiar to some nematode larvae, especially Heterodera and Meloidogyne spp., in which no internal bodies, structures or arrangements can he seen.

Tail Whorl The cuticular configurations in the perineal pattern of Meloidogyne spp., at the site of the tail terminus. .

Tandem A set of two or more things or objects arranged one behind another.

Tangoreceptor An organ of touch possessing a single sense cell.

Taper A gradual reduction in breadth of an elongate body.

Target Spot A relatively small, roundish area of stunted plants.

Taste Bud A sensilla.

Tautonym Where the same name is used to designate a genus and one of its species or subspecies.

Tautotype A genotype due to tautonymy. See type.

Taxis Movement by mobile organisms in response to a stimulus. Taxis implies a change of place rather than an orientation (tropism). See tropism, rheotaxic, stereotaxis, thigmotaxis, thermotaxis, geotaxis, hydrotaxis,chemotaxis, galvanotaxis.

Taxon A general term for a taxonomic group of any category.

Taxonomy The systematic grouping of organisms according to their natural relationships.

Teardrop Having an oblanceolate shape.

Tease To shred plant or animal material under magnification in search of nematodes.

Teeth Apparatus The valvular apparatus.

Telamon A cuticularized thickening of the ventral wall of the cloaca lying more or less between the spicules. See capitulum. .

Telocentric Pertaining to the centromere lying in a terminal position. See metacentric.

Telogonia A gonad where proliferation of primordial germ cells arises only at the distal end. See hologonia.

Telogonic Gonad A telogonia.

Telophase The final stage of mitosis or meiosis in which the nucleus reverts to the resting stage. See prophase, metaphase, anaphase.

Telorhabdions The walls of the telostom. .

Telostom A short valve which connects the stoma with the esophagus. See cheilostom, protostom. .

Terminal Adhesive Organ The spinneret.

Terminal Bodies Phagocytic structures of some coelomocytes. See end organs, cellules.

Terminal Bulb The basal bulb of the esophagus.

Terminal Cell The primordial germ cell of a gonad. See apical cell, cap cell.

Terminal Duct The excretory duct,

Terminal Pore The spinneret.

Terminal Proliferative Cell The terminal cell.

Terminal Spine In mermithids, the terminal portion of the larval tail. See caudal appendage, caudal papillae.

Terrestrial Soil inhabiting. See arenicolous, hypogaeic. aquatic, edaphic, marine, rheophile, terricolous, xerophilous, cavernicolous, fossorial, xiphilous.

Terricolous Soil inhabiting.

Tessellate Checkered. Patterned like a mosaic.

Testicle A gonad of the male. A testis.

Testis (pl. **Testes**) The reproductive gland of the male which produces the spermatozoa. See ovary.

Textural Class Categories for the kinds of soil material according to the proportions of sand, silt and clay. The principal textural classes in increasing order of the amount of silt and clay are: sand, loamy sand, sandy loam, loam, silt loam, silt, sandy clay loam, clay loam, silty clay loam, sandy clay, silty clay and clay. These class names are modified to indicate the size of the sand fraction or the presence of gravel, cobbles or stones i.e. loamy fine sand, very fine sandy loam, gravelly loam, stoney clay and cobbly loam.

Thelyotokous A form of parthenogenesis in which only female offspring are produced from unfertilized eggs. See parthenogenesis

Therapeutic Pertaining to, or belonging to, the arts and methods of curing diseases.

Therapy The treatment of diseases.

Thermal Death Point The time-temperature combination lethal to a nematode species, especially in hot water treatment.

Thermal Death Time The period of exposure prior to attaining the thermal death point.

Thermal Inactivation Point The thermal death point.

Thermotaxis A taxis where heat is the response initiating stimulus. See taxis, tropism.

Thermolabile Substances, structures, etc. which lose their characteristic properties on being heated.

Thigmotaxis Stereotaxis.

Thigmotropism Stereotropism.

Thorn-shaped See corniform, acanthiform.

Thread-like See filiform.

Threadworm A vernacular term for any individual of the phylum Nematoda, especially as applied to parasites of vertebrates. See eelworms, nema, nematode, roundworm.

Threshold The minimum point or level at which a response or affect is produced on an organism or system.

Tiling, Tiled Having the appearance of a tiled roof due to transverse and longitudinal striae on the cuticle.

Titilla (pl. **Titillae**) Small projections on either side of the distal end of the gubernaculum.

Tobra Virus A tobacco rattle or pea early-browning plant virus consisting of rod-like particles and transmitted by Trichodoridae.

Tolerance The capacity to endure nematode parasitism.

Tongue-shaped See linguiform, lingulate.

Tooth-like See dentiform, serrate.

Topomorph A morphologic variant of a cosmopolitan species developed in response to local environment.

Toponym The location or place of origin of a type species. Type locality.

Topotype A specimen taken from the locality of the original type. See type.

Torulose Knobby. Having knob-like swellings.

Torus A smooth, rounded protuberance, Having knob-like swellings.

Totomount A whole mount of a nematode.

Toxic Pertaining to a poison. Poisonous.

Toxicity The quality or degree of being poisonous,

Toxin A poisonous product of secretion derived through the metabolism of organisms.

Tract Some special purpose system of parts or organs, i.e. the digestive tract.

Transfer Ribonucleic Acid See tRNA.

Transformation The geneticalteration of a cell by direct uptake, incorporation and expression of exogenous DNA taken up or inserted through cell membranes. See DNA.

Transport Host A vector.

Transverse Fibrils Noncontractile or static fibers of the muscle sarcoplasm. See longitudinal fibrils.

Transverse Plane A plane or section perpendicular to the longitudinal axis. Cross section. See sagittal plane, longitudinal plane.

Transverse Striation In the nematode cuticle a circular groove or arc whose plane is perpendicular to the longitudinal axis. See longitudinal striation.

Trap Craps A control technique where a crop is destroyed after becoming infested by sedentary nematodes and before maturity of the nematode is reached. See enemy crops.

Trapeziform Having the form or shape of a trapezium.

Trapezium A plane four-sided figure in which no two sides are parallel.

Trehalose A disaccharide sugar, C12 H12 O11 stored in the nematode body in place of starch.

Triangle-shaped See deltoid.

Trichoid Hair-like.

Trichotomous Bearing three branches or forks. See trifurcate.

Tricuspid Divided into three points.

Tricuspidate Divided into three points. Tricuspid.

Tridentate Having three teeth or three tooth-like processes.

Trifid Having three clefts or three parts.

Trifurcate Bearing three branches or forks. See trichotomous.

Trilobate Bearing three lobes.

Tripartite Consisting of three parts.

Triploblastic Composed of three germ layers, the ectoderm, mesoderm and endoderm, from which develop complex organ-systems.

Triquetrous Having three angles or arms.

Triradiate Having three radiating arms or branches.

Triturate To rub, grind or otherwise reduce to minute particles. See comminute, macerate.

Trivial Name The second term of a species name under the system of binomial nomenclature. See species name.

Trixenic Pertaining to the rearing of an organism with three known species of associated organisms. See gnotobiotic, agnotobiotic, xenic, axenic, synxenic, monoxenic, dixenic, polyxenic.

tRNA Transfer ribonucleic acid. Molecules that adapt the 20 letter code of amino acids in proteins to the four letter (ACGU) genetic code of messenger RNA (See mRNA) with the 20 letter code of amino acids in proteins. tRNAs are essential components of the biological synthesis of proteins.

Trophosome A modification of the intestine in some nematode forms in which the walls and lumen of the intestine disappear resulting in a syncyteal structure.

Tropism Involuntary movement of turning or curvature (but not of position) of an organism or its parts. An orientation response induced automatically or by external stimuli. See taxis, rheotropism, thigmotropism, stereotropism, geotropism.

True Bulb A valvate muscular swelling of the esophagus in which the lumen widens forming a cavity, the lining is cuticularized and radiating muscle fibers extend to the periphery of the esophagus. See pseudobulb, basal bulb.

True Diversity A measure of the equality of abundance of the species present; the effective number of species. If all species are equally abundant, true diversity will equal species richness. See species richness, abunmdance.,

Truncate Having a squared terminus.

Tube-like See duct.

Tubercle A small knob-like protuberance of cuticle. See torus.

Tubes The marginal tubes of esophageal radii.

Tubiform Having the form of a tube.

Tubiform Organ The deManian system.

Tubiform Setae Ambulatory setae.

Tubule A minute tube.

Tuft A small cluster of outgrowths.

Tulip Root A disease of oats incited by Ditylenchus dipsaci.

Turgor The state of normal distension of the cell wall by fluid contents.

Tylenchoid Esophagus Having a narrow procorpus, a strongly formed median bulb, typical isthmus and terminating with a glandular basal bulb.

Type A group of things thought of as a category due to the likeness of the individuals in one or more characters.

Type Host The designated organism from which a type specimen has been collected. See type.

Type Locality The locale from which a type specimen has been collected. See toponym.

Type Slide A microscope mount which contains a type specimen.

Type Specimen A single example of a nematode designated as a type individual with the characteristics of the species description.

U

Ubiquitous Existing in all areas. Cosmopolitan.

Ufra Disease A disease of rice incited by Ditylenchus angustus.

Unarmed A nematode not possessing a stylet, tooth, cutting plate, etc..

Unciform Having the form of a hook.

Uncinate Hooked.

Underbridge The anterior bifurcation of the vagina composed of tough, brown material which may extend across the vulval cone and parallel to the vulval bridge. .

Undulate Having a wavy surface, margin or form.

Unilateral Sited or occurring on one side only.

Uninucleate Having one nucleus per cell.

Unispire A single turn of a spiral.

Unsegmented The condition of lacking Metameric Segmentation

Unsuitable Host An immune or resistant plant.

Uterine Pertaining to or belonging to the uterus.

Uterine Chamber That proportion of the reproductive tract connecting the uterus and the vagina.

Uterine Loop A posterior extension of the uterus.

Uterine Sac A thin-walled, muscular sac to which the uterus is joined and into which the vagina opens. It may be obscurely separated from the uterus by a slight constriction or merely by a differentiation of tissue.

Uterine Tract The uterus.

Uterine Vagina The vagina uterina.

Uterus (pl. **Uteri**) A region of the oviduct modified to function as a place of development (sometimes maturation) of the egg. .

Uvette A rosette or group of cells between the uterine afferent duct and efferent duct of the deManian system.

V

Vagina (pl. **Vaginae, Vaginas**) A canal lined with cuticle connecting the uterus or the ovijector with the genital orifice, the vulva. The vagina is considered to have three sections, proximal, intermediate and distal. See Pars Proximalis Vaginae, Pars Refringens Vaginae and Pars Distalis Vaginae.

Vagina Uterina An extension of the vagina inwards incorporating the distal part of the uterus which histologically resembles the vagina but lacks the cuticular lining. See Pars Proximalis Vaginae.

Vagina Vera The outermost part of the vagina and is lined with cuticle. See Pars Refringens Vaginae and Pars Distalis Vaginae

Vaginal Cuticle The cuticular lining of the vagina which differs in some respects from the external cuticle.

Vagrant Mobile nematodes in larval and adult stages and capable of existence away from their hosts for various lengths of time. See semivagrant.

Vagrant Endoparasites Nematodes which feed on cells and reproduce as they move about within roots and are capable of entering the soil and infecting new roots. See migratory, sessile.

Values A term used to denote the measurables of the de Man formula.

Valve The valvular apparatus.

Valve Apparatus The valvular apparatus.

Valve Flaps The valvular apparatus.

Valve Plates The valvular apparatus.

Valvular Apparatus A refractive structure situated in the median bulb or basal bulb of some nematode forms and probably functions as a pump during feeding.

Variant A departure from the normal form or condition. See topomorph.

Variation Any divergence from the normal form or condition.

Vas Deferens (pl. **Vasa Defentia)** A slender muscular tube extending from the ejaculatory duct to the testis.

Vector An organism or other agency which transmits or disseminates pathogens or other identities.

Velum A delicate euticular membrane on the inner side of the spicule.

Venter The belly. Undersidc. Any part corresponding to the belly. See dorsum.

Ventrad In the direction of the venter. See dorsal.

Ventral The bottom side aspect of a nematode. Pertaining to the lower side. See dorsal.

Ventral Cord The ventral line of the four longitudinal lines of thickening on the inner side of the hypodermis. See cords, dorsal cord, lateral cords. .

Ventral Gland The excretory gland.

Ventral Nerve The main body nerve which is actually a ganglionated cord.

Ventricular Esophagus An. esophagus with a glandular basal bulb.

Ventricnlus A glandular region of the esophagus.

Ventrolateral The position on the nematode body situated 300 laterally from the ventromedian line and perpendicular to the anteroposterior axis. Subventral. See position. .

Ventrolateral Connectives The anolumbar conriectives.

Ventrolateral Muscle Fields A segment of the ventrosubmedian muscle fields being delimited by a lateral cord and the subventral cord. See dorsosubmedian muscle fields, ventrosubmedian muscle fields.

Ventromedian The true middle line on the lower side of an individual.

Ventrosubmedian Muscle Fields The somatic muscles bounded by the ventral cord and the lateral cords. See dorsosubmedian muscle fields, subdorsal muscle fields, dorsolateral muscle fields, ventrolateral muscle fields, subventral muscle fields. .

Vermiform Having a worm-like shape.

Vermifuge. An anthelminthic.

Vertical Gene Transfer The transmission of genes from the parental generation to offspring via sexual or asexual reproduction. See horizontal gene transfer.

Vertical Resistance Resistance effective against certain races or biotypes of a nematode. Conferred by one or a few genes. See resistance.

Vesicula Seminalis A sac serving as a reservoir for the semen. A widening of the ejaculatory duct. See spermatheca.

Vesiculate Sack-like. Bladder-like, Having a swollen appearance.

Vessel-shaped See scaphiform.

Vestibule In the narrow sense: the cheilostom. In broad sense: the stoma. The anterior portion of the sylet guide in stylet-bearing nematodes. An expansible oral opening leading to the pharynx.

Vestibule Extension An extension of the axial cephalic framework into the body cavity forming the posterior part of the stylet guide. .

Vestibulum Var. of vestibule.

Vestigial A structure small or degenerate representing what formerly was more fully developed or functional. See rudimentary.

Vestigial Posterior Uterus The postvulvar uterine branch.

Viable Capable of growth and development.

Virulence The relative ability to cause disease.

Viscosity The resistance offered by a fluid to flow.

Vitelline Membrane The surface covering membrane of the oocyte prior to fertilization. See lipoid membrane.

Viviparous Producing living young from within the body in contrast to eggs. See oviparous, ovoviviparous, larviparous, fissiparous.

Voracious Ravenous. Rapacious in feeding. See feeding.

V-Spot A structure in microfilariae which represents the future excretory pore.

Vulva The female genital opening and its delimiting margins of cuticle.

Vulval Bridge In some species of Heterodera the vulval slit persists cross the fenestra as a bridge forming two semifenestrae. .

Vulval Body Diameter (Width) The diameter of the body at the level of the vulva; sometimes abbreviated as vbd or vbw.

Vulval Cone A relatively moderate protuberance at the posterior end of *Heterodera* cysts. .

Vulval Fenestra In *Heterodera* a thin-walled transparent region at the top of the vulval cone which may breakdown to form the hatching pore. .

Vulval Flap Cuticular membranes sited at both ends of the vulva. See epiptygma, lateral membrane.

Vulva Glands Glands emptying through the vulva probably functioning during copulation.

Vulval Muscles Specialized muscles of function for the dilation and constriction of the vulva. See dilator vulvae, constrictor vulvae.

Vulval Pad The subcuticular clear zone surrounding the vulva consisting of vulval musculature.

Vulval Slit The shape or border of the female external genital opening.

Vulviform Resembling the vulva.

W

Waddiform Having the shape of a club or staff.

Wart A cuticular excrescence somewhat potato-shaped.

Wavy See sinuate, sinuous, undulate.

Wedge-shaped See cuneiform, sphenoid.

White Bud A disease of strawberry incited by Aphelenchoides fragariae. See dwarf.

White Cysts Newly matured females of the genus Heterodera. See pearls.

White Tip A disease of rice incited by *Aphelenchoides besseyi.*

Wild Plant A disease of strawberry incited by *Aphelenchoides fragariae.* See dwarf.

Witchesâ€™ Broom A condition of excessive root branching induced by nematode feeding. See hexenbesen,

Wilt Loss of freshness and drooping of plant leaves due to inadequate water supply, excessive transpiration, a vascular disease or a toxin produce by an organism

Wilty Soil Light, sandy, well-drained soil which harbors the wilt disease organism of cotton and plant pathogenic nematodes.

Wing An anteriad configuration *Meloidogyne* perineal patterns into a lateral lingulate shape.

Wings See lateral wings, alae.

Wing-shaped See aliform.

Wool A mass of desiccated *Ditylenchus* larvae in a resistant state; frequently found in bulb crops. Nematode wool.

Worm Nematode in the very broad sense.

Worm-shaped See vermiform.

Wrinkle A small ridge or furrow on a surface. See rugose.

Wrinkled See rugose, rugosissimus, rugula, rugulose.

X

X-Bodies Binucleate, rounded and unbranched coelomocytes.

Xerophilous, Xerophytic Inhabiting dry places. See aquatic, arenicolous, edaphic, hypogaeic, marine, rheophile, terrestrial, terricolous,

Xiphiiform Sword-shaped.

Xiphoid Sword-shaped.

Y

Yellow Dwarf A disease of soybean incited by *Heterodera glycines.*

Yellows Disease A disease of black pepper vines incited by *Radopholus similis.*

Z

Z-Organ See Organ Z.

Zooparasitic Infective in a host organism of the animal kingdom.

Zygote A fertilized egg resulting from the union of an ovum and a spermatozoon. A diploid nucleus resulting from the union of two gametes of opposite kind. See synkaryon.

APPENDIX

Appendix I

STANDARD SIEVE SIZE

Mesh /inch	Pore aperture (micrometer)
4	4000
6	2800
8	2000
12	1400
16	1000
20	840
22	710
30	500
44	355
60	250
72	210
85	180
100	150
120	125
150	105
170	90
200	75
240	63
300	53
350	45
400	38
500	26

Appendix II

MAJOR JOURNALS IN NEMATOLOGY

African Plant Protection
Asian Journal of Nematology
CIH Descriptions of Plant Parasitic Nematodes
Egyptian Society of Agricultural Nematology Newsletter
Fungicide and Nematicide Tests
Helminthological abstracts
Indian Journal of Nematology
International Journal of Nematology
Japanese Journal of Nematology
Journal of Nematology
Nematologia Brasileira
Nematologia Mediterranea (Istituto di Nematologia Agraria of the C.N.R.)
Nematologica
Nematology
Nematology News
Nematology News
Nematology Newsletter
Nematropica
Organization of Nematologists of Tropical America Newsletter
Pakistan Journal of Nematology
Pakistan Society of Nematologists Newsletter
Revue de Nematologie
Russian Journal of Nematology
The Bibliography on Plant, Soil and Freshwater nematodes
The Egyptian Journal of Agronematology

MAJOR NEMATOLOGICAL SOCIETIES

Afro-Asian Society of Nematologists
American Phytopathological Society
Australasian Association of Nematologists
Brazilian Nematological Society
Cereal Nematode Network
Chinese Society of Plant Nematologists
Egyptian Society of Agricultural Nematology
Helminthological Society of Washington
International Federation of Nematology Societies
Japanese Nematological Society
Mediterranean Nematological Society
Nematological Society of India
Nematological Society of Southern Africa
Nematology Initiative in East and Southern Africa
Organization of Tropical American Nematologists
Pakistan Society of Nematologists
Society for Invertebrate Pathology
Society of European Nematologists
Society of Nematologists
West and Central African Nematology Network

Appendix III

1. ROOT-KNOT SCORING IN CARROT

(Adapted from Paul Speijer and Dirk De Waele-1997).

No galling Slight Mild Moderate. Severe

2. ROOT-KNOT SCORING IN OTHER CROPS

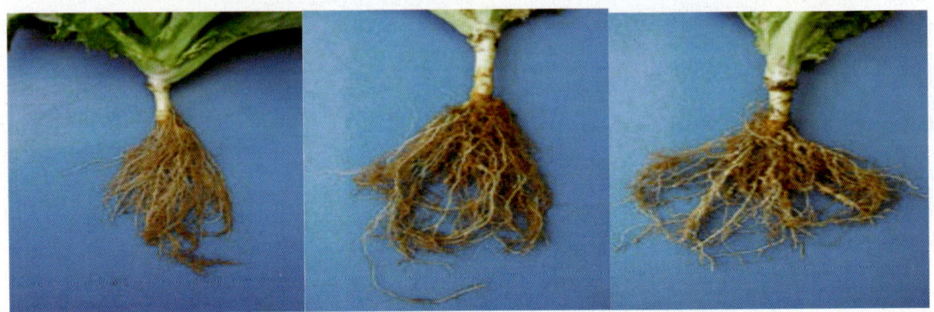

No galling Slight Mild

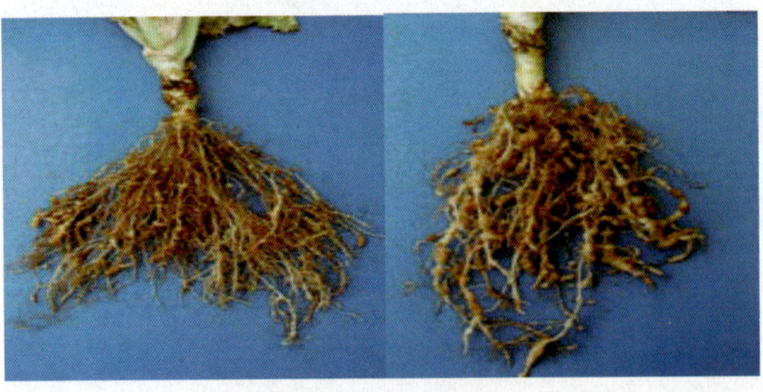

Moderate Severe

Appendix IV

3. LESION SCORING IN BANANA

(Adapted from Paul Speijer and Dirk De Waele -1997)

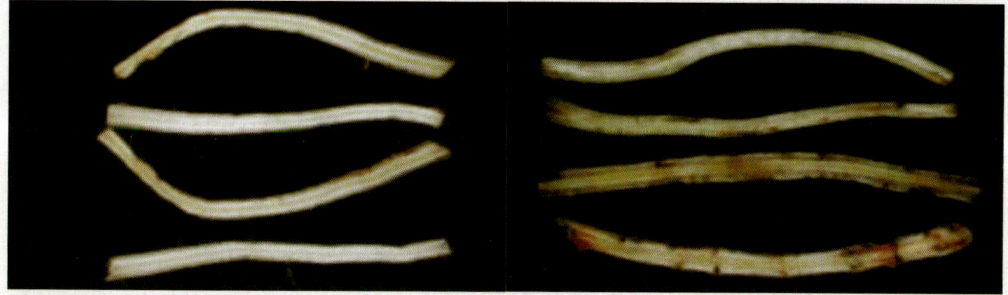

0%, clean 5%

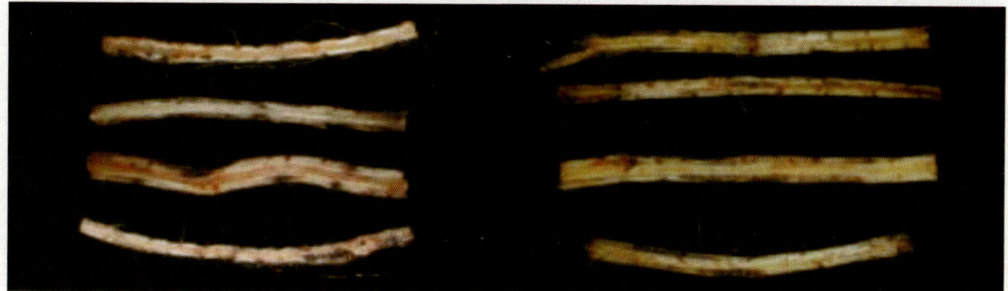

10% 25 %

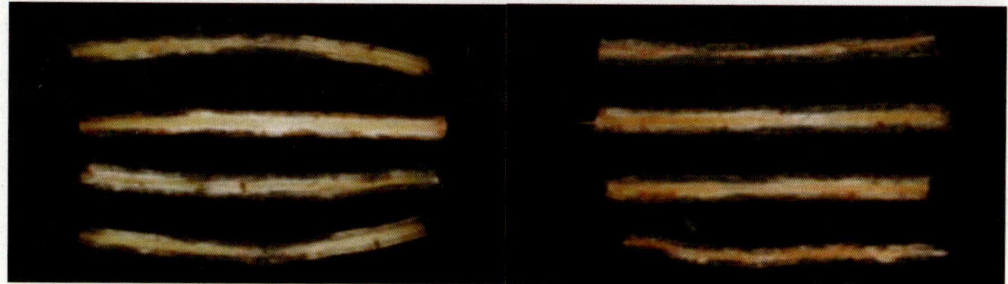

50 % 75 %

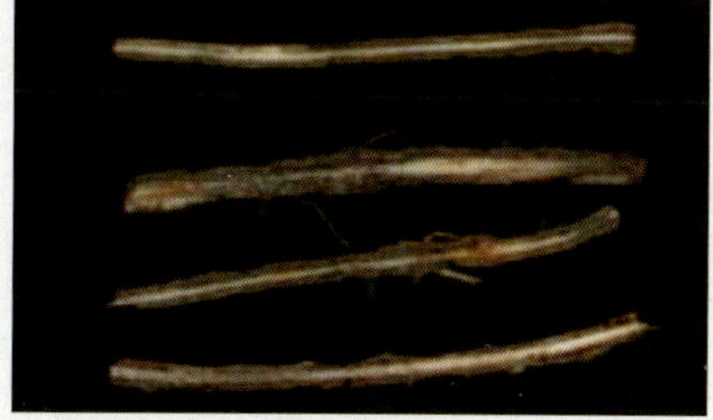

1200 %

Appendix V

CONVERSIONS

i. From Per cent to ppm:

1 per cent = 10,000 ppm = 10 g per litre of kg

0.1 per cent = 1000 ppm = 1000 mg per litre

0.01 per cent = 100 ppm = 100 mg per litre

0.001 per cent = 10 ppm = 10 mg per litre

0.0001 per cent = 1 ppm = 1 mg per litre

ii. Dose per ha to per acre, per kg soil and per sq.m.:

X kg per ha = 2X/5 kg per acre

X kg per ha = X/2 mg per kg soil

X kg per ha = X/10 g per sq.m

iii. Units of area and length:

Area:

1 sq. inch = 6.4516 sq. cm

1 sq. yard = 0.836 sq.m

1 acre = 4047 sq.m = 43,560 sq.feet

1 ha = 2.47 acres = 107593 sq.feet

1 sq. mile = 2.589 sq.km

1 sq. km = 0.386 mile

Length:

1 km = 1000 m

1m = 100 cm = 3.28 feet

1 furlong = 220 yards

1 foot = 12 inches = 30.48 cm

1 inch = 2.54 cm = 25.4 mm

1 yard = 0.9144 m

1 mile = 1.6093 km

1 km = 0.62137 mile

1 mm = 0.03937 inch

1 cm = 0.0328 foor

Appendix VI

COMMON NAMES OF MAJOR NEMATODES

Sl. No.	COMMON NAME	SCIENTIFIC NAME
1.	African spiral nematode	*Helicotylenchus africanus*
2.	Alfalfa root nematode	*Heterodera goettingiana*
3.	Almond cyst nematode	*Heterodera amygdali*
4.	Amaranth cyst nematode	*Cactodera amaranthi*
5.	American dagger nematode	*Xiphinema americanum*
6.	Amu-Darya nematode	*Heterodera oxiana*
7.	Apple cyst nematode	*Globodera mali*
8.	Apple root-knot nematode	*Meloidogyne mali*
9.	Awl nematode	*Dolichodorus spp.*
10.	Banana meadow nematode	*Pratylenchus coffeae*
11.	Banana nematode	*Pratylenchus musicola*
12.	Banana root-lesion nematode	*Pratylenchus coffeae*
13.	Banana spiral nematode	*Helicotylenchus multicinctus*
14.	Banana-root nematode	*Radopholus similis*
15.	Barley cyst nematode	*Heterodera hordecalis*
16.	Barley root-knot nematode	*Meloidogyne nassi*
17.	Beachgrass root-knot nematode	*Meloidogyne sasseri*
18.	Beer nematode	*Panagrellus silusiae*
19.	Beet nematode	*Heterodera schachtii*
20.	Beet stem nematode	*Ditylenchus dipsaci*
21.	Begonia leaf nematode	*Aphelenchoides fragariae*
22.	Bentgrass nematode	*Anguina agrostis*
23.	Bermudagrass cyst nematode	*Heterodera cardiolata*
24.	Birch cyst nematode	*Cactodera betulae*
25.	Black currant nematode	*Aphelenchoides ritzemabosi*
26.	Blueberry root-knot nematode	*Meloidogyne carolinensis*
27.	Boxwood spiral nematode	*Rotylenchus buxophilus*
28.	Brassica root eelworm	*Heterodera cruciferae*
29.	Brassica root nematode	*Heterodera cruciferae*
30.	Brazilian root-knot nematode	*Meloidogyne exigua*
31.	Brazilian root-knot nematode	*Meloidogyne inornata*
32.	British root-knot nematode	*Meloidogyne artiellia*

33.	British spiral nematode	*Scutellonema brachyurum*
34.	Buckwheat cyst nematode	*Heterodera graduni*
35.	Bud and leaf nematodes	*Aphelenchoides spp.*
36.	Bud and stem nematode	*Ditylenchus dipsaci*
37.	Bulb and stem nematode	*Ditylenchus spp.*
38.	Bulb nematode	*Ditylenchus dipsaci*
39.	Bulb or stem nematode	*Ditylenchus spp.*
40.	Burrowing nematode	*Radopholus similis*
41.	Burrowing nematode	*Radopholus spp.*
42.	Cabbage cyst nematode	*Heterodera cruciferae*
43.	Cabbage nematode	*Heterodera cruciferae*
44.	Cabbage root nematode	*Heterodera cruciferae*
45.	Cactus cyst nematode	*Cactodera cacti*
46.	Cajanus cyst nematode	*Heterodera cajani*
47.	California dagger nematode	*Xiphinema index*
48.	California meadow nematode	*Pratylenchus neglectus*
49.	California root-lesion nematode	*Pratylenchus neglectus*
50.	California sessile nematode	*Cacopaurus epacris*
51.	Camel thorn cyst nematode	*Heterodera oxiana*
52.	Camellia root-knot nematode	*Meloidogyne camelliae*
53.	Canadian root-knot nematode	*Meloidogyne microtyla*
54.	Carnation pin nematode	*Paratylenchus curvitatus*
55.	Carnation pin nematode	*Paratylenchus dianthus*
56.	Carolina spiral nematode	*Scutellonema brachyurum*
57.	Carrot cyst nematode	*Heterodera carotae*
58.	Carrot root nematode	*Heterodera carotae*
59.	Cereal cyst nematode	*Heterodera avenae*
60.	Cereal cyst nematode	*Heterodera latipons*
61.	Cereal root nematode	*Heterodera avenae*
62.	Cereal root-knot nematode	*Meloidogyne nassi*
63.	Cereals root eelworm	*Heterodera major*
64.	Cereals root nematode	*Heterodera avenae*
65.	Chamber's dagger nematode	*Xiphinema chambersi*
66.	Christie's spiral nematode	*Scutellonema christiei*
67.	Christie's stubby root nematode	*Trichodorus christiei*
68.	Chrysanthemum foliar nematode	*Aphelenchoides ritzemabosi*

69.	Chrysanthemum leaf nematode	*Aphelenchoides ritzemabosi*
70.	Chrysanthemum nematode	*Aphelenchoides ritzemabosi*
71.	Citrus nematode	*Tylenchulus semipenetrans*
72.	Citrus ring nematode	*Criconemoides citri*
73.	Citrus root nematode	*Tylenchulus semipenetrans*
74.	Citrus root-knot nematode	*Meloidogyne indica*
75.	Citrus spine nematode	*Criconema civellae*
76.	Clover cyst nematode	*Heterodera trifolii*
77.	Clover root nematode	*Heterodera trifolii*
78.	Clover stem nematode	*Ditylenchus dipsaci*
79.	Cobb's awl nematode	*Dolichodorus heterocephalous*
80.	Cobb's lance nematode	*Hoplolaimus galeatus*
81.	Cobb's meadow nematode	*Pratylenchus penetrans*
82.	Cobb's ring nematode	*Criconemoides simile*
83.	Cobb's root lesion nematode	*Pratylenchus penetrans*
84.	Cobb's root-knot nematode	*Nacobbus batatiformis*
85.	Cobb's spiral nematode	*Helicotylenchus multicinctus*
86.	Cobb's stubby root nematode	*Trichodorus primitivus*
87.	Coconut nematode	*Rhadinaphelenchus cocophilus*
88.	Coconut palm nematode	*Rhadinaphelenchus cocophilus*
89.	Cocopalm nematode	*Rhadinaphelenchus cocophilus*
90.	Coffee meadow nematode	*Pratylenchus coffeae*
91.	Coffee root-knot nematode	*Meloidogyne exigua*
92.	Coffee root-lesion nematode	*Pratylenchus coffeae*
93.	Columbia nematode	*Hoplolaimus columbus*
94.	Columbia root-knot nematode	*Meloidogyne chitwoodi*
95.	Corn cyst nematode	*Heterodera zeae*
96.	Corn meadow nematode	*Pratylenchus zeae*
97.	Corn root-lesion nematode	*Pratylenchus zeae*
98.	Cotton root-knot nematode	*Meloidogyne incognita acrita*
99.	Cowpea cyst nematode	*Heterodera vigni*
100.	Crown-headed lance nematode	*Hoplolaimus tylenchiformis*
101.	Currant nematode	*Aphelenchoides ribes*
102.	Cyperuscyst nematode	*Heterodera mothi*
103.	Cyst nematodes	*Globodera spp.*
104.	Cyst nematodes	*Heterodera spp.*

105.	Cyst-forming nematode	*Heterodera spp.*
106.	Cystoid body nematode	*Meloidoderita spp.*
107.	Cystoid nematode	*Meloidodera spp.*
108.	Dagger nematode	*Xiphinema spp.*
109.	De Man's meadow nematode	*Pratylenchus pratensis*
110.	De Man's root-lesion nematode	*Pratylenchus pratensis*
111.	Dock cyst nematode	*Heterodera rosii*
112.	Douglas Fir nematode	*Nacobbodera chitwoodi*
113.	Ear-cockle nematode	*Anguina tritici*
114.	Estonian cyst nematode	*Cactodera estonica*
115.	Eucalypt cystoid nematode	*Cryphodera eucalypti*
116.	European dagger nematode	*Xiphinema diversicaudatum*
117.	False root-knot nematode of sugar beet	*Nacobbus batatiformis*
118.	False root-knot nematode	*Nacobbus spp.*
119.	Fern nematode	*Aphelenchoides fragariae*
120.	Fern nematode	*Aphelenchoides olesistus*
121.	Fescue leaf gall nematode	*Anguina graminis*
122.	Ficuscyst nematode	*Heterodera fici*
123.	Fig cyst nematode	*Heterodera fici*
124.	Fig pin nematode	*Paratylenchus hamatus*
125.	Fig spine nematode	*Criconema decalineatum*
126.	Foliar nematodes	*Aphelenchoides spp.*
127.	Galeopsis cyst nematode	*Heterodera galeopsidis*
128.	Galeopsis root nematode	*Heterodera galeopsidis*
129.	Gall-forming nematodes	*Meloidogyne spp.*
130.	Godfrey's meadow nematode	*Pratylenchus brachyurus*
131.	Godfrey's root-lesion nematode	*Pratylenchus brachyurus*
132.	Gold-plated nematode	*Globodera rostochiensis*
133.	Golden nematode	*Globodera rostochiensis*
134.	Golden nematode of potato	*Globodera rostochiensis*
135.	Grass cyst nematode	*Punctodera punctata*
136.	Grass root-gall nematode	*Subanguina radicicola*
137.	Grass sheath nematode	*Hemicycliophora similis*
138.	Grass spiral nematode	*Helicotylenchus erythrinae*
139.	Great root nematode	*Heterodera avenae*
140.	Hairy-gall nematode	*Nacobbus batatiformis*

141.	Heart-shaped cyst nematode	*Heterodera cardiolata*
142.	Hop cyst nematode	*Heterodera humuli*
143.	Hop nematode	*Heterodera humuli*
144.	Hop root nematode	*Heterodera humuli*
145.	Horsenettle cyst nematode	*Globodera tabacum virginiae*
146.	Indian root-knot nematode	*Meloidogyne brevicauda*
147.	Iris nematode	*Ditylenchus destructor*
148.	Japanese cyst nematode	*Heterodera elachista*
149.	Javanese root-knot nematode	*Meloidogyne javanica*
150.	Kansas cyst nematode	*Heterodera longicolla*
151.	Kidney-shaped nematode	*Rotylenchulus reniformis*
152.	Kikuyu grass nematode	*Meloidogyne kikuyuensis*
153.	Knapweed nematode	*Mesoanguina picridis*
154.	Knawel cyst nematode	*Heterodera scleranthii*
155.	Knotweed cyst nematode	*Cactodera weissi*
156.	Kona coffee root-knot nematode	*Meloidogyne konaensis*
157.	Lance nematodes	*Hoplolaimus spp.*
158.	Lesion nematodes	*Pratylenchus spp.*
159.	Lespedeza cyst nematode	*Heterodera lespedezae*
160.	Lucerne cyst nematode	*Heterodera medicaginis*
161.	Maple root-knot nematode	*Meloidogyne ovalis*
162.	Meadow nematode	*Pratylenchus spp.*
163.	Mediterranean cereal cyst nematode	*Heterodera latipons*
164.	Milfoil cyst nematode	*Globodera millefolii*
165.	Motha cyst nematode	*Heterodera mothi*
166.	Mushroom nematode	*Aphelenchoides composticola*
167.	Mushroom spawn nematode	*Ditylenchus myceliophagus*
168.	Needle nematode	*Longidorous spp.*
169.	Nettle cyst nematode	*Heterodera urticae*
170.	Nigerian dagger nematode	*Xiphinema nigeriense*
171.	Northern root-knot nematode	*Meloidogyne hapla*
172.	Nutgrass cyst nematode	*Heterodera cyperi*
173.	Oak root-knot nematode	*Meloidogyne querciana*
174.	Oak sheathoid nematode	*Hemicriconemoides biformis*
175.	Oat cyst nematode	*Heterodera avenae*
176.	Oat cyst nematode	*Heterodera major*

177.	Oat nematode	*Heterodera avenae*
178.	Oat root nematode	*Heterodera avenae*
179.	Onion stem nematode	*Ditylenchus dipsaci*
180.	Osborne's cyst nematode	*Globodera tabacum solanacearum*
181.	Pacific dagger nematode	*Xiphinema radicicola*
182.	Pea cyst nematode	*Heterodera goettingiana*
183.	Pea root eelworm	*Heterodera goettingiana*
184	Pea root nematode	*Heterodera goettingiana*
185.	Peanut root-knot nematode	*Meloidogyne arenaria*
186.	Persian sessile nematode	*Cacopaurus pestis*
187.	Phlox stem nematode	*Ditylenchus dipsaci*
188.	Pigeon pea cyst nematode	*Heterodera cajani*
189.	Pin nematodes	*Paratylenchus spp.*
190.	Pine cystoid nematode	*Meloidodera floridensis*
191.	Pine sheathoid nematode	*Hemicriconemoides floridensis*
192.	Pine sting nematode	*Belonolaimus gracilis*
193.	Pine wood nematode	*Bursaphelenchus xylophilus*
194.	Potato cyst eelworm	*Globodera rostochiensis*
195.	Potato cyst nematode	*Globodera pallida*
196.	Potato cyst nematode	*Globodera rostochiensis*
197.	Potato nematode	*Globodera rostochiensis*
198.	Potato root eelworm	*Globodera rostochiensis*
199.	Potato root nematode	*Globodera rostochiensis*
200.	Potato rot nematode	*Ditylenchus destructor*
201.	Potato tuber eelworm	*Ditylenchus destructor*
202.	Potato tuber nematode	*Ditylenchus destructor*
203.	Pseudo root-knot nematode	*Hypsoperine graminis*
204.	Ramie pin nematode	*Paratylenchus elachistus*
205.	Red ring nematode	*Rhadinaphelenchus cocophilus*
206.	Reniform nematode	*Rotylenchulus reniformis*
207.	Reniform nematodes	*Rotylenchulus spp.*
208.	Rice blind root nematodes	*Hirschmanniella spp.*
209.	Rice cyst nematode	*Heterodera oryzae*
210.	Rice nematode	*Aphelenchoides oryzae*
211.	Rice root nematode	*Hirschmanniella oryzae*
212.	Rice root-knot nematode	*Meloidogyne graminicola*

213.	Rice stem nematode	*Ditylenchus angustus*
214.	Rice stunt nematode	*Tylenchorhynchus martini*
215.	Rice white-tip nematode	*Aphelenchoides besseyi*
216.	Rice-root nematode	*Radopholus oryzae*
217.	Ring nematodes	*Criconema spp.*
218.	Ring nematodes	*Criconemoides spp.*
219.	Root-gall nematode	*Meloidogyne spp.*
220.	Root-knot nematode	*Meloidogyne spp.*
221.	Root-lesion nematode	*Pratylenchus spp.*
222.	Round cystoid nematode	*Thecavermiculatus andinus*
223.	Rubber cyst nematode	*Heterodera fici*
224.	Rumex cyst nemtode	*Heterodera rumicis*
225.	Scribner's lesion nematode	*Pratylenchus scribneri*
226.	Scribner's meadow nematode	*Pratylenchus scribneri*
227.	Scribner's root-lesion nematode	*Pratylenchus scribneri*
228.	Sedge cyst nematode	*Heterodera cyperi*
229.	Seed gall nematode	*Afrina wevelli*
230.	Seed gall nematodes	*Anguina spp.*
231.	Seed-gall nematode	*Anguina tritici*
232.	Seinhorst's stubby root nematode	*Trichodorus pachydermis*
233.	Sessile nematodes	*Cacopaurus spp.*
234.	Seville root-knot nematode	*Meloidogyne hispanica*
235.	Sheath nematodes	*Hemicycliophora spp.*
236.	Sheathoid nematode	*Hemicriconemoides spp.*
237.	Shoot gall nematode	*Anguina spp.*
238.	Smooth-headed lesion nematode	*Pratylenchus brachyurus*
239.	Smooth-headed meadow nematode	*Pratylenchus leiocephalus*
240.	Smooth-headed nematode	*Pratylenchus brachyurus*
241.	Sorghum root-knot nematode	*Meloidogyne acronea*
242.	Sour paste nematode	*Panagrellus redivivus*
243.	South African pin nematode	*Paratylenchus curvitatus*
244.	Southern root-knot nematode	*Meloidogyne incognita*
245.	Sowthistle cyst nematode	*Heterodera sonchophila*
246.	Soybean cyst nematode	*Heterodera glycines*
247.	Spear nematodes	*Dorylaimus spp.*
248.	Spine nematodes	*Criconema spp.*

249.	Spiral nematodes	*Helicotylenchus spp.*
250.	Spiral nematodes	*Rotylenchus spp.*
251.	Spiral nematodes	*Scutellonema spp.*
252.	Spring crimp nematode	*Aphelenchoides fragariae*
253.	Spring dwarf nematode	*Aphelenchoides besseyi*
254.	Steiner's spiral nematode	*Helicotylenchus dihystera*
255.	Stem gall nematode	*Pterotylenchus cecidogenus*
256.	Stem nematode	*Ditylenchus dipsaci*
257.	Sting nematode	*Belonolaimus gracilis*
258.	Sting nematode	*Belonolaimus longicaudatus*
259.	Sting nematodes	*Belonolaimus spp.*
260.	Strawberry bud nematode	*Aphelenchoides besseyi*
261.	Strawberry bud nematode	*Aphelenchoides fragariae*
262.	Strawberry foliar nematode	*Aphelenchoides fragariae*
263.	Strawberry nematode	*Aphelenchoides fragariae*
264.	Stubby root nematode	*Trichodorus christiei*
265.	Stubby root nematode	*Trichodorus kurumeensis*
266.	Stubby root nematode	*Paratrichodorus spp.*
267.	Stubby root nematode	*Trichodorus spp.*
268.	Stunt nematode	*Tylenchorhynchus claytoni*
269.	Stunt nematodes	*Tylenchorhynchus spp.*
270.	Stylet nematodes	*Tylenchorhynchus spp.*
271.	Sugar beet cyst nematode	*Heterodera schachtii*
272.	Sugar beet nematode	*Heterodera schachtii*
273.	Sugar cane cyst nematode	*Heterodera sacchari*
274.	Sugar cane cyst nematode	*Heterodera schachtii*
275.	Sugar cane stylet nematode	*Tylenchorhynchus martini*
276.	Summer dwarf nematode	*Aphelenchoides fragariae*
277.	Sycamore root-knot nematode	*Meloidogyne platani*
278.	Tadzhik cyst nematode	*Heterodera tadshikistanica*
279.	Tadzhik cystoid nematode	*Meloidodera tadshikistanica*
280.	Tadzhik root-knot nematode	*Meloidogyne tadshikistanica*
281.	Tarjan's sheath nematode	*Hemicycliophora parvana*
282.	Tea root-knot nematode	*Meloidogyne brevicauda*
283.	Teasel nematode	*Ditylenchus dipsaci*
284.	Tesselate stylet nematode	*Tylenchorhynchus claytoni*

285.	Thames' root-knot nematode	*Meloidogyne thamesi*
286.	Thorne's cyst nematode	*Cactodera thornei*
287.	Thorne's lance nematode	*Rotylenchus uniformis*
288.	Thorne's meadow nematode	*Pratylenchus thornei*
289.	Thorne's needle nematode	*Longidorus sylphus*
290.	Thorne's root-lesion nematode	*Pratylenchus thornei*
291.	Tobacco cyst nematode	*Globodera tabacum*
292.	Tobacco stunt nematode	*Tylenchorhynchus claytoni*
293.	Tulip root nematode	*Ditylenchus dipsaci*
294.	Turf spiral nematode	*Rotylenchus christiei*
295.	Turkmen cyst nematode	*Heterodera turcomanica*
296.	Ustinov cyst nematode	*Heterodera ustinovi*
297.	Valentine cyst nematode	*Heterodera cardiolata*
298.	Vinegar eels	*Turbatrix aceti*
299.	Vinegar nematode	*Turbatrix aceti*
300.	Walnut meadow nematode	*Pratylenchus vulnus*
301.	Walnut root-lesion nematode	*Pratylenchus vulnus*
302.	Walnut sessile nematode	*Cacopaurus pestis*
303.	Wesson's sheathoid nematode	*Hemicriconemoides wessoni*
304.	West African spiral nematode	*Scutellonema blaberum*
305.	Wheat cyst nematode	*Heterodera latipons*
306.	Wheat gall nematode	*Anguina tritici*
307.	White-tip nematode	*Aphelenchoides besseyi*
308.	Willow cyst nematode	*Heterodera salixophila*
309.	Yam nematode	*Scutellonema bradys*
310.	Yarrow cyst nematode	*Globodera achilleae*
311.	Zimmerman's spiral nematode	*Helicotylenchus erythrinae*
312.	Zoysia spine nematode	*Criconema spinalineatum*